SMART CITIES FOR SUSTAINABILITY

ADVANCED SERIES IN MANAGEMENT

Series Editors: Miguel R. Olivas-Lujan and Tanya Bondarouk

Previous Volumes

Social Media in Human Resources Management
Eds. Tanya Bondarouk and Miguel R. Olivas-Luján

Social Media in Strategic Management
Eds. Miguel R. Olivas-Luján and Tanya Bondarouk

(Dis)honesty in Management: Manifestations and Consequences
Eds. Tiia Vissak and Maaja Vadi

Commercial Diplomacy and International Business: A Conceptual and Empirical Exploration
Ed. H. Ruël

Electronic HRM in Theory and Practice
Eds. T. Bondarouk, H. Ruël and J. C. Looise

Relational Practices, Participative Organizing
Eds. Chris Steyaert and Bart Van Looy

Autopoiesis in Organization Theory and Practice
Eds. Rodrigo Magalhaes and Ron Sanchez

Organizations as Learning Systems "Living Composition" as an Enabling Infrastructure
Ed. Marjatta Maula

Complex Systems and Evolutionary Perspectives on Organizations: The Application of Complexity Theory to Organizations
Ed. Eve Mitleton-Kelly

Managing Imaginary Organizations: A New Perspective on Business
Eds. Bo Hedberg, Philippe Baumard and A. Yakhlef

Systems Perspectives on Resources, Capabilities and Management Processes
Eds. John Morecroft, Ron Sanchez and Aimé Heene

Tracks and Frames: The Economy of Symbolic Forms in Organizations
Ed. K. Skoldberg

Human Resource Management, Social Innovation and Technology
Eds. Tanya Bondarouk and Miguel R. Olivas-Luján

Dead Firms: Causes and Effects of Cross-Border Corporate Insolvency
Miguel M. Torres, Virginia Cathro and Maria Alejandra Gonzalez Perez

New Ways of Working Practices: Antecedents and Outcomes
Jan de Leede

Age Diversity in the Workplace
Silvia Profili, Alessia Sammarra and Laura Innocenti

International Business Diplomacy: How Can Multinational Corporations Deal With Global Challenges?
Ed. Huub Ruel

Organisational Roadmap Towards Teal Organisations
Tanya Bondarouk, Anna Bos-Nehles, Maarten Renkema, Jeroen Meijerink and Jan de Leede

Indigenous Management Practices in Africa: A Guide for Educators and Practitioners
Eds. Uchenna Uzo and Abel Kinoti Meru

Diversity Within Diversity Management: Country-Based Perspectives
Eds. Andri Georgiadou, Maria Alejandra Gonzalez-Perez and Miguel R Olivas-Lujan

Diversity Within Diversity Management: Types of Diversity in Organizations
Eds. Andri Georgiadou, Maria Alejandra Gonzalez-Perez and Miguel R Olivas-Lujan

HRM 4.0 For Human-Centered Organizations
Rita Bissola and Barbara Imperatori

Sustainable Hospitality Management: Designing Meaningful Encounters With Talent and Technology
Huub Ruël, Angelique Lombarts and Jeoren A. Oskam

Agri-Food 4.0: Innovations, Challenges and Strategies
Rahul S. Mor, Dinesh Kumar and Anupama Singh

Smart Industry – Better Management
Tanya Bondarouk and Miguel R. Olivas

Management and Information Technology in the Digital Era: Challenges and Perspectives
Chemma Nawal, Mohammed El Amine Abdelli, Anjali Awasthi and Emmanuel Mogaji

Sustainable Management in COVID-19 Times
Aldo Alvarez-Risco, Marc A. Rosen and Shyla Del-Agulia-Arcentales

Technology, Management and Business: Evolving Perspectives
Rajnish Kumar Misra, Archana Shrivastava and Charu Sijoria

ADVANCED SERIES IN MANAGEMENT VOLUME 32

SMART CITIES FOR SUSTAINABILITY: APPROACHES AND SOLUTIONS

EDITED BY

MOHAMMED EL AMINE ABDELLI
University of Brest, France

ASMA SGHAIER
ISSCA School, University of Sousse, Tunisia

ATILLA AKBABA
Faculty of Tourism, İzmir Kâtip Çelebi University, Turkey

SAMIA CHEHBI GAMOURA
EM Strasbourg, University of Strasbourg, France

AND

HAMID DOOST MOHAMMADIAN
FHM University of Applied Sciences, Germany

emerald PUBLISHING

United Kingdom – North America – Japan – India – Malaysia – China

Emerald Publishing Limited
Emerald Publishing, Floor 5, Northspring, 21-23 Wellington Street, Leeds LS1 4DL

First edition 2024

British Library Cataloguing in Publication Data
A catalogue record for this book is available from the British Library

ISBN: 978-1-80455-903-1 (Print)
ISBN: 978-1-80455-902-4 (Online)
ISBN: 978-1-80455-904-8 (Epub)

ISSN: 1877-6361 (Series)

Printed and bound by CPI Group (UK) Ltd, Croydon, CR0 4YY

INVESTOR IN PEOPLE

CONTENTS

PART 1: SMART CITIES AND TECHNOLOGIES

Post-Pandemic Urban Planning and Use of Information Technologies in Smart Cities *3*
Seher Konak

Smart Destinations and Eco-friendly Practices *15*
Cenk Murat Koçoğlu, Burak Pinaroğlu and Emrah Yaşarsoy

An Alternative Model of Living: Smart Eco-cities *33*
Beyza Hatirnaz, Dilek Demirer and Emrah Özkul

Urban Transport Solutions for a Sustainable and Smart Mobility Future: Macro-Environmental Analysis *49*
Marwa Ben Ali and Ghada Boukettaya

Evaluation of Smart City Projects in Eco-Gastronomy Dimension: The Example of Turkey *71*
Fatih Varol, Merve Oksuz and Eren Yalcin

Blue-Green Smart U-Cities via Clean Technologies: Toward High Sustainable and Low Greenhouse Gas Emissions Urban Areas *83*
Hamid Doost Mohammadian

DRMM and Comprehensive Global Blue-Green Clean Sustainable Urban Mobility Risk Mitigation Plan for Mapping Future Smart Cities Through the 5th Wave Theory *103*
Hamid Doost Mohammadian

PART 2: DIGITALIZATION AND SUSTAINABILITY

Eco-City Tourism in Smart Cities for Sustainability *139*
Burcu Kıvılcım Zorba

Smart Tourism Destinations and Digitalization of Cultural Heritage for Sustainability *151*
Nil Sonuç and Seda Süer

Recent Developments in the Evaluation of Renewable Energy Resources in Tourism Businesses: A Bibliometric Analysis *169*
Metin Sürme and Dilara Bahtiyar Sari

Urban 6.0 and Utopia Concepts via Sustainable, Clean, Inclusive, Innovative, and U-Mobility Through the Theory of Comprehensive Everything *179*
Hamid Doost Mohammadian

Mapping Future Urban Plan – Toward Blue-Green Smart City and Mobility Through the 5th Wave, i-Sustainability Plus, and DCT Theories *211*
Hamid Doost Mohammadian

Index *233*

PART 1

SMART CITIES AND TECHNOLOGIES

POST-PANDEMIC URBAN PLANNING AND USE OF INFORMATION TECHNOLOGIES IN SMART CITIES

Seher Konak

ABSTRACT

Today, the rapid increase in the population living in cities and new developments in information and communication technologies (ICTs) increase the demand for smart cities day by day. It is thought that limited public resources and crowded cities will be managed better by making more use of the opportunities offered by smart technologies. At this point, many countries around the world are turning to smart city applications. Especially after the Covid-19 pandemic, local governments have started to give importance to smart city projects due to the advantages of smart cities. It is thought that urban planning against epidemics will gain more importance in the coming years.

Keywords: Smart city; information technologies; artificial intelligence; urban planning

1. INTRODUCTION

In cities, which are an important living space, it is important to include urban transformation models in this process for increasing social welfare levels and creating a more livable environment (Karayılmaz & Özker, 2020). According to 2016 United Nations World Urbanization report data, the period we are in is defined as "the most striking urban growth period in history". Accordingly, it is expected that 66% of the world population will live in urban areas by 2050.[1] The difficulty here is to provide this population with basic resources such as sufficient energy, clean water, and safe food while ensuring economic, social, and environmental sustainability (Lai et al., 2020). Also, cities which cover only 3% of the earth's surface consume 70% of energy production and produce 75% of carbon emissions.[2] Herewith, smart cities applications in the world are important.

[1]www.un.org.tr
[2]www.un.org.tr

Smart Cities for Sustainability
Advanced Series in Management, Volume 32, 3–14

Published under exclusive licence by Emerald Publishing Limited
ISSN: 1877-6361/doi:10.1108/S1877-636120230000033002

Smart cities, the changes after the Covid-19 epidemic (socioeconomic, demographic etc.), the Ukraine–Russia war, and the energy crisis that emerged in Europe are a reflection of the dramatic changes in the quality of life and are seen as a potential solution to the problems. It is necessary to develop effective solutions against globalization, urbanization, climate crisis, and epidemics in the twenty-first century. Therefore, urban planning is important against epidemics. It is possible to determine the needs of cities correctly and to meet these needs in a sustainable way with smart solutions (Örselli & Akbay, 2019). The smart cities approach will ensure the further development of cities by applying the Internet of Things (IoT) to the economy, management, and environment. It is important that safe and cost-effective services are applicable in these cities with information technology infrastructure.

Covid-19 is the disease caused by a new coronavirus called SARS-CoV-2, which first appeared in China on December 31, 2019 (WHO, 2021). According to the report published by United Nations Habitat in April 2020, a total of 1,430 cities in 210 countries were affected by the Covid-19 outbreak, with 95% of the cases occurring in urban areas (UN Habitat, 2021). Population density in cities may be higher than in rural areas due to the possibilities of people living in cities to encounter epidemics and inequalities in access to services, lack of infrastructure, migration, and urban poverty (Diez Roux, 2015). The Covid-19 epidemic has greatly affected and changed the way we behave in our living spaces (such as accommodation, work, entertainment, cultural and sports activities) (Şolt, 2021). As a result of the transmission of the disease through human-to-human contact in general, it has caused a faster spread, especially in urban areas. For this reason, urban areas are among the areas where measures should be taken to reduce the spread of epidemic crisis on a national and international scale (Budak & veSezgin, 2021). It is necessary to adopt the smart city approach in many parts of the world and to start using and adapting new technological tools in these cities with the developing technology (Allam & David, 2020). These technologies play an important role in controlling and effectively managing epidemics (Gretzel, Sigala, Xiang, & Koo, 2015).

Technological innovations with smart city thinking claim to transform urban space. Smart city applications that have reached technological maturity are highlighted with an aggressive marketing strategy. The emphasis is on energy efficiency, smart infrastructure, open data, open networks, seamless access, and transformative mobile services (Velibeyoğlu, 2019).

According to the European Union (2011), the concept of smart city supports the idea of environmental sustainability because the main purpose of smart cities is to reduce greenhouse gas emissions in urban areas through innovative technologies. The growing interest in the concept of the smart city and the need to solve the challenges of urbanization lead to various private and public investments in this area (Lee & Hancock, 2012). Today, it is seen that there are over 1,000 smart city projects around the world (Karayılmaz & Özker, 2020).

Singapore, Seoul, London, Barcelona, and Helsinki were at the top of the Top 50 Smart City Government rankings for 2020/2021 as published by the Eden Strategy Institute.[3] These smart cities offer great opportunities for businesses to collaborate with public authorities who are willing to leverage the private sector's expertise in systems design and strategic management. For instance, authorities in Moncton, Canada, announced a comprehensive green city project aimed at promoting public transport,

[3]www.smartcitygovt.com

cycling, and more recycling. LED traffic lights were installed throughout the city, and electric cars were provided for city services (Barrionuevo, Berrone, & Ricart, 2012). Following the devastating floods and landslides that hit Rio de Janeiro in 2010, a $14 million smart operations center equipped with cutting-edge disaster management technology was established in the city.

Due to this system, the city's traffic flow and public transportation systems can be controlled and power cuts can be managed. Rio de Janeiro has become one of the metropolises that uses smart resources to streamline their daily operations and improve the quality of life of their citizens (Barrionuevo et al., 2012). In this section, first of all, the definition of smart cities will be made, and the features of smart cities will be mentioned. Then, the information technologies that are the basis of smart cities will be mentioned, and examples of smart cities applications in the world will be given.

2. DEFINITION OF SMART CITY

The origin of the smart city idea is based on the smart growth movement at the end of the 1990s. This movement advocated creating new policies for urban planning. After the 2000s, the concept of smart city has been adopted by various technology companies (Siemens, IBM, etc.) and has been used to describe the integration of complex information systems into buildings, transportation, public safety, communication, and urban infrastructure and services such as electricity–water distribution level (Harrison & Donnelly, 2011).

The concept of smart city first emerged in 1994 (Dameri & Cocchia, 2013), and since 2010, smart city projects which have been encouraged by the European Union managed to stay on the agenda (Jucevicius, Patašienė, & Patašius, 2014). A few years later, Management Center at the University of Ottawa began to criticize the idea of smart cities for being too technical. There upon, researchers wanted to reveal the hidden aspects behind the term "smart city" (Hollands, 2008). Today, a smart city is generally defined as a city that uses information and communication technologies (ICTs), new urban infrastructure to improve service, and thus makes the life of local people (citizens) more comfortable, including improving the ecological situation (Dolinina, Pechenkin, Gubin, & Kushnikov, 2018, p. 3). In this context, smart city provides health benefits to a community with improved and educational options, greater security, social and political systems, economic prosperity, better housing, education, communication, energy, and other helpful resources (Pelton & Singh, 2019).

Similar concepts such as "intelligent city," "sustainable city," "green city," "knowledge city," "wired city," or "digital city" are used, which overlap with smart city (Albino, Berardi, & Dangelico, 2015; Camero & Alba, 2019). The concept of smart city is examined in three dimensions: technology, people, and society (Harrison et al., 2010). There is no generally accepted definition of the term of smart city. However, among the common elements frequently used in definitions, adaptation to innovative technologies, use of big data, and cooperation with urban actors are at the forefront (Köseoğlu & Demirci, 2018). The smart city is an urban development model that expresses the use of human, collective, and technological capital (Angelidou, 2014). In this context, the smart city is defined as "an ultra-modern urban area that caters to the needs of businesses, institutions, and especially citizens" (Khatoun & Zeadally, 2016). Planning a smart city, the accuracy of technology, its compatibility with changing demographics, education and health needs, employment opportunities, and adaptable infrastructure systems should be taken into consideration.

Studies such as intelligent data analysis, computer simulation, etc., allow the estimation of results (Pelton & Singh, 2019).

It is important to identify some infrastructure and superstructure factors in order to understand the projects implemented within the framework of the smart city concept. These factors can be divided into internal and external factors that have an impact on various stages within the scope of the attempt to intellectualize the urban space. Projects developed for smart cities should focus on building urban space infrastructure and organizational systems based on modern technologies that can solve emerging problems. In these projects, management and organization, technologies, policy, social communities, economy, infrastructure, and natural environment factors are taken into account (Chourabi et al., 2012).

Conceptually, while megacities research explains the reason for the growth of urban areas, the smart cities debate advocates sustainability. Considering the basis of smart cities research, namely the goal of ICT, the well-being of citizens, and sustainability, smart cities research is scalable, that is, it can cover a mega city, a regular city, or a village. For example, smart village research, which focuses on transferring the information obtained from smart city researches to rural areas, has just started (Visvizi & Lytras, 2018).

3. CHARACTERISTICS OF SMART CITY

In a study aiming to rank medium-sized cities in Europe according to smart city factors, six characteristics and 33 factors were identified. According to this, the features of smart cities have six: smart economy, smart people, smart governance, smart mobility, smart environment and smart living (Center of Regional Science, 2007, p. 11).

The "Smart Cities Wheel (SCW)" method of Cohen (2014), which is one of the approaches to smart cities, has been accepted by the European Union and consists of six components. These are:

(1) Smart Mobility: It includes comprehensive transport systems for passengers, drivers, and operators as well as integrated environmentally friendly and modern and sustainable transport systems, especially for disadvantaged individuals.
(2) Smart Living: It includes various aspects of the quality of life of the citizens such as culture, health, safety, and housing.
(3) Smart Governance: Political participation, services for citizens, and practices aimed at providing transparency in public administration are established.
(4) Smart Environment: It refers to attractive natural conditions such as climate, green space, pollution level, resource management, and efforts to protect the environment.
(5) Smart Economy: It includes all the elements of economic competitiveness such as innovation, entrepreneurship, trademarks, productivity and flexibility of the labor market, and integration in the international market.
(6) Smart People: This feature, together with the qualifications and educational level of the citizens, refers to the quality of social interactions related to integration and public life and openness to the "outside" world.

According to Nam and Pardo (2011), a smart city has three components: technology, people, and institutions. Based on the link between these factors, a smart city can emerge when investments in human capital and information technology infrastructure support

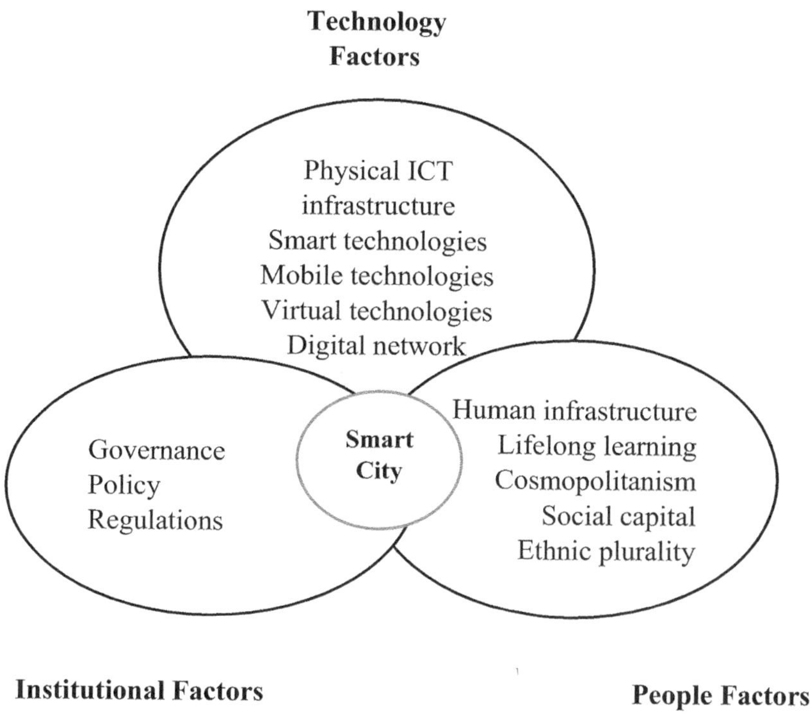

Fig. 1.1. Key Components of a Smart City. *Resource:* Samih (2019, p. 6).

sustainable growth and positively affect quality of life through the institutional factor. Fig. 1.1 shows the connections and subcomponents between these three components.

4. INFORMATION TECHNOLOGIES AT THE FOUNDATION OF SMART CITIES

Smart cities are people-oriented, have strong social services, have good governance, and have developed service areas and living standards. It is seen that these cities are not based solely on technological developments, but the concept is evaluated from a broad perspective based on the information and communication qualities of the society (Çelikyay, 2008).

Three key trends have emerged in smart city technologies: (1) the transition to wireless communication; (2) cloud computing, machine learning (ML), and smart data analytics; and (3) the development of the concept of smart technology in city security systems (Jameel, Ali, & Ali, 2019).

With the effect of digital transformation experienced today, innovative technologies that are at the core of smart cities come to the fore. Especially in recent years, big data, data mining, IoT, cloud computing, and Industry 4.0 are widely used in smart cities to solve urban problems (Köseoğlu & Demirci, 2018). Fig. 1.2 shows the relationship of smart technologies, including big data and cloud computing, with smart city applications.

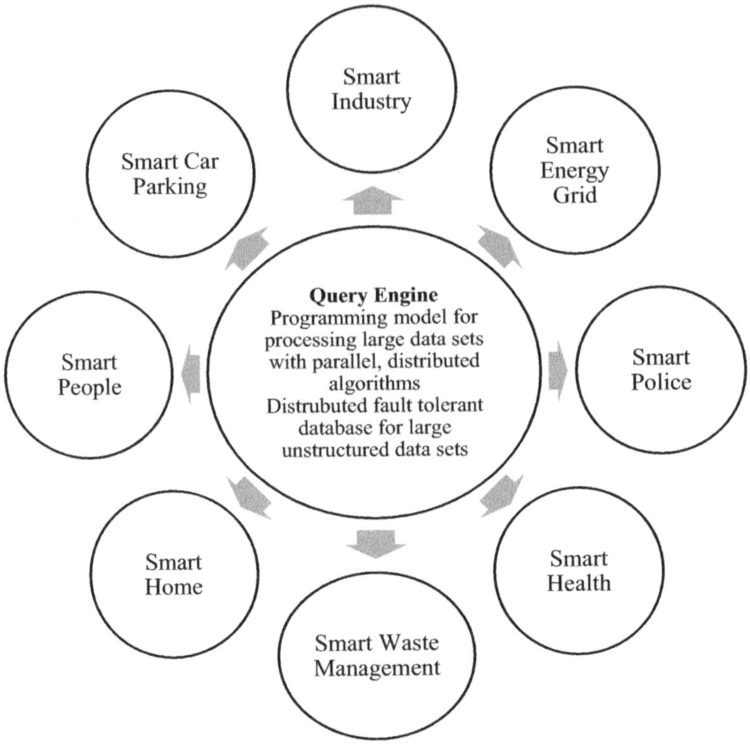

Fig. 1.2. Smart City Paradigms. Based upon: Hashem et al. (2016, p. 749).

According to Fig. 1.2, various smart applications share information by using sensors and other tools integrated with the cloud computing infrastructure to generate large amounts of unstructured data. This large amount of unstructured data is collected and stored in the cloud or data centers using fault tolerant databases. Thus, parallel algorithms and programming models are used to process large data sets, and data analytics are used to generate values from stored data (Hashem et al., 2016, p. 748). At this point, data mining technologies are used. In addition, due to the IoT, a variety of data is produced by many different actors such as individual users, private companies, city residents, and public institutions (Köseoğlu & Demirci, 2018). In Section 4.1, smart technologies used in smart cities will be explained.

4.1 Big Data and Data Mining

E-commerce on the internet, messaging, location sharing, all kinds of interaction, and communication through e-applications digitally produce and record high volumes of data without the knowledge or awareness of users (McKinsey, 2013). Everything that is recorded by modern ICTs, such as network sensors, smart objects and devices, and web and social media, constitutes big data (Rabari & Storper, 2015).

Data mining, on the other hand, is the application of ML algorithms with the help of various software packages on a smaller data set to help decision-makers make future-oriented decisions and provide services on behalf of their institutions by making use of the large data sets they have (Köseoğlu & Demirci, 2018, p. 48).

4.2 Internet of Things (IoT)

IoT notionally means "a worldwide network of uniquely addressable, interconnected objects based on standard communication protocols" (Atzori, Iera, & Morabito, 2010). The concept of the IoT gained notoriety through the work of the Center for Auto-Identification at the Massachusetts Institute of Technology which began designing and deploying the company radio frequency identification (RFID) infrastructure in 1999 (Sarma, Brock, & Ashton, 2000). This technology is a part of sustainable and smart cities infrastructure, especially in the industrial sector called industrial IoT. In IoT, industrial devices perform better production, performance, and maintenance functions (Qureshi, Rana, Ahmed, & Jeon, 2020).

There are many IoT-based smart solutions implemented worldwide. However, the most widely used solutions in smart cities are smart grids, smart homes, transportation and traffic management, e-health, waste management, public Wi-Fi, and so on. A smart grid is a modernized electricity grid that uses ICTs to automatically manage information between utilities and consumers (Ejaz & Anpalagan, 2019, p. 6).

4.3 Cloud Computing Technology

Cloud computing is a common concept that includes the delivery of services over the internet. The idea of cloud computing was put forward by John McCarty in 1961. Although services like webmail and YouTube have been used by many organizations for some time, companies didn't start using the cloud structure until 2009 to meet their broader information technology needs (Lin & Chen, 2012). In this context, Amazon modernized its data centers and enabled Amazon S3, which has an important place in the foundation of cloud computing, which had been put into service in 2006 (Bulut, 2018). Cloud computing provides services to perform large-scale computing tasks such as mining large social network data generated through smartphone applications (Chang, Bacigalupo, Wills, & Roure, 2010).

In today's technology, there are storage capacity problems because users want to store more personal data. As a solution to these problems, cloud technology refers to software applications accessed over the internet, data storage service, and processing capacity (Örselli & Akbay, 2019).

4.4 Artificial Intelligence (AI)

AI facilitates the processing and analysis of data generated from machine-to-machine communication in smart city installation (Batty, 2018). AI is also widely applied in healthcare, public safety, manufacturing, and retail. For example, it can be used to manage large sources of raw data such as ticket sales in public transport, police reports, sensors on the roads, and weather stations (Navarathna & Malagi, 2018).

AI can detect objects (vehicles, pedestrians, etc.), read license plates, and track their speed. It has been determined that drivers spend approximately 107 hours a year searching for a parking place. AI, which provides real-time data, can be used to solve this problem. Due to public transport technology, passengers can travel on trackable buses, and individuals know when they will arrive at their destination. AI is also used in waste management, which is an important solution in the development of smart cities (Mathur & Modani, 2016).

4.5 Industry 4.0

The term Industry 4.0 was first used at the Hannover Fair in 2011 (Hermann, Pentek, & Otto, 2015, p. 45). Industry 4.0 allows businesses to exchange information, take actions, and control themselves independently. Cyber-physical systems (CPS), IoT, and Internet of Services (IoS) are the key technology providers of Industry 4.0 (Chauhan, Jakhar, & Chauhan, 2021).

Industry 4.0 provides real-time smart services to minimize resources and maximize the use of existing infrastructures. Another feature of Industry 4.0 is that it is a CPS. CPS can provide autonomous information exchange and initiate necessary action in response to current conditions and mutually independent controls; sensors, machines, components, and IT systems will be interconnected (Lom, Pribyl, & Svitek, 2016).

5. SMART CITY APPLICATIONS IN THE WORLD

Considering the applications of smart cities in the world, it can be said that smart cities are not only based on technology but are evaluated more comprehensively based on the information and communication qualities of the society. The smart city is equipped with applications that use technology, are people-oriented, care about social services, have good governance, and have advanced living standards. Smart cities care about the participation and well-being of their citizens (Batty et al., 2012).

Countries in the world such as China, South Korea, the United States, and Singapore have developed many applications to combat the Covid-19 epidemic due to smart city applications. Reducing the distances between people with noncontact applications, more effective use of robotic technologies in the field of health, using drones or robots in the field of law enforcement, data-based monitoring methods, and isolating Covid-19 patients or contacts are some of the smart city applications used (Budak & veSezgin, 2021).

When the examples of resource savings and efficiency obtained from smart city applications are examined, with smart transportation systems in Los Angeles, it is seen that there is a 35% reduction in stops, 20% in waiting rates at intersections, 13% in travel time, and a 12.5% reduction in fuel consumption (Deloitte, 2016, p. 10). South Korea saved nearly $40 billion in 2010 by digitizing its entire supply cycle. Indonesia, on the other hand, used e-procurement applications in the purchasing processes of local governments between the years 2008 and 2013 for stationery, transportation, etc. It has saved 26 billion dollars in costs in transactions (Collin, Caramazana, & Habeau, 2016).

Considering the common features of smart cities (quality of life, respect for the environment, and smart use of information technologies), some smart cities in the world are as follows:

- Copenhagen, Denmark: It is a city at the forefront with its efforts to zero carbon emissions and investing in renewable energy until 2025 (Civelek, 2018). Copenhagen administration has produced projects for many years to make cycling a tradition in line with its clean environment policies. As a result of policies and investments that have been planned for half a century, 41% of Copenhagen's people cycle today and 63% of children cycle on the way to school. In addition, more than 5000 public charging stations have been put at the disposal of citizens to charge electric vehicles (Karayılmaz & Özker, 2020).

- San Francisco, ABD: The San Francisco administration aims to be a smart city that eliminates all the factors that cause traffic accidents, along with its vision of zero waste and zero emissions.[4] In San Francisco, the SFpark project was developed to better utilize existing spaces. Known as the smart parking meter, SFpark is one of the pilot initiatives of the city of San Francisco. It provides information to citizens about empty parking spaces and occupancy rates in parking lots in real time via wireless communication (SFMTA, 2018).
- Barcelona, Spain: In Barcelona, the garbage level is remotely monitored, and garbage collection services are optimized through sensors placed in the garbage cans. In addition, sensors in the street lamps of the city adjust the light intensity according to daylight.
- Amsterdam, Holland: A smartphone application called Mobypark is used to monitor the parking lots in the city and rent the empty parking spaces (Örselli & Akbay, 2019).
- Dubai, United Arab Emirates: Four different sustainable cities are planned in Dubai, which is the first smart city modeling in the world. There are schools, hospitals, commercial areas, and residential residences in this region. The region provides its energy needs entirely from solar energy, and only electric vehicles are allowed to reach the region. In this context, solar panels have been placed on the residential parking lot that individuals have to charge their vehicles.[5]
- Tokyo, Japan: Efficient devices are used wherever electricity is used in the city, and 100% LED bulbs are used in street and home lighting (Örselli & Akbay, 2019).

6. CONCLUSION

Today, in addition to globalization, urbanization, and climate change problems, urgent and effective solutions need to be found against epidemics. Therefore, urban planning against epidemics will become even more important in the coming years. Needs of cities, face measurements, demographic and economic conditions, global warming, and the developing and changing demands of individuals shape many factors. With limited resources, changing and rapidly increasing day by day meeting these needs in a sustainable way seems possible only with innovative and smart solutions (Örselli & Akbay, 2019). In this context, preferences should be made for the protection of natural resources; materials and methods that will not cause climate change should be focused on. In urban planning, decisions should be made by paying attention to sustainability in housing type choices and population densities. More attention should be paid to the amount of green space per person. A balanced distribution of urban equipment should be made in the physical space. Natural disasters, epidemics, and similar situations should be taken into account in urban design planning (Henden Şolt, 2021).

Smart technologies enable the institutions that provide urban services to work more efficiently, effectively, and economically, in solving the local common problems of the people living in the cities, increasing the quality of urban services, performing participatory management, and providing urban services (Khan, Anjum, & veKiani, 2013, p. 381). These developments, beyond the traditional e-municipal services offered in cities, provide the opportunity to make effective decisions by storing, processing, and transforming data

---

[4]www.sfmta.com

[5]www.thesustainablecity.ae

into information. Integration of smart technologies into urban services makes smart city applications widespread. In this context, solutions based on innovative technologies are used in many urban service areas such as reducing environmental pollution, traffic, and transportation problems; rationalizing resource use; and giving importance to renewable energy sources (Köseoğlu & Demirci, 2018). As a result, the creation of smart cities primarily requires a governance approach based on the cooperation of many actors consisting of the informatics sector, the state, local governments, and the city's stakeholders. In addition, while developing smart city policies, many factors such as technological problems, data security, confidentiality of private data, high costs, qualified manpower, service users, and the level of technology adoption by stakeholders should be considered.

REFERENCES

2020/21 Top 50 Smart City Government Rankings. Retrieved from https://www.smartcitygovt.com. Accessed on August 20, 2022.

Albino, V., Berardi, U., & Dangelico, R. M. (2015). Smart cities: Definitions, dimensions, performance, and initiatives. *Journal of Urban Technology*, *22*(1), 3–21.

Allam, Z., & David, S. (2020). On the coronavirus (COVID-19) outbreak and the smart city network: Universal data sharing standards coupled with artificial intelligence (AI) to benefit urban health monitoring and management. *Healthcare*, *8*, 46.

Angelidou, M. (2014). Smart city policies: A spatial approach. *Cities*, *41*(S1), S3–S11.

Atzori, L., Iera, A., & Morabito, G. (2010). The internet of things: A survey. *Computer Networks*, *54*(15), 2787–2805.

Barrionuevo, J. M., Berrone, P., & Ricart, J. E. (2012). Smart cities, sustainable progress. *IESE Insight*, *14*, 50–57.

Batty, M. (2018). Artificial intelligence and smart cities. *Environment and Planning B: Urban Analytics and City Science*, *45*(1), 3–6.

Batty, M., Axhausen, K. W., Giannotti, F., Pozdnoukhov, A., Bazzani, A., & Wachowicz, M. (2012). Smart cities of the future. *European Physical Journal: Special Topics*, *214*, 481–518.

Budak, S., & ve Sezgin, S. (2021). COVID-19 ile Mücadelede Akıllı Kent Uygulamalarının Önemi: Balıkesir Büyükşehir Belediyesi Örneği. *Tesam Akademi Dergisi*, *8*(2), 521–552.

Bulut, C. (2018). *Bulut Bilişim (cloud computing) Nedir?* Retrieved from https://www.endustri40.com/bulut-bilisim-cloud-computing-nedir/. ErişimTarihi August 19, 2022.

Camero, A., & Alba, E. (2019). Smart city and information technology: A review. *Cities*, *93*, 84–94.

Çelikyay, H. H. (2008). Teknoloji Girdabından Akıllı Şehre Dönüşüm: İstanbul Örneği. In *II. Türkiye Lisanüstü Çalışmaları Kongresi- Bildiriler Kitabı V, ss.* (pp. 1315–1328).

Center of Regional Science. (2007). Retrieved from http://www.smart-cities.eu/download/smart_cities_final_report.pdf. Accessed on August 10, 2022.

Chang, V., Bacigalupo, D., Wills, G., & Roure, D. D. (2010). A categorisation of cloud computing business models. In *Paper presented at the proceedings of the 2010 10th IEEE/ACM international conference on cluster, cloud and grid computing*.

Chauhan, A., Jakhar, S. K., & Chauhan, C. (2021). The interplay of circular economy with industry 4.0 enabled smart city drivers of healthcare waste disposal. *Journal of Cleaner Production*, *279*, 123854.

Chourabi, H., Nam, T., Walker, S., Gil-Garcia, J. R., Mellouli, S., Nahon, K., … Scholl, H. J. (2012). Understanding smart cities: An integrative framework. In *45th Hawaii international conference on systems science (HICSS-45 2012), proceedings, 4–7 January 2012, GrandWailea, Maui, HI, USA* (pp. 2289–2297).

Civelek, A. (2018). Smart cities and smart tourism: Smart city projects and applications in turkey. In *Proceedings of MAC 2018* (pp. 323–332). MAC Prague Consulting Ltd Prague.

Cohen, B. (2014). *Smart city index master indicators survey*. Retrieved from https://smartcitiescouncil.com/resources/smart-city-index-master-indicatorssurvey

Collin, M. N., Caramazana, D. L., & Habeau, J. F. (2016). The impact of smart technologies in the municipal budget: Increased revenue and reduced expenses for better services. In *Uraía Workshop, Lefkoşa*.

Dameri, R., & Cocchia, A. (2013). Smart city and digital city: Twenty years of terminology evolution. In *X conference of the Italian chapter of AIS, ITAIS 2013, Università Commerciale Luigi Bocconi, Milan (Italy)* (pp. 1–8).

Deloitte. (2016). *Akıllı Şehir Yol Haritası*. Retrieved from https://www.sehirsizin.com/Documents/DeloitteVodafone-Akilli-Sehir-Yol-Haritasi.pdf. Accessed on July 5, 2023.

Diez Roux, A. V. (2015). Health in cities: Is a systems approach needed? *Cadernos de Saúde Pública*, *31*(Supp), S9–S23.

Dolinina, O., Pechenkin, V., Gubin, N., & Kushnikov, V. (2018, October). An approach to the garbage collection's simulation in the "smart clean city" project. In *The proceedings of the third international conference on smart city applications* (pp. 3–14). Cham: Springer.

Ejaz, W., & Anpalagan, A. (2019). *Internet of things for smart cities: Technologies, big data and security* (pp. 1–15). Berlin/Heidelberg: Springer International Publishing.

European Union. (2011). *Cities of tomorrow. Challenges, visions, ways forward.* Brussels. Retrieved from http://ec.europa.eu/regional_policy/sources/docgener/studies/pdf/citiesoftomorrow/citiesoftomorrow_final.pdf. Accessed on August 20, 2022.

Gretzel, U., Sigala, M., Xiang, Z., & Koo, C. (2015). Smart tourism: Foundations and developments. *Electronic Markets*, *25*(3), 179–188.

Harrison, C., & Donnelly, I. A. (2011). *A theory of smart cities.* IBM Corporation.

Harrison, C., Eckman, B., Hamilton, R., Hartswick, P., Kalagnanam, J., Paraszczak, J., & Williams, P. (2010). Foundations for smarter cities. *IBM Journal of Research and Development*, *54*(4).

Hashem, I. A. T., Chang, V., Anuar, N. B., Adewole, K., Yaqoob, I., Gani, A., ... Chiroma, H. (2016). The role of big data in smart city. *International Journal of Information Management*, *36*(5), 748–758.

Hermann, M., Pentek, T., & Otto, B. (2015). *Design principles for Industrie 4.0 scenarios: A literature review.* Dortmund: Technische Universität Dortmund. Accessed on August 20, 2022.

Hollands, R. G. (2008). Will the real smart city please stand up? *City*, *12*(3), 303–320.

Jameel, T., Ali, R., & Ali, S. (2019, March). Security in modern smart cities: An information technology perspective. In *2019 2nd international conference on communication, computing and digital systems (C-CODE)* (pp. 293–298). IEEE.

Jucevicius, R., Patašienė, I., & Patašius, M. (2014). Digital dimension of smart city: Critical analysis. *Procedia Social and Behavioral Sciences*, *156*, 146–150.

Karayılmaz, C., & Özker, A. N. (2020). Kamusal Nitelikli Özel Malların Sunumunda Akıllı Şehirler Olgusu: Akıllı Şehir Uygulamalarında Küresel Değişimler. *KMÜ Sosyal ve Ekonomik Araştırmalar Dergisi*, *22*(38), 82–100.

Khan, Z., Anjum, A., & ve Kiani, S. L. (2013), Cloud based big data analytics for smart future cities. In *IEEE/ACM 6th international conference on utility and cloud computing, Dresden, Germany, (9–12 December)*.

Khatoun, R., & Zeadally, S. (2016). Smart cities: Concepts, architectures, research opportunities. *Communications of the ACM*, *59*(8), 46–57.

Köseoğlu, Ö., & Demirci, Y. (2018). Akıllı Şehirler ve Yerel Sorunların Çözümünde Yenilikçi Teknolojilerin Kullanımı. *Uluslararası Politik Araştırmalar Dergisi*, *4*(2), 40–57.

Lai, C. S., Jia, Y., Dong, Z., Wang, D., Tao, Y., Lai, Q. H., ... Lai, L. L. (2020). A review of technical standards for smart cities. *Cleanroom Technology*, *2*(3), 290–310.

Lee, J. H., & Hancock, M. G. (2012). *Towards a framework for smart cities: A comparison of Seoul, San Francisco and Amsterdam.* Stanford Program on Regions of Innovation and Entrepreneurship. Retrieved from http://iis-db.stanford.edu/evnts/7239/Jung_Hoon_Lee_final.pdf. Accessed on August 20, 2022.

Lin, A., & Chen, N. C. (2012). Cloud computing as an innovation: Perception, attitude, and adoption. *International Journal of Information Management*, *32*(6), 533–540.

Lom, M., Pribyl, O., & Svitek, M. (2016, May). Industry 4.0 as a part of smart cities. In *2016 Smart Cities Symposium Prague (SCSP)* (pp. 1–6). IEEE.

Mathur, S., & Modani, U. S. (2016). Smart city- a gateway for artificial intelligence in India. In *2016 IEEE students' conference on electrical, electronics and computer science (SCEECS), Bhopal, India.*

McKinsey Company. (2013). Bilgi Toplumu Stratejisinin Yenilenmesi Projesi. In *Bilgi ve İletişim Teknolojileri Destekli Yenilikçi Çözümler Ekseni Mevcut Durum Raporu, T. C. Kalkınma Bakanlığı Bilgi Toplumu Dairesi Yayını, Ankara.*

Nam, T., & Pardo, T. A. (2011). Smart city as urban innovation: Focusing on management, policy, and context. In *Proceedings of the 5th international conference on theory and practice of electronic governance* (pp. 185–194). Estonia: Tallinn.

Navarathna, P. J., & Malagi, V. P. (2018). Artificial intelligence in smart city analysis. In *2018 international conference on smart systems and inventive technology (ICSSIT)* (pp. 44–47).

Örselli, E., & Akbay, C. (2019). Teknoloji ve kent yaşamında dönüşüm: akıllı kentler. *Uluslararası Yönetim Akademisi Dergisi*, *2*(1), 228–241.

Pelton, J., & Singh, I. (2019). *Smart cities of today and tomorrow.* Cham: Springer International Publishing.

Qureshi, K. N., Rana, S. S., Ahmed, A., & Jeon, G. (2020). A novel and secure attacks detection framework for smart cities industrial internet of things. *Sustainable Cities and Society*, *61*, 102343.

Rabari, C., & Storper, M. (2015). The digital skin of cities: Urban theory and research in the age of the sensored and metered city, ubiquitous computing and big data. *Cambridge Journal of Regions, Economy and Society*, *8*(1), 27–42.

Samih, H. (2019). Smart cities and internet of things. *Journal of Information Technology Case and Application Research*, *21*(1), 3–12.

Sarma, S., Brock, D. L., & Ashton, K. (2000). *The networked physical world.* TR MIT-AUTOID-WH-001, MIT Auto-ID Center.

SFMTA. (2018). *SFpark pilot project evalution: The SFMTA's evaluation of the benefits of the SFpark pilot Project, San Francisco.*

Şolt, H. B. H. (2021). Covid-19 pandemisinde kentlerde yaşanan değişimler ve kentsel hizmetler. *Uluslararası Yönetim ve Sosyal Araştırmalar Dergisi*, *8*(15), 2148–1415.

UN. Habitat. (2021). *UN-Habitat COVID-19 response plan.* Retrieved from https://unhabitat.org/sites/default/files/2020/04/final_un-habitat_covid-19_response_plan.pdf. Accessed on December 10, 2022.

Velibeyoğlu, K. (2019). Akıllı şehirler dönüşümün neresinde? *Kent Dergisi.* Retrieved from https://www.researchgate.net/publication/331802415_Akilli_Sehirler_Donusumun_Neresind. ErişimTarihi September 15, 2022.

Visvizi, A., & Lytras, M. D. (2018). Rescaling and refocusing smart cities research: From mega cities to smart villages. *Journal of Science and Technology Policy Management*, *9*(2), 134–145.

WHO. (2021). Retrieved from https://www.who.int/. Accessed on November 10, 2022.

https://www.un.org/sustainabledevelopment/cities/

https://www.sfmta.com/projects/electric-mobility-san-francisco

https://www.thesustainablecity.ae/inside-the-city/

SMART DESTINATIONS AND ECO-FRIENDLY PRACTICES

Cenk Murat Koçoğlu, Burak Pinaroğlu and Emrah Yaşarsoy

ABSTRACT

Globalization, along with its technological developments, has brought important changes and developments in all sectors and led to innovation and modernization. In the tourism industry, the concept of smart destination, derived from the concept of smart city, has become widespread to meet the changing needs of tourists. In this context, smart technologies have started to be used in destinations, but more than technology in destinations is needed in terms of sustainability. To ensure sustainability in tourism, these smart technologies must also be eco-friendly. Smart destinations and eco-friendly practices are important for the sustainable development of the tourism industry. The use of technology in destinations can increase sustainability and efficiency, while using eco-friendly practices can protect and preserve the natural environment. In this study, first of all, smart destinations and the characteristics of smart destinations are discussed. Then, smart tourism and related eco-friendly practices are argued. Following this, information about the future of smart destinations is given, and assumptions are mentioned. Afterward, problems and solutions are discussed. Finally, the study has been completed with the conclusion part. The findings of the study revealed that the concept of green especially comes to the fore in smart destinations, that there are studies to prevent all kinds of pollution (air, noise, etc.), and applications such as smart parking systems, smart lighting systems, augmented reality (AR), and virtual reality (VR) are given importance.

Keywords: Tourism; smart tourism; smart destinations; eco-friendly practices; green; sustainability

1. INTRODUCTION

Smart destinations are innovative concepts rapidly spreading in the tourism sector and aiming to improve the experience of tourists. *Smart destinations* are technologically advanced tourism centers that offer the information and services tourists need when planning their trips. These destinations reduce the difficulties that tourists face when

Smart Cities for Sustainability
Advanced Series in Management, Volume 32, 15–32

Published under exclusive licence by Emerald Publishing Limited
ISSN: 1877-6361/doi:10.1108/S1877-636120230000033003

making their travel plans more enjoyable and efficient. (For example, smart destinations provide tourists with information and services on transportation, accommodation, food, excursions, and activities) This way, tourists can plan their travels more easily and use their time more efficiently. Smart destinations also provide economic growth by increasing investments in the tourism sector (Buhalis & Amaranggana, 2013).

Eco-friendly practices are technological solutions that emphasize the importance of tourism activities in protecting nature and the environment. These practices support the tourism industry to cause less damage to the environment. While these practices reduce the impact of the tourism industry on the environment, they also increase the tourists' experience of discovering and recognizing nature (Lamsfus, Martin, Alzua-Sorzabal, & Manzenera, 2015).

This chapter will examine smart destinations and eco-friendly practices in terms of the tourism sector. In particular, information will be given on how smart destinations are applied in the tourism sector and which technological solutions are used. Likewise, how eco-friendly practices are applied to tourism activities and which technological solutions are used will also be explained. The place of these practices in the tourism sector and the opportunities and challenges they present will also be discussed. Likewise, how eco-friendly practices reduce the impact of tourism activities on the environment and how tourists increase their experience of discovering and recognizing nature will also be examined.

To sum up, firstly, general information about smart destinations and eco-friendly practices will be given. Then, the relationship between smart tourism and eco-friendly practices will be explained. After that, the development of information and communication technologies (ICTs) and some applications widely used in the tourism sector will be mentioned. Finally, in light of these applications, some predictions will be made about the future of smart destinations, and some of the problems and solutions they bring will be mentioned.

2. SMART DESTINATIONS AND THE CHARACTERISTICS OF SMART DESTINATIONS

The tourism industry is the perfect strategy for starting and accelerating development in developed and developing countries. Despite the recent difficulties, the contribution of travel and tourism to gross domestic product (GDP) increased by USD 1 trillion (+21.7%) to reach USD 5.8 trillion in 2021, while the share of the sector in the entire economy increased from 5.3% in 2020 to 6.1% in 2021. According to World Travel and Tourism Council (WTTC) forecasts, strong growth is expected over the next decade. Travel and tourism GDP is expected to grow by an average of 5.8% annually between 2022 and 2032, outpacing the growth of the general economy (2.7% per year). It is also estimated that travel and tourism GDP could return to 2019 by the end of 2023 (WTTC, 2022).

Through a carefully prepared and implemented tourism strategy, natural resources in a country can be transformed into much-needed economic capital (UNWTO, 2015). Furthermore, if the result is successful in the carefully prepared and implemented strategy, a cyclical relationship can be developed between growth rates in the overall economy and growth in the tourism industry. Therefore, countries actively develop destination management techniques and strategies to create a positive perception of themselves and increase rates of tourism-related revenues (Lee, Rodriguez, & Sar, 2012).

ICTs have deeply influenced the interaction, and the way businesses interact with stakeholders of institutions and organizations in the tourism sector (Buhalis & Law, 2008). Technological developments cause major changes by affecting tourism markets, management styles, and competitive strategies. With the development of information technology, tourists receive support in various ways, such as easy access to information, obtaining information, decision-making, travel experience, and planning (Neuhofer, Buhalis, & Ladkin, 2012). E-tourism is just one of the consequences of the tourism industry intertwining with technology. Due to the widespread use of ICTs, e-tourism has brought innovations to tourism services and led to various changes in the use of services (Buhalis & Amaranggana, 2013).

With the intensive use of information in the tourism sector and considering the ICT dependency of the sector (Benckendorff, Sheldon, & Fesenmaier, 2014; Gretzel, Sigala, Xiang, & Koo, 2015), it has become inevitable to use the term "smart" for tourism as well. There are various global institutional supports to develop and support smart tourism. For example, South Korea and China have provided significant funds to initiatives aiming to contribute to smart tourism to form smart tourism infrastructure (Hwang, Park, & Hunter, 2015). Smart applications aim to anticipate the needs of tourists who visit the destination and improve the tourist experience by providing location-based and customized services with a variety of recommendations to tourists. Additionally, thanks to smart applications in destinations, tourists' travel experiences are shaped significantly, which can impact other potential tourists by increasing satisfaction (Yoo, Sigala, & Gretzel, 2015).

Smart tourism generally refers to using ICT to improve the quality of life and the experience (Buhalis & Amaranggana, 2013). The main focus of smart tourism is tourists, even if they travel for different purposes. Hence, using ICT and data collection, processing, and exchange, a smart business ecosystem, smart tourism mobility, and smart structuring of destinations allow tourists to benefit from smart experiences (Gretzel et al., 2015). Furthermore, smart tourism affects and develops the search for the technology-based tourism market day by day (Li, Hu, Huang, & Duan, 2017).

When it comes to smart destinations and smart mobility, since the principles adopted and applied in rural areas, urban areas and smart destinations take into account not only the residents living in the city but also the tourists, the concept of smart city and smart destination are closely intertwined concepts (Buhalis & Amaranggana, 2013). From this point of view, technology in cities encourages mobility in tourism.

There are fundamental differences between smart tourism and e-tourism in the context of destination. While e-tourism only exists in the digital area, smart tourism applications, which have an important place in smart destinations, are more in the physical area. While applications such as smartphones, sensors, and new technologies form the basis of a smart tourism destination, it is seen that websites form the basis of e-tourism (Gretzel et al., 2015).

Research on smart destinations shows that data provided by the Internet of Things (IoT) and cloud databases allow the identification of tourists' needs and desires and the intelligent analysis of tourist preferences. It demonstrates that these activities can improve tourist experiences by affecting them positively (Huang, Yuan, & Shi, 2012). In addition, integrating technological applications into destinations can increase tourist satisfaction by providing tourists with various information about the destination to be traveled to (Yoo, Goo, Huang, Nam, & Woo, 2017). In general, using and disseminating various technologies in smart destinations benefit tourism and tourists differently.

Smart and Sustainable Destinations (SSD) have significant potential for tourists in terms of providing innovative and high-end amenities in various ways. Governments and various tourism organizations encourage technological initiatives to develop tourism destinations and become smart destinations (Li et al., 2017). Additionally, destinations transformed by technology should be considered in terms of sustainability. For this reason, sustainability should be considered while making efforts to transform tourism destinations into smart tourism destinations.

Sustainability in tourism generally has five dimensions (economic, social, environmental, political, and cultural). When these dimensions are considered, separate policies and plans should be made and carried out in each destination for each dimension. As in many sectors, sustainability should be highly important to gain a competitive advantage in tourism. These principles, which are at the foundation of sustainable tourism, are a holistic approach to justice, equality, and a long-term tourism future in the global ecosystem. From this point of view, while providing various conveniences to tourists in destinations transformed with technology, sustainability should be considered (Ritchie & Crouch, 2003; Soares, Ruiz, & Baidal, 2022).

3. SMART TOURISM AND ECO-FRIENDLY PRACTICES

Rapid population growth in cities has created some problems in terms of management and resource utilization. There is a need for some innovations that will provide convenience to people to meet the needs of the increasing population. Today, web-based applications are used to find solutions to these problems. With the spread of the internet, information about cities has been put into service with artificial intelligence (AI) and web-based applications to meet human needs more easily and increase cities' efficiency. In parallel with these services, the concept of a smart city has emerged (Kulualp & Sarı, 2020). The implementation and spread of these services in cities and the acceleration and spread of the process in various sectors have led to radical changes. The tourism sector is one of the sectors most affected by these changes because, with the emergence of new services, the expectations and needs of the region's people and tourists have changed simultaneously.

Some innovative practices have been introduced in tourism destinations as a solution to the changing needs and demands of tourists and local people. These innovative changes appear as smart tourism. Smart tourism has also brought innovations to cover all tourism stakeholders. It has become mandatory to consume sociocultural and environmental resources, especially important for tourism, by considering the balance of protection and use. The use of resources by considering the protection–use balance also contributes to understanding sustainable tourism. In brief, thanks to the concept of smart tourism, an integrated system was established to cover all tourism sector stakeholders.

Today, environmental problems have revealed important dimensions that affect every moment of our lives. Negativities such as increasing environmental pollution, decreasing biodiversity, areas facing the danger of extinction, etc., prevent future generations from living in a safer environment. The most important factor in such negative events is being seen as human beings. Unconscious consumption and irresponsible behaviors have created and continue to create serious problems in the environment associated with the increasing population. For example, sea coasts and forested areas were opened for construction, and historical and natural resources were unconsciously consumed under the name of tourism activities. Reduction in water resources, soil fertility, and food resources has begun to be seen parallel with global warming. All these and similar problems are seen as a serious threat to human life and national economies. It is thought that understanding sustainability in all areas of our lives will solve these problems (Çetin & Karataş, 2022).

While explaining "eco-friendly tourism practices" in the context of sustainable tourism, the concepts of smart tourism and ecotourism should be taken into account because sustainable tourism means protecting and using destinations with their environmental, social, and economic dimensions (Janusz & Bajdor, 2013, p. 524), can be confused with the concept of ecotourism. In sustainable tourism, much more is explained than protecting nature, while in ecotourism, this situation expresses a narrower content compared to sustainable tourism (Aydın & Yılmaz, 2013, p. 10). Smart tourism, on the other hand, creates a link between physical and digital, takes the personalization of touristic movements to the next level, and tries to increase sustainability and efficiency effectively (Gretzel, Reino, Kopera, & Koo, 2016).

Ensuring sustainability is as important as attracting tourists. In this context, sustainable and eco-friendly smart tourism practices significantly contribute to the protection of destinations. Smart and eco-friendly practices that are popularly used in the tourism industry are the IoT, AI (robot technology, chatbot, travel bot), virtual reality (VR), Wi-Fi connection, wearable devices, mobile applications, web technologies, VR, augmented reality (AR), audio technology, waste management systems, smart parking systems, air control systems, energy monitoring systems, and smart lighting systems. These practices speed up processes while increasing quality and performance. In addition, it also reduces costs and environmental damage in the long run. Furthermore, thanks to these smart and eco-friendly practices, customer satisfaction can be increased by providing more personalized services to customers (Avci, 2019; Kečkeš & Tomičić, 2017).

When creating sustainable smart destinations, it is generally difficult to understand what is effective in a destination and why tourists prefer it. For this reason, while increasing efficiency with technological applications' support in destinations, efforts should be made to simultaneously make the destination more livable. The negative consequences arising from the interaction in the destinations should be minimized, and a protective approach should be taken (Encalada, Boavida-Portugal, Ferreira, & Rocha, 2017). Based on the smart destination paradigm, smart tourism applications can address both tourists' environmental behavior and consumption. While creating these applications, destination carrying capacity, sustainability, and extreme tourism should also be considered (Fistola, Anna, & Rocca, 2013). Briefly, integrated eco-friendly practices should be used and expanded to ensure sustainability in smart tourism destinations.

3.1 Information Communication Technologies

ICTs form the basis of the digital economy. The development and progress of this technology have prepared a favorable environment for the development of cities (Komninos, 2002). The concept of "smart city," based on people's welfare and sustainable development, has emerged in conjunction with integrating these technologies into cities. Technological applications, which were initially applied in urban environments, started to be applied to tourism destination management after the smart city approach and the "smart tourism destination" was derived (Law, Buhalis, & Cobanoglu, 2014). These developments and the spread of the internet toward the end of the 1990s, the increase in the use of mobile devices and social media have created a digital revolution in the tourism sector (Benckendorff et al., 2014; Buhalis, 2003; Buhalis & Law, 2008; Xiang, Tussyadiah, & Buhalis, 2015). Adopting ICT applications by tourists in a short time has affected tourism marketing and management and caused changes (Gretzel, Yuan, & Fesenmaier, 2000). These applications have become the main competitive factor for tourism destinations (Buhalis & Matloka, 2013). While ICT applications are being implemented in destinations, reducing environmental damage has started to be taken into consideration, apart from tourist satisfaction and income generation. In this context, it has become mandatory to use eco-friendly practices to reduce the environmental damage caused by tourism. Otherwise, the depletion of resources is an inevitable end.

Additionally, environmental values, sustainability of resources, and service quality affect tourists' purchasing preferences more and more and are seen as an important factor in destination choice (Swarbrooke & Horner, 2001, p. 36). As a result of increasing information sharing along with technological progress, there has been an increase in the expectation level, culture level, and consciousness level of tourists. Besides, environmental sensitivity has developed against tourism activity, where the environment is at the forefront. With the increasing environmental awareness, it is thought that the development and implementation of eco-friendly practices will become more important day by day at the point of demand and sustainability of the environment (Ayaş, 2007).

The smart city concept is still in its growth phase, so there is no concrete concept formulation and standardization yet (Dohler, Vilajosana, Vilajosana, & Llosa, 2011; Schaffers et al., 2011). Since there is no clear standardization and specific limitation, and eco-friendly practices in tourism destinations are discussed in this section, the smart applications most associated with the tourism sector will be listed and explained under subheadings.

3.1.1 Traffic Control Systems

Due to the increasing population in cities, transportation problems have also been growing significantly. Urban transportation problems, especially in big cities, are vital to people. Transportation disruptions affect social and economic life negatively. Hence, it is necessary to solve the road transportation problems in dense cities and to manage them more smartly. Problems such as insufficient infrastructure, lack of technology, intersection problems, and traffic signaling that cause these transportation problems can be solved by developed traffic control systems. More information on traffic flow, density, delays, inadequacies, etc., can be gathered about the problems using these technologies. Easier solutions can be found with this information obtained. Ultimately, this situation can positively contribute to city life (Gordon et al., 1996; Mandhare, Kharat, & Patil, 2018).

3.1.2 Air Quality Monitoring Systems

Today, air pollution is one of the biggest problems facing the world. The most important reason for the increase in air pollution, which negatively affects the ecosystem, climate, and human health, is the increased rate of pollutants in the atmosphere. The air quality of the cities is adversely affected by harmful gases and particulate matter. These negative situations also cause health problems. The factors that threaten air quality should be determined and eliminated or reduced for the living standards and health of people living in environments and areas with poor air quality. Various air quality monitoring systems have been developed with technological developments. Thanks to these systems, the factors that affect the air quality of the cities negatively, which are people's living spaces, can be determined and eliminated by taking precautions for them. When the same problems are in question for an attractive destination, the decrease in air quality, which is the basic product of human life, can affect destinations negatively (Marinov, Topalov, Gieva, & Nikolov, 2016; Üçgün, Gömbec, Yüzgeç, & Yalçın, 2020). When the same problems come into question for an attractive destination, the decrease in air quality, the basic life product of human beings, can affect destinations negatively.

3.1.3 Noise Monitoring Systems

There are discussions on detecting noise pollution within the scope of smart cities. Zappatore, Longo, and Bochicchio (2016) and Rana, Chou, Kanhere, Bulusu, and Hu (2010) focused on monitoring systems for investigating noise pollution. As smart city systems, the noise inspection system measures the noise that occurs in an area and at a certain time in the city. This measurement is made based on a certain period by analyzing the data. The noise measurement measures and analyzes the accidents that occur in smart destinations and the noise outside the normal flow that occurs in different city conditions in real-time. When noise is evaluated in terms of tourism, a noisy tourism environment was found to impact tourist experiences (Mace, Corser, Zitting, & Denison, 2013; Merchan, Diaz-Balteiro, & Solino, 2014; Taff et al., 2014).

3.1.4 Energy Monitoring Systems

In smart cities, the system of monitoring the energy consumed by individual houses is the equivalent of the smart energy monitoring system (Al-Hader, Rodzi, Sharif, & Ahmad, 2009; Zhou, Wu, & Yan, 2017). This smart energy consumption monitoring system involves using resources such as electricity, water, and gas, distributed via pipelines to every citizen in a city. Thanks to the sensors placed in these pipelines that deliver resources to citizens, the amount of resources transported is monitored. The amount of energy resources consumed by individuals is measured with the measured resource amount data with the support of sensors. In addition to these data, available systems such as traffic lights, heating systems, and security cameras of buildings in the city are monitored within the city's general lighting systems. In the light of the data collected on the axis of this system, the authorities in the cities can plan the resource production of the resource-generating facilities and the circulation of this production to create the possibilities of creating a better future with energy sustainability as well as saving energy (Krishnamurthi, Nayyar, & Solanki, 2019).

3.1.5 Waste Management Systems

It enables the waste management system to monitor waste bins in certain areas in smart cities (Anagnostopoulos et al., 2017). It is also very useful to ensure the cleanliness of the cities in an economical and eco-friendly manner by implementing smart waste management in this way. Thanks to the sensors placed in the garbage cans in the cities, the weight of the garbage cans is detected, the garbage trucks in charge of emptying the garbage cans are warned, and the garbage is transported for disposal. In addition, waste management systems streamline the collection path for emptying bins, and the cost of all these services is greatly reduced. These management systems prevent odor and visual pollution in cities and minimize environmental damage (Tomar & Kaur, 2019).

3.1.6 Smart Parking Systems

In today's existing parking systems, manual systems depend on the experience of the drivers and the chance of free parking space (Pham, Tsai, Nguyen, Dow, & Deng, 2015). Considering the cities with high population density, the existing parking system is time-consuming. Recent research on this topic has provided information on the use of different wireless technologies such as radio frequency identifiers (RFID), near frequency communication (NFC), and the internet for solving parking problems and smart parking solutions (Lee & Chang, 2008). The necessary information, such as the GPS locations of the vehicles, the distance of the vehicle from the parking areas, and the parking costs, are available in the smart parking system. In the smart parking system, when drivers need to park their vehicles, smart parking services can be easily accessed via phones or tablets. With this service, drivers can reserve a parking space for their vehicles near the location they want to arrive using phone applications to reduce parking difficulty. In this way, the reduction of parking time with the smart parking system and the reduction of the amount of CO_2 released into the air in direct proportion to this are also supported. With the parking service, the vehicles create less traffic congestion, and as a result, the satisfaction of the individuals in the city is another advantage. In addition, this smart parking system will shorten parking times and minimize the negative effects of the parking problem in the city, especially for disadvantaged individuals (e.g., elderly and disabled) in the community, by preserving parking space (Fahim, Hasan, & Chowdhury, 2021; Khanna & Anand, 2016; Krishnamurthi et al., 2019).

3.1.7 Smart Lighting Systems

While the share of lighting systems in global energy consumption is 19%, 6% of greenhouse gas emissions and environmental pollution are caused by lighting systems. In addition to increasing energy savings through smart wireless sensor networks (WSNs) and smart light-emitting diode (LED) luminaires to be used in lighting systems, potential benefits can be provided for cities with continuous monitoring of the system and controlled detection services (Shelby, Hartke, & Bormann, 2014). Smart lighting systems offer different options according to the current situation of cities. These options provide smart system possibilities in the smart city concept of smart lighting systems that dim the public lights according to the current conditions in different weather conditions and the density of people in the city. ZigBee network is seen as the most suitable network to be used in the lighting system due to low power usage, easier use, and a more secure network in light of recent research. Smart lighting systems in public areas can be monitored remotely, controlled, and directed by authorized staff through interactive software interfaces and luminaires connected with the ZigBee network mesh most suitable for the system (Budhiputra & Putra, 2016; De Paz, Bajo, Rodríguez, Villarrubia, & Corchado, 2016). Energy saving can be achieved by using smart lighting systems in tourist areas (hotels, recreation areas, airports, food and beverage businesses, etc.). This can provide both environmental and economic benefits.

3.2 Green Energy

Climate change and the effects of global warming are one of the most important problems to be solved for sustainable development, including the tourism sector. These problems have important impacts in terms of tourism development and a sustainable economy. The energy demand is increasing at the same rate as the increasing demand for tourism destinations. Although there are seasonal densities in destinations, the demand, in general, is increasing regularly. These problems can be tried to be solved by the use of green energy in destinations. Green energy is an essential element for the sustainability and development of destinations. Furthermore, green energy transformation in destinations is considered necessary to protect the environment, provide maximum benefit for many years, and meet the increasing demand (Bojnec & Papler, 2009; Tomar & Kaur, 2019). When the potential consumer demands of destinations are examined, it is seen that the destinations implementing green practices are one step ahead of other destinations with the increasing awareness of the environment. From this point of view, green energy and the eco-friendly transformation of destinations are considered important.

3.3 Green Infrastructure

Green infrastructure is a system that focuses on the sustainable development of cities with an environmentally friendly approach. The sustainability of destinations depends on the use of smart and energy-efficient applications. Power supply is very important for uninterrupted continuation of life. The use of renewable energy sources such as sun, water, and wind alleviates the burden on conventional energy sources. Renewable energy is more environmentally friendly and cheaper than other energy sources. The main purpose of green infrastructure is to contribute to ecological sustainability while providing economic benefits. Besides, these applications provide positive benefits to smart destinations and cities. In general, the idea of green infrastructure is to provide a more sustainable future by making practical contributions to the development of a destination or city (Pan et al., 2018).

Green infrastructure applications also have positive social effects such as providing new employment areas for individuals, obtaining greener and more attractive areas, improving human welfare and health, making energy and transportation solutions more integrated, and improving tourism and recreation opportunities (Parlak & Atik, 2020).

3.4 Green Applications

The negative effects of climate and global warming are seen in destinations. Sustainable tourism activities in destinations have gained importance to reduce these negative effects and gain a competitive advantage (Dinç, Kömür, & İlban, 2021). In this direction, destinations have made significant progress toward becoming green destinations by seeking innovative and eco-friendly practices. It is aimed to reduce the damage caused by tourism to the environment, thanks to these developments (Bayar, 2022). To achieve this goal, an organization called Green Destinations (GD), based on the principles of nonprofit and sustainability, was established within the framework of the criteria standardized by the United Nations, and is present in more than 80 countries. This organization gives certificates and awards to destinations that meet the criteria set worldwide (Green Destinations, 2022). There are also some labels and certificates issued by various institutions:

- *Blue Flag:* The Blue Flag eco-label for marinas and beach destinations is an international environmental award.
- *Green Globe:* It is an eco-label program widely used in the world. It is given to tourism and travel businesses that meet certain environmental standards.
- *Green Key:* This practice is based on the training of the hotel and its customers to minimize the damage to the environment and to raise awareness about the environment. It shows that businesses with this application are in harmony with nature.
- *European Union Eco-Label:* It is a label that is valid in all EU member countries, based on products and services offered with a high standard and environmental focus. It has a wide range of products, covering buildings, businesses, cleaning products, and lighting systems.

- *International Environmental Management Standard - ISO 14001:* All environmental issues related to waste management, soil pollution, efficient use of resources, water and sewerage problems, climate change, and environmental adaptation are taken into account.
- *National Labels:* Labels or practices that may be different in each country, but are country-specific as environmental standards. For example; in Turkey, there are applications such as "green star and white star" that are valid throughout the country.

In addition to the applications and examples listed above, there are others. All of these applications have a common basic purpose. This common purpose can be expressed as reducing the damage to the environment, increasing environmental awareness, and encouraging positive practices toward the environment.

3.5 Low Carbon

Human beings leave a trace in the world in the course of their life with all their production and consumption activities. Considering the waste created by all the resources we consume throughout life (food, clothing, resources used for transportation and heating, and all consumer goods), it is seen that the trace left by human beings in the world is not very small (Han & Kaya, 2013). Different indicators are used for more sustainable development. One of them is known as an ecological footprint.

Greenhouse gas emissions from energy sources used today constitute two-thirds of total emissions. From this point of view, it is vital to pass to low-carbon economies within the scope of combating climate change. The transition to a low-carbon economy refers to a comprehensive process such as transforming energy resources into green energy, changing current consumption habits, preventing waste, and accelerating the forestation process. Low-carbon programs and practices are carried out in different ways in different countries for a more sustainable future. In this context, it is aimed to minimize the effects of human beings on the environment by reducing carbon emissions, and it is aimed to use green, renewable energy sources (Topçu, 2018).

4. THE FUTURE OF SMART DESTINATIONS

It is no longer possible to talk about a smart destination without technology and the IoT. When it comes to smart destinations, IoT is one of the applications that has attracted the most attention recently and will increase its popularity in the future. Construction companies incorporate IoT devices and solutions into their architectural plans to create cost-effective, qualitative, and self-sustainable infrastructure. The governments of both developed and developing countries are working on public–private partnerships to bring IoT solutions to the creation of smart destinations (Aydın, 2021). Every smart city should have the technology (SmartCity, 2017). Transmission technologies and infrastructure are key technologies supporting the framework of a smart city and are considered to be the most important issues in the future. It is difficult to transition to a smart city without a technology infrastructure. IoT devices can collect and analyze data to make smart destinations work better. This makes IoT a data collection technology. Smart destinations can use these data to improve infrastructure, public services, etc. (Zigurat, n.d.). IoT refers to a rapidly growing network of computing devices with built-in sensors and software to connect and share data. It enables billions of devices and objects equipped with smart sensors to connect, collect, and send information in real time. These data are transferred to central control systems via wireless communication. These, in turn, manage traffic, reduce energy use and improve a wide range of urban operations and services (ITU, 2019). The IoT may have some answers. Created as part of the smart technology movement, IoT enables various objects and entities to communicate with each other over the internet. It reveals a wide range of technological innovations that can improve public transport, provide accurate traffic reports, or provide real-time energy consumption data by creating a network of objects capable of intelligent interactions (Ellsmoor, 2019). Smart cities IoT technology will affect every application area related to human life. Similar to the famous industrial revolution that has had a positive impact on everyone's life, IoT will impact humanity in several different ways. First of all, the application of IoT is not limited to any industry and the fact that it can provide wireless connectivity to all physical media. Thus, it makes it possible for all physical entities to communicate with each other without the need for any third-party intervention. We can talk about human-to-human interaction, human-to-machine, or machine-to-machine transfer of information. Thanks to IoT technology, all physical and natural beings in the world will show more comfortable and intelligent leadership (iotWorm, 2015). The basic infrastructure, various sensors, support technologies, and environmental background of a smart city are essential. IoT is considered one of the most important elements for successful implementation (Jin, Gubbi, Marusic, & Palaniswami, 2014). IoT is an indispensable technology for smart city initiatives. Objects of IoT devices, sensors, and applications collect data that makes technology solutions effective. For example, smart water meters report water quality, and usage, show leaks or possible contamination at the water company. Big data analytics are necessary for smart city initiatives to work. IoT generates large datasets that must be analyzed and processed to implement smart city services. Big data platforms, which are part of the city's ICT infrastructure, have to sort, analyze, and process the data collected from the IoT. Smart destinations around the world are implementing initiatives to promote greener and safer urban environments with cleaner air and water, better mobility, and efficient public services. These initiatives are supported by technologies such as the IoT and big data analytics, which form the basis of the smart city model (Here, 2021). IoT refers to devices connected to the network or the internet, such as devices, thermostats, monitors, sensors, and portables that can measure, store, and transmit information (Tech Republic, 2021). Big data and business analytics, IoT, collects large amounts of data with its built-in sensors and transfers them to business intelligence and analytics tools. These data are then used to solve problems and provide value-added services to customers (Kim, 2016). Large amounts of data are needed to protect, retrieve, store, and analyze information. Hence, big data includes comprehensive information generated by IoT technology (Bhatt, Dey, & Ashour, 2018). Considering approximately 50 billion devices, it is necessary to store, recall, and analyze data (Botta, de Donato, Persico, & Pescapé, 2016). The networking of people, things, and data is known as the IoT, and home automation systems, among others, are used as tools for people's well-being through smart parking (Batalla, Mastorakis, Mavromoustakis, & Pallis, 2017).

There are significant developments in the fields of innovation, technology, sustainability, and accessibility, which are the cornerstones of a smart destination. For example, in March 2021, Málaga published the Tourism Plan for 2021–2024, targeting post-pandemic restructuring with more than 50 projects (Karakaş & Atay, 2021). In this context, the main smart applications planned to be implemented in the future are as follows (Malaga Turismo, 2021, p. 40):

- creating a map that identifies the current information needs and the resources available to meet these needs,
- establishing a customer relationship management (CRM) infrastructure to collect data for the relevant market segment,
- improvement of data provided by the tourism observatory,
- establishing a new mechanism (technological surveillance, planning and opportunity identification, and follow-up activities) to identify and implement new technologies suitable for the region,
- málaga City Council's nomination to enter the UNWTO International Network of Sustainable Tourism Observatories (INSTO) and making Málaga a world benchmark in tourism sustainability,
- developing a reliable "tourism big data system" that can be integrated with other projects, collected through social media, point-of-sale system (POS), mobile devices, sensors, and applications,
- designing a smart vehicle with a digital instrument panel that offers Málaga simultaneous decision-making.

In addition to the tourism plan, another smart destination application, which is intended to be developed by Málaga in the coming years, is a VR game similar to Pokemon Go that allows people to explore the city and its innovations (Independent, 2021). Thus, it is aimed to provide tourists with an unforgettable experience both for the past and the present (Karakaş & Atay, 2021).

Cities of the future should remember about recycling processes for liquid and solid wastes beyond being just smart destinations by taking into account the logic of green technology. Additionally, sewerage, water, electricity, gas, and communication networks should be considered to be integrated with roads. While roads are considered together with the subway and light rail system as a whole, bicycle and walking paths should not be neglected in this system. All kinds of roads and vehicles related to transportation of cities and buildings should be designed appropriately for the disabled and the elderly. In the future, it should be planned that the disabled tracking areas in the city and buildings will also be used as a way for service robots. Starting from the center, the architecture of the buildings should be designed with a floor easement descending outward. These designs should be created with a structure in the logic of a spider web by taking into account the science of biomimetics, which is the science of imitating nature. The interior of the circle at the most central point should be left as a completely green area. The settlement parcels immediately following should be created as green areas and residential areas with the logic of circular checkers. Thus, the residential areas' front, back, and sides will be green. The parcels within the settlement parcels and consecutively should be created with a floor servitude decreasing from the center to the outside, as well (Alkan, 2015).

Some recent studies indicated to which point smart destinations should pay more attention in the future. Radojević, Lazić, and Cimbaljević (2020) stated that in the future, the full integration of various sensing technologies along with strategies for gathering and transforming data, and contextual rescaling will help us discover all the critical relationships and good and bad causalities that exist between tourists and fully sensor-equipped smart tourism destination.

Jeong and Shin (2020) emphasized that regardless of tourists' age, the majority of tourists used at least two smart tourism technologies (STTs) from their smartphones while traveling to the destination. Tourists' usage rate of STTs at destinations expects to increase in the future. To accommodate tourists' needs and their preferred environment for STT usage, destination marketers should ensure high performance on at least three key STT attributes, namely, interactivity, personalization, and informativeness.

5. PROBLEMS AND SOLUTIONS

The problems and difficulties that arise in the destinations are also increasing in parallel with the increasing population and business fields. Since the existing transportation infrastructure cannot handle the increasing number of vehicles with the population, the time spent in traffic is increasing. Job losses increase, productivity decreases, transportation costs increase, and as a result of traffic accidents, loss of life and property and health costs increase (Alkan, 2015). It has become the main problem of many destinations where energy is expensive and water sustainability cannot be fully achieved. Also, air pollution, problems in health services, unplanned urbanization, infrastructure and superstructure problems, and security problems are among the main problems. At this point, smart destination applications have solving features. The readiness of the technologies to be used in establishing the cities of the future, on the other hand, is hopeful and encouraging. Some of these technologies that are ready to use can be listed as: "Geographic Information System (GIS), City Digital Maps, Transportation Information System, Smart Signaling, Smart Meter Systems, Information Kiosks, Fast Internet Infrastructure, Wireless Internet Infrastructure 3G – 4G GSM Technology, Smart Building Architecture, Citizen Address, and Population Information System, Waste Management System, Biological, Mechanical, and Rail System". Renewable energy, which is expected to meet a large part of the energy of the cities in the future, needs to be developed in the storage phase. However, it is thought that this technology, which is developing rapidly, will come to the storage capacity to meet the needs in the medium term. In this case, considering that Industry 5.0 has started to become widespread, applications with more advanced technology such as autonomous transportation, intravehicular communication, and water monitoring system which improve efficiency and productivity by improving the collaborative interactions between humans and machines may be needed.

Karayilmaz and Özker (2020) mentioned the difficulties encountered in smart city formation in their research. In this context, the participants mostly stated that there was financial insufficiency. They argued that the weak interinstitutional cooperation also affected this situation negatively. It has been determined that institutions do not want to share this application with other institutions, especially when they have technological applications. Furthermore, GIS infrastructure, expertise in ICTs, citizen's involvement in the process, knowledge, and experience of smart cities, the existence of standards, political elements, legal regulations on the subject, integrated and innovative approaches, visionary leadership, and prediction of implementation risks smart city stated as the most important difficulties in their implementation (Komninos, 2006).

Alawadhi et al. (2012) revealed the common features and challenges of smart city initiatives, with the findings they obtained from interviews with government officials and managers within the scope of smart city initiatives. Firstly, it was determined that budget constraints emerged as the most fundamental challenge in the sustainability of smart city initiatives. Secondly, it has been determined that the development of social media applications and mobile communication technologies offers new opportunities within the scope of inclusion of society in smart enterprise applications (Barutçu, 2021).

Moreover, a key challenge lies in evaluating the effectiveness of smart destinations and measuring their level of "intelligence" (Femenia-Serra, Neuhofer, & Ivars-Baidal, 2019;

Rucci, Moreno-Izquierdo, Perles-Ribes, & Porto, 2021). Williams, Rodriguez, and Makkonen (2020) advocated more thorough understanding of smart destinations by stating that "An imprecise understanding of smart tourism as a development tool (both in conceptual and practical terms) can lead to ambiguity in its use in tourism strategies for destinations" (Sorokina et al., 2022).

While the verbal signs hinted at residents and tourists, the visual signs (logos, videos, infographics) completely neglected residents and communicated a clean and idealized version of a destination with faceless, lonely tourists in the periphery. Substantially, smart destinations might run the risk of alienating residents and tourists alike and remain unable to address challenges such as overtourism and anti-tourism sentiment (Gretzel & de Mendonça, 2019). Joss, Sengers, Schraven, Caprotti, and Dayot (2019) suggest that public discourse surrounding smart cities indicates that they are facing critical junctures. Thus, it is important to add an academic voice to this discourse to provide alternatives or shape existing pathways to smart development.

Although smart sensors play an increasing role in the acquisition of data to enable smart service optimization, the understanding of their spatial capabilities to facilitate the tourism process as they interfere with and influence the tourism system has received only minimal attention (Radojević et al., 2020). In essence, smart tourism destinations face the challenge of the lack of geospatial understanding, which is primarily related to the different spatial coverage of smart technologies. This implies the need to upgrade existing and develop new urban innovations where smart systems help strengthen social and physical infrastructure to improve the provision of tourism, public, and social services (González-Reverté, 2019).

6. CONCLUSION

The development of technology has affected all areas of life, as well as the tourism sector. In the classical concept of tourism, while meeting basic needs such as food and beverage, accommodation, entertainment, and security was sufficient for tourists, different expectations have emerged in today's world. At the top of these are traffic control systems, which are included in eco-friendly practices, such as air quality monitoring systems, noise monitoring systems, energy monitoring systems, waste management systems, smart parking systems, green energy, and green infrastructure. In addition, guests staying in hotels, for example, expect the room lamps to turn on and off automatically and adjust the room temperature automatically. Such situations have revealed the so-called "smart destination" phenomenon.

A smart city invests in social capital and people, manages resources, collects, stores, and analyzes real-time data using smart technology systems, creates smart development policies using the findings, and thus provides services to improve the quality of life such as public safety, education, and health (Harrison et al., 2010). The smart destination is based on the four basic information technologies that exist in the smart city which means that IoT, mobile communication, cloud communication, and AI technology (Guo, Liu, & Chai, 2014). While building smart destinations, minimizing the damage to nature is of great importance in terms of sustainability. At this point, eco-friendly practices such as zero waste, low carbon, waste management, and green energy should be implemented. In the future, such environmentally friendly practices will be an important part of smart destinations.

In this study, first of all, smart destination concept and the characteristics of smart destinations were examined. In this context, it was demonstrated that implementations such as ICT, sustainability, and AR are important for smart destinations. Then, smart destinations, which are the main theme of the study, and eco-friendly practices associated with these destinations are included. As a result of the examinations, it was determined that applications such as traffic control systems, air quality monitoring systems, noise monitoring systems, energy monitoring systems, smart parking systems, green energy, green infrastructure, green applications, and low carbon are the most common applications.

It is thought that IoT applications in smart destinations will become even more popular in the future. Companies will pay more attention to IoT applications to reduce costs and be more sustainable. Besides, the concept of ecology will be more prominent. Eco-friendly practices will make themselves felt in almost every aspect of life. It is thought that applications such as developing different technologies according to regions, creating CRM according to different market segments, social media, POS, mobile devices, and sensors will be used more frequently in tourism.

REFERENCES

Al-Hader, M., Rodzi, A., Sharif, A. R., & Ahmad, N. (2009, September). Smart city components architecture. In *International conference on computational intelligence, modelling and simulation* (pp. 93–97).

Alawadhi, S., Aldama-Nalda, A., Chourabi, H., Gil-Garcia, J. R., Leung, S., Mellouli, S., . . . Walker, S. (2012, August). Building understanding of smart city initiatives. In *International conference on electronic government* (pp. 40–53).

Alkan, T. (2015). Akıllı kentler ya da 21. Yüzyıl şehirleri. *Bilişim Dergisi, 182*, 70–77.

Anagnostopoulos, T., Zaslavsky, A., Kolomvatsos, K., Medvedev, A., Amirian, P., Morley, J., & Hadjieftymiades, S. (2017). Challenges and opportunities of waste management in IoT-enabled smart cities: A survey. *Transactions on Sustainable Computing, 2*(3), 275–289. doi:10.1109/tsusc.2017.2691049

Avci, E. (2019). Enhancing the cultural tourism experience through augmented reality. In C. Çobanoglu, M. Çavuşoğlu, & A. Çorbacı (Eds.), *Advances in global business and economics*. South Florida: M3 Center Publishing.

Ayaş, N. (2007). Çevresel sürdürülebilir turizm gelişmesi. *Gazi Üniversitesi İktisadi ve İdari Bilimler Fakültesi Dergisi, 9*(1), 59–69.

Aydın, N. (2021). Akıllı Şehirlerde IoT: Gelecekteki Kentsel Gelişim Planı. *Balkan ve Yakın Doğu Sosyal Bilimler Dergisi, 7*(4), 44–51.

Aydın, T. Ş., & Yılmaz, İ. A. (2013). Ekoturizm. In Ş. T. Aydın & M. Boz (Eds.), *Turizmde Güncel Konu ve Eğilimler*. Ankara: Detay Yayıncılık.

Barutçu, B. (2021). *Akıllı şehirler üzerine sistemik bir literatür taraması ve akıllı şehirlerde endüstri mühendisliği uygulama alanları* (Unpublished master thesis), University of Başkent.

Batalla, J. M., Mastorakis, G., Mavromoustakis, C., & Pallis, E. (2017). *Beyond the Internet of things: Everything interconnected.* Cham: Springer.

Bayar, S. B. (2022). Destinasyonlarda sürdürülebilirlik ve yeşil uygulamalar: Green destinations top 100 sürdürülebilir hikayeler yarışmasına katılan destinasyonların incelenmesi. *Turizm ve İşletme Bilimleri Dergisi, 2*(2), 170–198.

Benckendorff, P., Sheldon, P. J., & Fesenmaier, D. R. (2014). *Tourism information technology*. Wallingford: CABI International.

Bhatt, C., Dey, N., & Ashour, A. (2018). *Internet of things and big data analytics toward next-generation intelligence*. Cham: Springer.

Bojnec, Š., & Papler, D. (2009). Tourism demand, green energy supply, and sustainable development. *Academica Turistica, II*, 11–16.

Botta, A., de Donato, W., Persico, V., & Pescapé, A. (2016). Integration of cloud computing and internet of things: A survey. *Future Generation Computer Systems, 56*, 684–700.

Budhiputra, P. M., & Putra, K. P. (2016, July). Smart city framework based on business process re-engineering approach. In *2016 international conference on ICT for smart society, Surabaya, Indonesia* (pp. 69–73).

Buhalis, D. (2003). *eTourism: Information technology for strategic tourism management*. London: Pearson PLC.

Buhalis, D., & Amaranggana, A. (2013). Smart tourism destinations. In Z. Xiang & I. Tussyadiah (Eds.), *Information and communication technologies in tourism 2014*. Springer.

Buhalis, D., & Law, R. (2008). Progress in information technology and tourism management: 20 years on and 10 years after the internet-The state of eTourism research. *Tourism Management*, *29*(4), 609–623.

Buhalis, D., & Matloka, J. (2013). Technology-enabled tourism destination management and marketing. In C. Costa, E. Panyik, & D. Buhalis (Eds.), *Trends in European tourism planning and organisation* (pp. 339–350). Buffalo, NY: Channel View Publications.

Çetin, B., & Karataş, A. (2022). Sürdürülebilirlik kapsamında akıllı turizm ve çevre dostu uygulamalar. In Ü. Aydın (Ed.), *Turizm ve Destinasyon Araştırmaları I*. Çanakkale: Paradigma Akademi.

De Paz, J. F., Bajo, J., Rodríguez, S., Villarrubia, G., & Corchado, J. M. (2016). Intelligent system for lighting control in smart cities. *Information Sciences*, *372*, 241–255.

Dinç, Y., Kömür, T., & İlban, M. O. (2021). Sürdürülebilir destinasyon yönetimi için bir model önerisi: ayvalık örneği. *OPUS International Journal of Society Researches*, *17*(37), 4376–4399. doi:10.26466/opus.896321

Dohler, M., Vilajosana, I., Vilajosana, X., & Llosa, J. (2011, December). *Smart cities: An action plan* (pp. 1–6). Barcelona: Barcelona Smart Cities Congress.

Ellsmoor, J. (2019, May 01). Smart cities: The future of urban development. *Forbes*. Retrieved from https://www.forbes.com/sites/jamesellsmoor/2019/05/19/smart-cities-the-future-of-urban-development/#7f62acef2f90

Encalada, L., Boavida-Portugal, I., Ferreira, C., & Rocha, J. (2017). Identifying tourist places of interest based on digital imprints: Towards a sustainable smart city. *Sustainability*, *9*(12), 2317. doi:10.3390/su9122317

Fahim, A., Hasan, M., & Chowdhury, M. A. (2021). Smart parking systems: Comprehensive review based on various aspects. *Heliyon*, *7*(5).

Femenia-Serra, F., Neuhofer, B., & Ivars-Baidal, J. A. (2019). Towards a conceptualisation of smart tourists and their role within the smart destination scenario. *Service Industries Journal*, *39*(2), 109–133.

Fistola, R., Anna, R., & Rocca, L. (2013, September). Smart city planning: A systemic approach. In *Proceedings of the 6th knowledge cities World summit 2013, Istanbul, Turkey* (pp. 520–529).

González-Reverté, F. (2019). Building sustainable smart destinations: An approach based on the development of Spanish smart tourism plans. *Sustainability*, *11*(23), 6874. doi:10.3390/ su11236874

Gordon, R. L., Reiss, R. A., Haenel, H., Case, E., French, R. L., Mohaddes, A., & Wolcott, R. (1996). *Traffic control systems handbook*. Washington, DC: Federal Highway Administration.

Green Destinations. (2022). *Our story*. Retrieved from https://greendestinations.org/about/

Gretzel, U., & de Mendonça, M. C. (2019). Smart destination brands: Semiotic analysis of visual and verbal signs. *International Journal of Tourism Cities*, *5*(4), 560–580.

Gretzel, U., Reino, S., Kopera, S., & Koo, C. (2016). Smart tourism challenges. *Journal of Tourism*, *16*(1), 41–47.

Gretzel, U., Sigala, M., Xiang, Z., & Koo, C. (2015). Smart tourism: Foundations and developments. *Electronic Markets*, *25*(3). 179–188. doi:10.1007/s12525- 015-0196-8

Gretzel, U., Yuan, Y. L., & Fesenmaier, D. R. (2000). Preparing for the new economy: Advertising strategies and change in destination marketing organizations. *Journal of Travel Research*, *39*(2), 146–156. doi:10.1177/004728750003900204

Guo, Y., Liu, H., & Chai, Y. (2014). The embedding convergence of smart cities and tourism Internet of Things in China: An advance perspective. *An International Journal of Akdeniz University Tourism Faculty*, *2*(1), 54–69.

Han, E., & Kaya, A. A. (2013). *Kalkınma Ekonomisi, Teori ve Politika*. Ankara: Nobel Yayıncılık.

Harrison, C., Eckman, B., Hamilton, R., Hartswick, P., Kalagnanam, J., Paraszczak, J., & Williams, P. (2010). Foundations for smarter cities. *IBM Journal of Research and Development*, *54*(4), 1–16.

Here. (2021, May 02). *Smart city technologies: Role and applications of big data and IoT*. Retrieved from https://mobility.here.com/learn/smart-city-mobility/smart-city-technologies-role-and-applications-big-data-and-iot#pgid-1771

Huang, X., Yuan, J., & Shi, M. (2012). Condition and key issues analysis on the smarter tourism construction in China. In F. L. Wang, J. Lei, R. W. H. Lau, & J. Zhang (Eds.), *Multimedia and signal processing*. Berlin: Springer.

Hwang, J., Park, H. Y., & Hunter, W. C. (2015). Constructivism in smart tourism research: Seoul destination image. *Asia Pacific Journal of Information Systems*, *25*(1), 163–178. doi:10.14329/apjis.2015.25.1.163

Independent. (2021, April 3). How Malaga is using 'smart tourism' to rebound from Covid. *The Independent*. Retrieved from https://www.independent.co.uk/travel/europe/malaga-smart-city-tourism-covid-apps-technology-b1764077.html

iotWorm. (2015, May 05). *Smart cities Internet of Things (IoT) examples and applications*. Retrieved from https://iotworm.com/smart-cities-internet-things-examples-applications/

ITU. (2019, December). *Smart sustainable cities*. Retrieved from https://www.itu.int/en/mediacentre/backgrounders/Pages/smart-sustainable-cities.aspx

Janusz, G. K., & Bajdor, P. (2013). Towards to sustainable tourism–framework, activities, and dimensions. *Procedia Economics and Finance*, *6*, 523–529. doi:10.1016/S2212-5671(13)00170-6

Jeong, M., & Shin, H. H. (2020). Tourists' experiences with smart tourism technology at smart destinations and their behavior intentions. *Journal of Travel Research*, *59*(8), 1464–1477.

Jin, J., Gubbi, J., Marusic, S., & Palaniswami, M. (2014). An information framework for creating a smart city through Internet of Things. *IEEE Internet of Things Journal*, *1*(2), 112–121.

Joss, S., Sengers, F., Schraven, D., Caprotti, F., & Dayot, Y. (2019). The smart city as global discourse: Storylines and critical junctures across 27 cities. *Journal of Urban Technology*, *26*(1), 3–34.

Karakaş, E., & Atay, L. (2021). COVID-19 Sürecinde Geliştirilen Akıllı Destinasyon Uygulamaları: Malaga Örneği. *Avrasya Turizm Araştırmaları Dergisi*, *2*(1), 33–41.

Karayilmaz, C., & Özker, A. N. (2020). Kamusal nitelikli özel malların sunumunda akıllı şehirler olgusu: Akıllı şehir uygulamalarında küresel değişimler. *Karamanoğlu Mehmetbey Üniversitesi Sosyal ve Ekonomik Araştırmalar Dergisi*, *22*(38), 82–100.

Kečkeš, A. L., & Tomičić, I. (2017). Augmented reality in tourism–Research and applications overview. *Interdisciplinary Description of Complex Systems*, *15*(2), 157–167.

Khanna, A., & Anand, R. (2016, January). IoT based smart parking system. In *International conference on Internet of Things and applications* (pp. 266–270).

Kim, G.-H. (2016). What is the appropriate policy direction to develop Internet of things? *Journal of Multimedia Services Convergent with Art, Humanities, and Sociology*, *6*(2), 89–100.

Komninos, N. (2002). *Intelligent cities: Innovation, knowledge systems, and digital spaces*. New York, NY: Spon Press.

Komninos, N. (2006). The architecture of intelligent cities. *Intelligent Environments*, *6*, 53–61.

Krishnamurthi, R., Nayyar, A., & Solanki, A. (2019). *Green and smart technologies for smart cities*. Boca Raton, FL: CRC Press.

Kulualp, H. G., & Sarı, Ö. (2020). Smart tourism, smart cities, and smart destinations as knowledge management tools. In *Handbook of research on smart technology applications in the tourism industry* (pp. 371–390). Hershey, PA: IGI Global.

Lamsfus, C., Martin, D., Alzua-Sorzabal, A., & Manzenera, E. T. (2015). Smart tourism destinations: An extended conception of smart cities focusing on human mobility. In Tussyadiah & Inversini (Eds.), *Information and communication technologies in tourism*. Springer.

Law, R., Buhalis, D., & Cobanoglu, C. (2014). Progress on information and communication technologies in hospitality and tourism. *International Journal of Contemporary Hospitality Management*, *26*(5), 727–750. doi:10.1108/IJCHM-08-2013-0367

Lee, C. C., & Chang, C. P. (2008). Tourism development and economic growth: A closer look at panels. *Tourism Management*, *29*(1), 180–192.

Lee, S., Rodriguez, L., & Sar, S. (2012). The influence of logo design on country image and willingness to visit: A study of country logos for tourism. *Public Relations Review*, *38*(4), 584–591.

Li, Y., Hu, C., Huang, C., & Duan, L. (2017). The concept of smart tourism in the context of tourism information services. *Tourism Management*, *58*, 293–300. doi:10.1016/j.tourman.2016.03.014

Mace, B. L., Corser, G. C., Zitting, L., & Denison, J. (2013). Effects of overflights on the national park experience. *Journal of Environmental Psychology*, (35), 30–39. doi:10.1016/j.jenvp.2013.04.001

Malaga Turismo. (2021, April 9). *Plan Estrategico 2021 – 2024 Turismo De Malaga*. Retrieved from http://s3.malagaturismo.com/files/933/933/plan-estrategico-turismo-2021-2024-def.pdf

Mandhare, P. A., Kharat, V., & Patil, C. Y. (2018). Intelligent road traffic control system for traffic congestion: a perspective. *International Journal of Computer Science and Engineering*, *6*(7).

Marinov, M. B., Topalov, I., Gieva, E., & Nikolov, G. (2016, May). Air quality monitoring in urban environments. In *39th international spring seminar on electronics technology (ISSE)* (pp. 443–448).

Merchan, C. I., Diaz-Balteiro, L., & Solino, M. (2014). Noise pollution in national parks: Soundscape and economic valuation. *Landscape and Urban Planning*, *123*(1), 1–9. doi:10.1016/j.landurbplan.2013. 11.006

Neuhofer, B., Buhalis, D., & Ladkin, A. (2012). Conceptualising technology enhanced destination experiences. *Journal of Destination Marketing & Management*, *1*(1), 36–46.

Pan, S. Y., Gao, M., Kim, H., Shah, K. J., Pei, S. L., & Chiang, P. C. (2018). Advances and challenges in sustainable tourism toward a green economy. *The Science of the Total Environment*, *635*, 452–469.

Parlak, E., & Atik, M. (2020). Dünyadan ve ülkemizden mavi–yeşil altyapı uygulamaları. *Peyzaj*, *2*(2), 86–100.

Pham, T. N., Tsai, M. F., Nguyen, D. B., Dow, C. R., & Deng, D. J. (2015). A cloud-based smart-parking system based on Internet-of-Things technologies. *IEEE Access*, *3*, 1581–1591. doi:10.1109/ACCESS.2015.2477299

Radojević, B., Lazić, L., & Cimbaljević, M. (2020). Rescaling smart destinations: The growing importance of smart geospatial services during and after COVID-19 pandemic. *Geographica Pannonica*, *24*(3), 221–228.

Rana, R. K., Chou, C. T., Kanhere, S. S., Bulusu, N., & Hu, W. (2010, April). Ear-phone: An end-to-end participatory urban noise mapping system. In *International conference on information processing in sensor networks* (pp. 105–116).

Ritchie, J. B., & Crouch, G. I. (2003). *The competitive destination: A sustainable tourism perspective*. CABI International.

Rucci, A. C., Moreno-Izquierdo, L., Perles-Ribes, J. F., & Porto, N. (2021). Smart or partly smart? Accessibility and innovation policies to assess smartness and competitiveness of destinations. *Current Issues in Tourism*, *25*(8), 1270–1288. doi:10.1080/13683500.2021.1914005

Schaffers, H., Komninos, N., Pallot, M., Trousse, B., Nilsson, M., & Oliveira, A. (2011). *Smart cities and the future internet: Towards cooperation frameworks for open innovation*. Berlin: Springer.

Shelby, Z., Hartke, K., & Bormann, C. (2014). *The Constrained Application Protocol (CoAP)*. Internet Engineering Task Force (IETF). Retrieved from https://www.rfc-editor.org/rfc/rfc7252

SmartCity. (2017, July 18). *IOT in smart cities: Future urban development plan of building one community around wireless and internet connectivity*. Smart City Press. Retrieved from https://www.smartcity.press/internet-of-things-smart-cities/

Soares, J. C., Ruiz, D. T. C., & Baidal, I. J. A. (2022). Smart destinations: A new planning and management approach? *Current Issues in Tourism*, *25*(17), 2717–2732. doi:10.1080/13683500.2021.1991897

Sorokina, E., Wang, Y., Fyall, A., Lugosi, P., Torres, E., & Jung, T. (2022). Constructing a smart destination framework: A destination marketing organization perspective. *Journal of Destination Marketing & Management*, *23*, 1–17.

Swarbrooke, J., & Horner, S. (2001). *Business travel and tourism*. Burlington, MA: Butterworth-Heinemann.

Taff, D., Newman, P., Lawson, S. R., Bright, A., Martin, L., Gibson, A., & Archie, T. (2014). The role of messaging on acceptability of military aircraft sounds in Sequoia National Park. *Applied Acoustics*, *84*(1), 122–128. doi:10.1016/j.apacoust.2013.09.012

Tech Republic. (2021). Internet of Things policy. *Tech Republic*. Retrieved from https://www.techrepublic.com/resource-library/downloads/internet-of-things-policy/

Tomar, P., & Kaur, G. (2019). *Green and smart technologies for smart cities*. Boca Raton, FL: CRC Press.

Topçu, F. H. (2018). Düşük karbon ekonomisine geçme(me): İklim değişikliği ve enerji politikaları bağlamında bir bakış. *Akdeniz İİBF Dergisi*, *18*, 115–154.

Üçgün, H., Gömbeci, F., Yüzgeç, U., & Yalçın, N. (2020). IoT tabanlı platform ile gerçek zamanlı iç ortam hava kalitesi izleme sistemi. *Bilecik Şeyh Edebali Üniversitesi Fen Bilimleri Dergisi*, *7*(1), 370–381.

UNWTO Tourism Highlights. (2015). *UNWTO tourism highlights 2015*. Retrieved from https://www.e-unwto.org/doi/pdf/10.18111/9789284416899

Williams, A. M., Rodriguez, I., & Makkonen, T. (2020). Innovation and smart destinations: Critical insights. *Annals of Tourism Research*, *83*, 102930.

WTTC. (2022). *Travel & tourism economic impact 2022*. Retrieved from https://wttc.org/Portals/0/Documents/Reports/2022/EIR2022-Global%20Trends.pdf

Xiang, Z., Tussyadiah, I., & Buhalis, D. (2015). Smart destinations: Foundations, analytics, and applications. *Journal of Destination Marketing & Management*, *4*(3), 143–144. doi:10.1016/j.jdmm.2015.07.001

Yoo, C. W., Goo, J., Huang, C. D., Nam, K., & Woo, M. (2017). Improving travel decision support satisfaction with smart tourism technologies: A framework of tourist elaboration likelihood and self-efficacy. *Technological Forecasting and Social Change*, *123*, 330–341. doi:10.1016/j.techfore.2016.10.071

Yoo, K. H., Sigala, M., & Gretzel, U. (2015). Exploring Tripadvisor. In R. Egger, I. Gula, & D. Walcher (Eds.), *Open tourism – Open innovation, crowdsourcing and collaborative consumption challenging the tourism industry*. Heidelberg: Springer.

Zappatore, M., Longo, A., & Bochicchio, M. A. (2016, July). Using mobile crowd sensing for noise monitoring in smart cities. In *2016 international multidisciplinary conference on computer and energy science, Split, Croatia* (pp. 1–6).

Zhou, Y., Wu, Q., & Yan, T. (2017, June). A hierarchical system for energy consumption monitoring and information management. In *2017 12th IEEE conference on industrial electronics and applications, Siem Reap, Cambodia* (pp. 449–454).

Zigurat. (n.d.). *New technologies in smart cities*. E-Zigurat. Retrieved from https://www.e-zigurat.com/innovation-school/blog/new-technologies-smart-cities/

AN ALTERNATIVE MODEL OF LIVING: SMART ECO-CITIES

Beyza Hatirnaz, Dilek Demirer and Emrah Özkul

ABSTRACT

Changes and transformations experienced today have created alternative lifestyles in cities. Many life concepts try to find solutions to environmental problems, so much so that this section focuses on smart eco-city concept, which is one of the trending city initiatives of the last century and is expected to become more popular each day. In the chapter, firstly, information about the concepts of eco-city and smart city is given and then what is meant by the concept of smart eco-city is discussed. Next, smart eco-cities have been tried to be explained with application examples. Finally, the chapter emphasizes the importance of ecological cities blended with technology for the sustainability of living spaces and offers implications for future research.

Keywords: Eco-city; smart city; smart eco-city; information and communication technologies; sustainability; sustainable living

1. INTRODUCTION

Our generation has gone down in history as the first generation to see the concrete traces of the global environmental crisis and to struggle with it. Today, deforestation, desertification, loss of fresh water, disappearing biodiversity, pollution and deforestation of the oceans, and the climate crisis due to global warming are affecting the daily life of humanity for the first time in history. As a result of sustainable life and transformation meetings and agreements signed in the international arena for years, optimum use of resources and ecological transformations in all areas of life are maintained, albeit in different dimensions and qualities. The special need that brings together the concepts of sustainable, ecological, and smart has emerged due to this evolutionary transformation.

Although the definition of smart plays a savior role for cities that are trying to maintain populations above their capacities, the conservative attitude in transforming consumption habits has contributed to ecological characteristics as a necessity. At this point, smart solutions are interpreted as indispensable for sustainability for cities, albeit at different levels. Cities have become one of the common denominators of the transition to

Smart Cities for Sustainability
Advanced Series in Management, Volume 32, 33–48

ISSN: 1877-6361/doi:10.1108/S1877-636120230000033004

sustainable development during the transition to society where ecological practices can be applied. In today's world, where the speed of movement of the world is faster than ever, technological developments, ecological solutions and recipes, and the concept of smartness are evolving day by day. In order to determine suitable living conditions and to increase the duration of our existence in the world, all kinds of peaceful solutions are tried to be integrated with each other, and a way is sought to create alternative solutions that are suitable for different living spaces. Especially since the Rio Conference in 1992, economies have laid the perceptual foundations of the effort to develop special techniques to offset the carbon they create. Definitions such as zero carbon, low carbon, carbon neutral, smart, ecological, and especially sustainability have become a common discourse for the sustainability of cities and companies. One of the concepts that emerged as a result of this movement is smart eco-cities. Cities are living spaces where ecological requirements are left behind human necessities, and the instrumental role of nature deepens in terms of their priorities. The meeting of technological developments with the conditions for cities to be ecological has also paved the way for the beginning of a new era in urban planning and for cities to take on the quality of supporting the world. In this way, sustainable living will be possible without sacrificing the benefits of city life, such as efficient production, safety, and comfort. On the other hand, blending eco-cities with intensive technology gives rise to different ethical and ecological debates arising from the definition and scope of the concepts of "natural" and "ecological."

Smart eco-cities have emerged due to integrating two concepts with different purposes and offering an alternative example to sustainable urban planning. The definition of smart eco-city, which paves the way for cities to gain an ecological character based on technology, has evolved by preserving its continuity in sociocultural and economic dimensions. However, in the face of the undeniable potential of the features that make a city smart to increase the ecological level of that city, the question of how well a city life built based on technology is compatible with the ecological and social features frequently discussed in sustainable development and ecological ethics has not yet been answered.

Based on all these, in this section, the eco-city and smart city are discussed in detail, and information is given about the basic factors that will make a city a smart eco-city. The authors first present a multifaceted approach that includes both concepts in detail and then explain what "smart eco-city" means. Thus, a framework for smart eco-cities, an emerging concept that has not been studied much, will be developed. This section differs because it includes an effort to present a critical interpretation and evaluation in the context of ecological ethical approaches in which human and nature relations are questioned deeply. It also highlights the importance of smart eco-cities for environmental sustainability while discussing implications and recommendations for further research.

2. ECO-CITY

The concept of eco-city was first explained by Register and his friends in 1975 with the title of "rebuild cities in balance with nature" with Urban Ecology, a nonprofit organization (principles such as transportation opportunities and alternatives, restoration, housing construction, social justice, local agriculture, recycling, waste management, raising local environmental awareness). The concepts of green and sustainability have been associated with the "eco-city," and it has developed over time with the understanding of environmental and economic effects. It is aimed to achieve sustainable urban development in the future by taking into account the macro-environmental factors in which the production model, consumption behaviors, and decision-making tools change (Cao & Li, 2011; Cheng & Hu, 2010; Dong et al., 2016; Register, 1993; Urban Ecology, 1990; Zhou et al., 2015). Particularly, world urban problems were discussed (Engwicht's contributions as an urban planner), suggestions were presented, and principles for structuring cities according to ecological principles were determined in the international conferences held in Berkeley, Adeliade, and Yoff between 1990 and 1996. Themes such as "eco-neighborhoods," "urban eco-village," and "eco-communities" were also mentioned (Roseland, 1997).

Many pioneers are mentioned based on the subject of the new western civilization that emerged with the industrial revolution. In the nineteenth century, Olmsted, Geddles, and Howard included concerns for preserving ecological values (Welter & Lawson, 2000) in planning with survey, analysis, plan, and decentralization concepts. In the 20th century, "unsustainability" and a "more organic world" concepts were defined by the ideas expanded by Mumford and Stein. In this process, social problems have occurred with the increasing urban sprawl. A new understanding emerged in the 1980s with McHarg's theory of ecology, which he put forward by integrating human and natural environments and planners who applied it differently. Different solutions have been produced for a system that emphasizes cultural values, encourages urban development, and provides ecological integrity. Thus, it is associated with the concept of today's eco-city (Wong & Yuen, 2011). An urbanization system in which natural resources are used the least, renewable energy is preferred, green areas and biodiversity are protected, air, water, and soil pollution are reduced, safe drinking and utility water are provided, and environmentally friendly agricultural practices are maintained at the forefront. On the other hand, it is mentioned that a system in which unemployment is reduced economically, economic inequality is resolved, the population is evenly distributed, and public transportation becomes widespread. In social terms, they are the cities where the public can access the health and education system without any problems, support the planning process, and ensure equality, security, and justice (De Jong, Yu, Chen, Wang, & Weijnen, 2015; Öztekin & Çolak, 2022).

When examined as a whole, it is seen that the titles of "healthy communities" where traditional conditions are provided, "appropriate technology" defined as medium-sized by Schumacher and in cooperation with the local, "social-economic development" where solutions to common economic problems are found, and local employment is encouraged, "social ecology," which affects the relationship of social society with nature as a whole, "bioregionalism," which also includes an ecological footprint, to answer how natural systems work and how they should be developed, "world view," which separates societies from each other in terms of production, behavior, and use, "green movement" in harmony with nature and diversity, "sustainable development" where a certain basis is formed with our common future and agenda 21 is examined (Roseland, 1997).

Whether it is on the effort of ecological ethical approaches to make sense of life, nature, and nature-human interaction or it is within the scope of sustainable development that has been discussed in the international arena for years, it is necessary to accept that ecological problems and solutions cannot be independent of social issues and situations. Otherwise, we both misinterpret the current situation and put possible solutions into a dead end in vicious circles. One of the most important issues here is the internalization of the ecologicalization process beyond technical planning and the abandonment of the human-centered perspective, which has been reflected in all economic systems and, accordingly, in planning in all sectors for years. At the most basic point, the issue is to reveal what we call "*ecological.*"

"*It has to be ecological*" (Wong & Yuen, 2011). It is clear, but what is ecological? Does it contain nonhuman species' rights to live or exist? Do we have to think about their welfare? What is our position in nature, and what is nature? The main issue here is how naturally the systems shaped by our view of the world and the rules created to execute these systems are. At this point, it is necessary to mention the necessity of reversing the biased evolution of the naturally accepted issue for centuries, which has developed in a way that supports our current consuming lifestyle. It should be emphasized that filling in the definition of "nature-friendly" presented for the buildings that need to be established today is necessary for the right approach and planning, so much so that we now consider adopting "sustainable thinking" as the basis of success in order for the sustainable development process to be successful (Van Antwerp & Heun, 2022). In other words, first of all, we need to start by accepting that being able to talk about ecologicalization is a holistic process that includes mental and social characteristics, current situations, and transformation. The adoption of this perception has also led to the inclusion of the concept of "socio-ecological" and "socio-technical" in eco-city studies and to be taken into account as features to be considered in eco-city planning (Hes & Coenen, 2018).

Naess (1989) and Bookchin (1974) presented ecological life and living space practices with their ecological ethical forms. Accordingly, it is possible to say that the perception of "ecologicalization" differs in every approach. The lack of a clear answer to this definition is related to the main reason why a global recipe cannot be created today. Namely, the differences in societies and geographies worldwide make it impossible to create a single correct answer for the world. Head and Lam (2011) discussed these differences in income level in their studies. They emphasized that the deficiencies are similar in the direction of the economic deficit and mentioned the necessity of shaping the plans accordingly. Although this is a correct approach regarding shedding light on the diversity of societies, it should be emphasized that developing ideas that individuals and societies can adopt for permanent solutions is vital.

At this point, it goes through ecological space designs suitable for societies shaped around the basic, indispensable practices that must be accepted and united. Register (1987), on the other hand, associated lifestyles with ecosystems, similar to their ecological ethical approaches. It is not known whether he handled it in this sense, but the uniqueness of the ecosystems also reflects the uniqueness of the eco-cities to be created. Such approaches talked about the necessity of transferring the complementary and supportive aspects of ecosystems to living spaces and created his living space practices with this perception beyond the "big fish swallows small fish" hypothesis, are the misinterpretation of ecology and nature and created the basis of social Darwinism and capitalism today. Similarly, Roseland (1997) reemphasized the ecologicalization processes of living spaces and their link to sociological and psychological factors within bioregion approach. The issue, which has been discussed for years, has been redefined with the same qualities under sustainable cities with the spread of the idea of sustainability (Kline, 2000; Tosun, 2013).

Roseland (1997) claimed that the theme "eco-cities" is contained in a complex series of related variations; it must be imagined as a hologram and defined eco-city feature parallel with sustainable city features. UNEP defines sustainable cities with Sustainable Development Goal (SDG), including a special focus on sustainable production and consumption necessities and interventions to reduce and manage pollution and waste (unep.org).

As they come today, eco-cities are known for similar qualities to sustainable cities. Kavcar (2017) defines the eco-city concept in environmental, social, and economic dimensions, similar to sustainable development. The eco-city standards determined by the Eco-city Builders Foundation, established by the Register, also support a holistic perception and lifestyle change beyond nature-friendly cities (Register, 2012; ecocitystandards.org). The determined eco-city standards are as follows:

- *Urban Design*: access by proximity, safe and affordable housing, green building, environmentally friendly transport;
- *Bio-Geo-Physical*: clean air, clean and safe water, healthy soil, responsible sources, clean and renewable energy, healthy and accessible food;
- *Sociocultural*: healthy culture, community capacity/governance, healthy and equitable economy, lifelong education, quality of life;
- *Ecological*: healthy biodiversity, earth's carrying capacity, ecological integrity.

As an urban reflection of the sustainability movement, Bibri (2021a, b) examined eco-cities in three types:

- Type 1 emphasizes passive solar design (eco-village, solar city, solar village, cohousing, sustainable housing).
- Type 2 combines passive solar design and greening (eco-city, eco-zone, environmental city, green city, garden city, sustainable neighborhood, sustainable community, sustainable urban living, living machines, techno-city, new city).
- Type 3 focuses on green energy technologies and/or smart energy and environmental technologies (symbiotic city, carbon neutral city, zero energy city, zero carbon city, net zero carbon community, low carbon city, energy efficient city, ubiquitous eco-city, smart eco-city, data-driven smart eco-city).

This list, which allows a holistic view of eco-city models, reveals that each type has gradually developed within itself and when compared to other types. Eco-cities took on new characters over time with the effect of both sustainability and global changes and transformations by maintaining the goal of ecological sustainability.

Although these and similar solutions are indispensable for first-stage contingency plans, all educational networks, from daily educational practices to institutional training given at school, should primarily be transformed to support this lifestyle to ensure continuity. With the idea that our cultural and educational past shapes our present (Hes & Coenen, 2018), the ecologicalization of educational and sociocultural transformation is vital for developing ecological thinking skills to shape tomorrow. Once you start thinking ecologically, all the requirements accepted as a rule today will take shape automatically. In deep ecology, this is expressed as the discovery of ties with nature and self-realization (Des Jardins, 2013; Naess, 1995). As a result, it is possible to say that it is not possible to deal with the ecologicalization adventure in a one-dimensional way. Although our definition of eco-city is new, it is possible to say that the issue of ecologicalization and examining ecological lifestyles are not newly discussed issues in the literature. The success of the ecological structures established today will be possible if either social or individual transformation can take on a quality suitable for the ecologicalization process.

3. SMART CITY

The concept of smart city, which has started to be discussed since the 2000s, is a concept that develops in line with the pace of cities. This concept is developing and diversifying more and more. In fact, due to each city's unique identity, cities follow different paths depending on the conditions they have while transforming into smart cities (European Parliament, 2014). This situation causes many perspectives about the "smart city."

There is no common definition for the concept of smart city that fits all situations (Kondepudi & Kondepudi, 2015). The lack of a generally agreed definition has caused many researchers, institutions, and organizations to express the concept in different ways. Thus, many definitions have been made for the smart city (European Parliament, 2014, p. 17).

While the concept of smart city is within the scope of the 2020–2023 National Smart Cities Strategy and Action Plan: "More livable solutions that are implemented through cooperation between stakeholders, that use new technologies and innovative approaches, that are justified based on data and expertise, and that anticipate future problems and needs, add value to life. And sustainable cities" (T.C. Ministry of Environment and Urbanization, 2019), it is expressed in Mapping Smart Cities in the European Union as cities that try to solve the problems of the city by using information and communication technologies (ICTs) based on stakeholders and management (European Parliament, 2014, p. 9).

Kondepudi and Kondepudi (2015), on the other hand, approached the concept with a holistic/comprehensive perspective and defined a smart city as a city that uses ICTs to (i) Improve the living standards of city residents, (ii) Lay the foundations for economic growth, (iii) Protect the well-being of its citizens, (iv) Developing and implementing an environmentally friendly and sustainable policy in line with the definition of sustainable development, (v) Optimize the physical and service infrastructure, (vi) To produce the necessary solutions for all kinds of disasters, (vii) To carry out an open, fair, and effective management process.

As can be seen in the definitions, there are two important criteria for smart cities: ICTs and sustainability. Here, what is meant by ICT is the establishment of an effective communication infrastructure in the city, while sustainability refers to a structure in which environmental problems are reduced, and the living standards of the people are improved (Kayapınar, 2017; Laleoğlu, 2021). These two criteria complement each other and are based on ensuring sustainability through the use of ICTs.

On the other hand, although what smart city means is perceived as a digitalized city, the situation is not only about this. City management and the public are also part of the smart city process. Therefore, their actions or decisions are effective in the success of the smart city system (Kloss, 2022).

4. COMPONENTS OF THE SMART CITY

The most widely accepted process of transformation of cities into smart cities proceeds through some components, the structure defined by Boyd Cohen in the "Smart Cities Wheel, SCW" (Laleoğlu, 2021). Cohen (2014) also showed these components in the Smart City Index, and categorized the components of the smart city concept as 18 work areas, indicators, and definitions under six main dimensions; "smart people, smart life, smart environment, smart transportation, smart governance, and smart economy." Indicators in these components are used to look at the smartness of a city:

4.1 Smart Governance

Smart governance includes online services, infrastructure, and open government. Open government creates a transparent environment between management, stakeholders, and residents, while citizens have easy access to data through online services. In addition, holistic management is provided by establishing a communication infrastructure throughout the city.

4.2 Smart Economy

Smart economy includes entrepreneurship and innovation, productivity, and local and global connectivity. Cities should encourage entrepreneurship and innovation. Production should be increased, exports should be made by establishing connections, and participation in international events should be ensured.

4.3 Smart Mobility

This component includes efficient transportation, multimodal access, and technology infrastructure. Smart mobility promotes clean energy and multimodal public transport. Transportation technology infrastructure should be improved in cities, and smart card usage and access to real-time information should be provided. Thus, both environmental pollution will be prevented and affordable quality service will be provided.

4.4 Smart Environment

Smart environment includes environmentally friendly solutions. Cities must achieve environmental sustainability through smart buildings, resource management, and sustainable urban planning.

4.5 Smart People

What is meant by smart people is a society that is educated, open to innovation, and has the skills to use technology. For this, the intelligent human component includes inclusivity, education, and creativity.

4.6 Smart Living

Intelligent life includes culture, welfare, security, and health. Smart life offers an environment where citizens can feel culturally rich, safe, and healthy by improving living conditions and quality with ICTs, facilitating access to medical records, and preventing crimes.

5. SMART ECO-CITY

In recent years, innovative solutions brought by the period in the face of the problems experienced by cities have led to the formation of many new city types, and some existing types have been combined and developed. One of these converging city initiatives is smart eco-cities.

Smart eco-cities are included in Type 3, which focuses on green and smart technologies among many eco-city models that can be divided into three types (Bibri, 2021a,b). Some authors think that the smart eco-city is an inclusive term for the types of cities aiming at sustainable life (Xie, Tan-Mullins, & Cheshmehzangi, 2019; Yiğitcanlar & Lee, 2014) and claim that zero- and low-carbon city and ubiquitous eco-cities can be smart eco-cities (Yigitcanlar & Lee, 2014, p. 101). Others state that eco-cities have transformed into data-driven smart eco-cities with the change they have experienced in the light of technological developments (Bibri, 2021a). As an example of this transformation in eco-cities, the results of the study examining the eco-city development in China can be shown. Eco-city development in China occurs over three periods: eco-cities in the 1990s, low-carbon cities in the 2000s, and smart eco-cities in the 2010s (Cai & Tang, 2021).

A smart eco-city is defined as an experimental city that functions as a potential niche where sustainable economic and environmental reforms/innovations can be tested, both in the immediate environment and internationally (Caprotti, Cowley, Flynn, Joss, & Yu, 2016). In smart eco-cities, combining two trending concepts in recent years, such as smartness and ecology, is being tested for urban development. While one of the concepts is software for urban sustainability, the other is hardware (Xie et al., 2019).

The following three technologies are used when proposing smart solutions to city problems (Bogunovich, 2002): (i) environmental technologies (energy, water, and waste technologies), (ii) information technologies (computer-based hardware and software technologies, environmental sensing technologies), and (iii) communication technologies (CT) (a set of technologies that enable the transfer of environmental data, information, information, and decisions). An eco-city that integrates with these technologies can transform into a smart eco-city and realize its sustainable development vision.

Huang, Gao, and Liu (2019) tried to determine the main success factors of smart eco-city development with a survey conducted on the Shanghai sample. As a result of the study, four basic factors (smart technology, smart citizen, smart environment, smart governance) and 16 evaluation indicators were determined to help the state in the effective planning of smart eco-cities in order to emphasize the superiority of the city and increase its urban competitiveness, and briefly, the following suggestions were presented:

- The public should have e-skills. The government should develop strategies to achieve smart eco-city goals.
- Cultural change should be ensured. At this point, a progressive and sustainable city should improve life and the environment and strike a balance between the urban development it creates and ecology. In addition, initiatives and people with talents, knowledge, and skills suitable for this culture should be attracted to the region.
- Citizens in smart eco-cities should support the city's positive developments, innovations, and synergies; they should not be afraid of innovation and technology. Smart eco-cities should improve the quality of life, sustainable development, and the city–nature balance.

5.1 Smart Eco-City Project Examples

Many smart eco-cities in the world today have existed for years. Technology-based, ecologically prioritized city ideas lead cities to develop different projects and practices within themselves and to build smarter ecological cities. Recently, reports on a research project covering the United Kingdom, France, Germany, the Netherlands, and China have been published under "Smart Eco-Cities for a Green Economy: A Comparative Study of Europe and China." In the reports, the smart eco-development of the selected cities is explained through city profiles and projects, giving an overview of the development and diversity of smart eco-cities. Below are examples of several smart eco-city profiles briefly described in these reports.

5.2 Munich, Germany

Munich stands out, especially with its energy production and cooling systems investments. EFRE, the European Fund for Regional Development, is used in Munich to fund green investments and rural planning. New legislation on the digitalization of energy transmission has been published, and SMART METERING has emerged. In this way, energy consumption began to be read regularly and automatically. The OKOPROFIT program supports the private sector in creating and developing environmental and climate protection measurements. In the city, where an action plan for climate protection has been created, many different issues, from traffic control to lighting, are controlled with this scale created within the framework of the action plan.

Again, the M.O.V.E. project has been developed to enable access to both government and public data, and thus transparency in government and governance principles is also fulfilled. Furthermore, action plans were created to support e-mobility and reduce car dependence for Munich within the scope of The H2020 Smarter Together Project (2016–2020). Because traffic is one of the major problems in Munich due to the increasing population, sustainable transportation opportunities or ways to develop such a system are needed in the city. So, the sustainable mobility-focused project CIVITAS ECCENTRIC has been developed (Späth, 2017).

5.3 Amsterdam, Netherlands

Amsterdam has an initiative called the Amsterdam Smart City platform that supports Smart City projects. Other smart city initiatives include City-Zen, which shows how to integrate innovations in green energy solutions into life; Vehicle to Grid (Vehicle2Grid), offering solutions for the storage of renewable electricity; Amsterdam Arena Innovation Center, which focuses on themes such as transportation, experience, safety, and sustainability (Sengers, 2016).

5.4 Bristol, UK

Bristol, one of the European Green Capitals (2015), aims to be a smart city with the aim of environmental sustainability and produces new smart city strategies in this direction. Bristol has many projects focused on digital infrastructure (e.g., Gigabit Bristol, Bristol is Open, Bristol Data Dome). In addition, 3-E Houses, which encourages the use of smart meters that will reduce energy consumption in residences, and Knowle West Media Center (KWMC), a community center that aims to strengthen and facilitate the life of citizens in many areas, are among the initiatives implemented by Bristol to transform into a smart and eco-city (Caprotti, 2016).

5.5 London, UK

The challenges of London as an international city have led it to take advantage of the opportunities that smart cities bring. London's desire to become a smart city was supported by the Smart London Plan in 2013. London's smart city approach is based on collaborative and creative governance, including local government, experts, community stakeholders, and citizens. Many digital platforms such as Talk London, Tech Londoners, and Smart London Innovation Network have been established for these management collaborations. For example, the public's reactions to all kinds of problems concerning the city can be monitored via Talk London. Another application is Rapid

Volunteering (Joss, 2016), which pairs volunteers with help seekers for various social and environmental activities.

5.6 Milton Keynes, UK

Milton Keynes also has many smart and eco-focused initiatives for urban sustainability. Among its projects, there are MK: Smart, where data obtained from various fields such as infrastructure, education, and health are presented by MK Data Hub; Low Carbon Urban Transport Zone, where driverless cars are tested; Anaerobic digestion plant built to produce energy from waste; Combined Heat and Power Plant, which aims to protect the environment by reducing fuel use and carbon dioxide emissions and is designed to produce electricity and heat; Falcon Smart Grid, which is looking for ways to reduce power cuts, Smart Parking designed to manage the use of parking spaces; Tattenhoe Park, an environmentally friendly neighborhood project to be built with smart measurement technology (Yu, 2016).

5.7 Chengdu, China

Chengdu has significant advantages for smart eco-development with its well-developed infrastructure and facilities. It is one of the major projects planned and designed in line with smart eco ideas in Tianfu New District. Emphasizing an environmentally friendly approach, this project includes six main environmental monitoring and mitigation programs while importance is given to low-carbon transportation, expansion, and improvements in public transportation. The Tianfu New District plan has also adopted the sponge city concept to absorb the city's water and reduce floods. Another project is Chengdu-Bonn Low Carbon City Partnership. This project's scope, which was started in 2012 with the China-Europe Partnership, focuses on urban sustainability (Chien, 2017a).

5.8 Kunming, China

Kunming, known as the "Spring City" with its air, collaborates and prepares plans and programs in line with its smart eco ambitions. Kunming is a testing ground for the national "Broadband Village" project, which aims to provide internet access and set up 4G infrastructure in rural villages. This project benefits villagers and administrators by developing databases on basic services in rural villages. Chenggong New District is an award-winning eco-city project designed according to the principles of low-carbon development. Another project is Flavors Orchard. In this project, villas with sustainable features are built in light of smart eco-initiatives (Chien, 2017b).

5.9 Shanghai, China

Both eco and smart city projects such as Smart Yangpu, Dongtan Eco-City, and Pudong are being implemented in Shanghai. While the Yangpu project focuses on reinforcing an old region with new ICT facilities, the Pudong project, implemented in 2015, provides smart infrastructure to meet the needs of citizens (Tan-Mullins, 2017).

5.10 Tianjin, China

Sino-Singaporean Tianjin Eco-City (SSTEC) was an eco-enterprise rather than a smart one (Cheshmehzangi, 2017). However, in 2017, the city turned into a smart eco-city example by adding smart elements such as developing a smart city master plan (Caprotti, 2020).

5.11 Lyon, France

There are many smart and eco initiatives in Lyon. The Confluence District Project leads them all. The buildings in the Confluence district, located between the Rhone and Saone rivers, are structured by environmental sustainability principles. The Lyon Smart Community (LSC) project also operates in the Confluence region and aims to establish an environmentally oriented smart city. The project is built on four steps to achieving its goals. These goals are (Jolivet & Bond, 2018):

(1) Construction of passive energy-smart buildings with smart grid system (Hikari Project),
(2) Putting electric vehicles charged by solar panels into service (SUNMOOV Project),
(3) Installation of the system where energy consumption can be monitored (ConsoTab project),
(4) Development of the system for monitoring and managing all data (Community Management System Project).

6. CONCLUSION AND IMPLICATIONS

Although there is no generally accepted definition yet, smart eco-cities refer to cities where technological infrastructure supports ecological sustainability. Accordingly, a smart eco-city means the mixed structures that emerge from the blending of dimensions such as smart environment, smart economy, smart life, smart people, smart management, and smart mobility with ecological priorities such as renewable energy, efficient buildings, mixed land use, green transportation, water cycle management, waste management, social sustainability, and biodiversity to have effective, sustainable urban planning (Fig. 3.1: Smart eco-city formation). Projects related to smart green environment, smart green economy, smart green transportation are carried out in these areas, which emerged from the coexistence of smart and eco-urbanism and offered smart solutions to urban environmental problems.

Smart eco-cities are one of the most important and effective opportunities of our age, developed to rule out or minimize the negative effects of city life. These cities, which have been developed with different characters and priorities all over the world, have the potential to be a footnote to today's capitalist world. And the right-minded initiative examples from different cities give us the hope that we can change the consuming human lifestyle to the supporting one. The efficiency and sustainability of these initiatives, the importance of which cannot be ignored, depends on the right answers to the right questions and the right areas to be created according to these answers, as in every subject, and, of course, with the reminder that the world is a whole and living organism. Now is the best time to think about these real-world problems because smart eco-cities have the potential to change the accepted city life and change the way the Planet Earth dragged along.

The concepts of smart and eco represent two important trends that complement each other quite effectively. However, the accustomedness and familiarity created by the sustainable development debates, which have been going on for years, include a shallow evaluation of the issue in the establishment of the eco-city (Naess, 1989, 2005a), and the implementations are not the root solution of the fundamental problems, but the papering the results over the cracks. Although it is mostly mentioned in the literature that these cities are managed with corporate logic, one of the main points to be discussed is how natural these cities are or can stay.

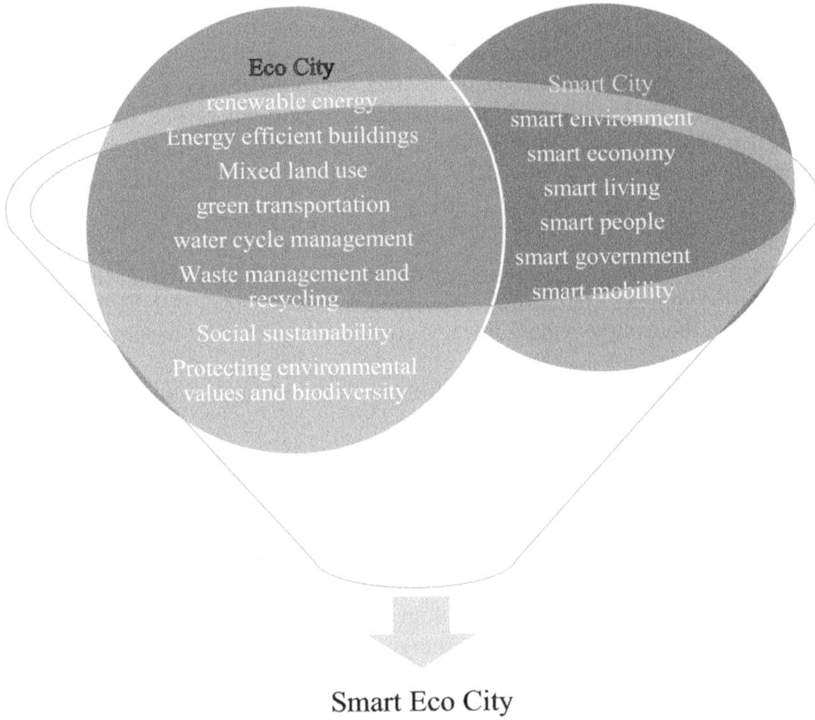

Fig. 3.1. Smart Eco-City. *Source:* Created by the authors.

At this point, it is our responsibility to consider whether new living spaces are being created where realities such as alienation from nature and ecophobia, which are some of the main reasons for humanity's destruction of nature, will deepen. In the struggle of humanity with nature, it is possible that reducing human life to technological, individual lives, completely isolated from nature through smart eco-cities, will lead to generations in which the interaction with nature is completely lost, and alienation from nature deepens in the long run. At this point, to take into account the questions of what is "nature," "natural," and "ecological," which are discussed in every "ecological" themed discussion and discussed for years within the scope of ecological ethics, an event to listen to the human-nature interaction, which is the main debate of ecological ethics, would be appropriate. It should not be overlooked that the basis of all ecological, sociological, and even psychological problems experienced today emerge from the practices of detachment from nature that alienate people from themselves, their species, and the planet they live on (Naess, 2005b). Accordingly, the necessity of leaving the naturalness of the cities established with the technological infrastructure open to discussion stems from this. Ecophobia, which has developed with thousands of years of experience, and the walls built between nature and humans as a reflection of this, have eliminated all our empathy for nonhuman life and made it easier for the damage to appear acceptable. One of the most significant threat potentials of smart eco-cities is that modern eco-cities built with technology raise this wall instead of removing it. The discovery of the real relationship with nature lies at the heart of ecological ethical approaches, whether it is an approach that calls for ecological communities to be shaped by individual enlightenment, such as deep ecology, or social ecology argues that social transformation will transform the individual. The prerequisite for this is abandoning the habit of being isolated from nature. Approaches that do not reject technology especially emphasize that technology and modernization should not blunt our ability to see the real nature, including all ecosystems, harmony, and symbiotic relations. Smart eco-city concept does not yet have globally accepted standards, that is, under development. At the formation stage, it is possible to transform the ecological part into a consideration of real issues by considering ecological and ethical principles beyond the generally accepted technical requirements. It should be more than possible for smart eco-cities to have a place in the ecologicalization adventure of cities.

Due to their intense technological structure, smart eco-cities are parallel to the second nature idea of social ecology, which accepts the naturalness of artificial areas and advocates the positive and improving effects of manufactured areas on nature (Bookchin, 1994, 2012, 2013). However, at this point, it should not be ignored that being able to talk about the existence of a social-ecological area depends on the existence of just and democratic structures and the fact that the ecology of the established cities should also include not breaking the human-nature interaction. In this context, the correct determination of the priority of cities is the prerequisite for the development of correct planning and practices. The criticisms of commercialization and company-style management developed for smart eco-cities in the literature bring over the prediction that these areas' "ecological" adventures will not be real and sustainable.

In the research and interpretations on the subject, it has been concluded that smart eco-cities are formed and carried out by coalitions of powerful actors (Hollands, 2015; McNeill, 2015; Ren, 2017; Wiig, 2015). Putting different goals other than ecologicalization based on the ecologicalization process, the fact that the reason that brings people together is not the effort of ecologicalization, and, as criticized, evaluating the efficiency of smart eco-cities in line with commercial outputs will bring the commercialization of the goals in the long run. This fact will bring about the softening, minimizing, and even tolerating ecological targets that should not be deviated from. And again, it's also likely to root problems (Caprotti, 2020). Therefore, it's necessary to have ecological ethical reasons appropriate for the culture and character of each region on its basis and develop all economic arrangements to be created by these objectives to ensure the ecological sustainability of exemplary projects such as smart eco-cities.

All the living spaces, building designs, landscaping and planning studies, energy production and consumption conditions, and production and consumption patterns should be aimed to comply with principles of ecological ethics, and all breakthroughs should be advanced toward this goal. Although it may seem that this can be achieved within strict rules, the main issue is internalization. In this context, the level of awareness of all actors involved in smart eco-cities, their internalization, their active participation, and their volunteering in adopting the lifestyle are important factors that directly affect the success of the areas (Bookchin, 2015; Naess, 1989).

It is also imperative to clearly state whether the priority of smart eco-cities is smartness or ecologicalness. Will technology play an intermediary role in the ecologicalization process in smart eco-cities, or will it be transformed into a control mechanism for individual freedoms, one of the red lines of every ecological ethical approach? In this context, the content of security and diagnostic systems should be determined and created by considering ethical dimensions.

REFERENCES

Bibri, S. E. (2021a). Data-driven smart eco-cities and sustainable integrated districts: A best-evidence synthesis approach to an extensive literature review. *European Journal of Forest Research*, *9*(16), 1–43.

Bibri, S. E. (2021b). Data-driven smart sustainable cities of the future: An evidence synthesis approach to a comprehensive state-of-the-art literature review. *Sustainable Futures*, *3*, 1–23.

Bogunovich, D. (2002). Eco-tech cities: Smart metabolism for a green urbanism. *WIT Transactions on Ecology and the Environment*, *54*.

Bookchin, M. (1974). Toward an ecological society. *Philosophica*, *13*.

Bookchin, M. (1994). *Ecology of freedo* (Çev. M. K. Coşkun). İstanbul: Ayrıntı.

Bookchin, M. (2012). Ecological crisis and the need to remaking the society. *Dipnot Sosyal Bilim Dergisi* (Çev. Aziz Adıgüzel), 8.

Bookchin, M. (2013). *Toward an ecologic society* (Çev. A. Yılmaz). İstanbul: Sümer.

Bookchin, M. (2015). *The ecology of freedom: The emergence and dissolution of hierarchy* (2. Baskı). (Çev. M. K. Coşkun). İstanbul: Sümer.

Cai, Z., & Tang, Y. (2021). Toward a sustainable city: A scoping review of eco-cities development and practices in China. *Chinese Urban Planning and Construction: From Historical Wisdom to Modern Miracles*, 179–199.

Cao, S., & Li, C. (2011). The exploration of concepts and methods for low-carbon eco-city planning. *Procedia Environmental Sciences*, *5*, 199–207.

Caprotti, F. (2016). Bristol. In F. Caprotti, R. Cowley, A. Flynn, S. Joss, & L. Yu (Eds.), *Smart-eco cities in the UK: Trends and city profiles 2016* (pp. 10–19). SMART-ECO Project. Exeter: University of Exeter.

Caprotti, F. (2020). Smart to green: Smart eco-cities in the green economy, In K. S. Willis & A. Aurigi (Eds.), *The routledge companion to smart cities* (pp. 200–209). New York, NY: Routledge.

Caprotti, F., Cowley, R., Flynn, A., Joss, S., & Yu, L. (2016). *Smart-eco cities in the UK: Trends and city profiles 2016*. Exeter: University of Exeter (SMART-ECO Project).

Cheng, H., & Hu, Y. (2010). Planning for sustainability in China's urban development: Status and challenges for Dongtan eco-city project. *Journal of Environmental Monitoring*, *12*(1), 119–126.

Cheshmehzangi, A. (2017). Tianjin (Sino-Singaporean Tianjin Eco-City). In M. Tan-Mullins, A. Cheshmehzangi, S.-S. Chien, & L. Xie (Eds.), *Smart-eco cities in China: Trends and city profiles 2016* (pp. 53–57). SMART-ECO Project. Exeter: University of Exeter.

Chien, S. S. (2017a). Chengdu. In M. Tan-Mullins, A. Cheshmehzangi, S. S. Chien, & L. Xie (Eds.), *Smart-eco cities in China: Trends and city profiles 2016* (pp. 6–11). Exeter: University of Exeter (SMART-ECO Project).

Chien, S. S. (2017b). Kunming. In M. Tan-Mullins, A. Cheshmehzangi, S. S. Chien, & L. Xie (Eds.), *Smart-eco cities in China: Trends and city profiles 2016* (pp. 19–26). Exeter: University of Exeter (SMART-ECO Project).

Cohen, B. (2014). *Smart city index master indicators survey*. Retrieved from www.smartcitiescouncil.com/resources/smart-city-index-master-indicators-survey. Accessed on March 30, 2023.

De Jong, M., Yu, C., Chen, X., Wang, D., & Weijnen, M. (2013). Developing robust organizational frameworks for Sino-foreign eco-cities: Comparing Sino-Dutch Shenzhen low carbon city with other initiatives. *Journal of Cleaner Production*, *57*, 209–220.

Des Jardins, J. R. (2013). *Environmental ethics*. Boston, MA: Cengage Learning.

Dong, H., Fujita, T., Geng, Y., Dong, L., Ohnishi, S., Sun, L., … Fujii, M. (2016). A review on eco-city evaluation methods and highlights for integration. *Ecological Indicators*, *60*, 1184–1191.

Eco-city Standarts. (2023). *Assessing and guiding progress towards ecologically healthy cities*. Retrieved from https://ecocitystandards.org/. Accessed on January 04, 2023.

Engwicht, D. (1992). *Towards an eco-city: Calming the traffic*. Sydney: Envirobook.

European Parliament. (2014). *Mapping smart cities in the EU*. Retrieved from https://www.europarl.europa.eu/RegData/etudes/etudes/join/2014/507480/IPOL-ITRE_ET(2014)507480_EN.pdf. Accessed on July 12, 2022.

Head, P., & Lam, D. (2011). How cities can enter the ecological age. In T. C. Wong & B. Yuen (Eds.), *Eco-city planning*. Dordrecht: Springer. doi:10.1007/978-94-007-0383-4_2

Hes, D., & Coenen, L. (2018). *Regenerative development and transitions thinking. Enabling eco-cities: Defining, planning, and creating a thriving future*. Singapore: Palgrave Pilot.

Hollands, R. G. (2015). Critical interventions into the corporate smart city. *Cambridge Journal of Regions, Economy and Society*, *8*(1), 61–77.

Huang, Z., Gao, J., & Liu, R. (2019). Discussion on the key success factors of developing smart eco-city based on analytic hierarchy process. *Ekoloji*, *28*(107), 1651–1657.

Jolivet, E., & Bond, A. (2018). *Smart-eco cities in France: Trends and city profiles 2017*. Toulouse: University of Toulouse I, Capitol.

Joss, S. (2016). London. In F. Caprotti, R. Cowley, A. Flynn, S. Joss, & L. Yu (Eds.), *Smart-eco cities in the UK: Trends and city profiles 2016* (pp. 30–38). Exeter: University of Exeter (SMART-ECO Project).

Kavcar, B. (2017). Eco-city urban model [Eko-Kent Kent Modeli]. Retrieved from http://sbpturkiye.com/eko-kent-modeli.html. Accessed on December 14, 2022.

Kayapınar, Y. E. (2017). Smart cities and application examples. *İTÜ Vakfı Dergisi*, *77*, 14–19.

Kline, E. (2000). Planning and creating eco-cities: Indicators as a tool for shaping development and measuring progress. *Local Environment*, *5*(3), 343–350.

Kloss, P. (2022). The interaction between resilience and intelligence of cities. In N. Minaei (Ed.), *Smart cities, critical debates on big data, urban development, and social environmental sustainability* (pp. 23–34). Boca Raton, FL: CRC Press.

Kondepudi, S., & Kondepudi, R. (2015). What constitutes a smart city? In A. Vesco & F. Ferrero (Eds.), *Handbook of research on social, economic, and environmental sustainability in the development of smart cities* (pp. 1–25). Hershey, PA: IGI Global.

Laleoğlu, B. (2021). *Smart cities, changing city management and Turkey*. İstanbul: SETA Siyaset, Ekonomi ve Toplum Araştırmaları Vakfı. Retrieved from https://setav.org/assets/uploads/2021/01/R179.pdf. Accessed on July 12, 2022.

McNeill, D. (2015). Global firms and smart technologies: IBM and the reduction of cities. *Transactions of the Institute of British Geographers*, *40*(4), 562–574.

Naess, A. (1989). *Ecology, community, lifestyle*. Cambridge: Cambridge University Press.

Naess, A. (1995). Self-realization. An ecological approach to being in the world. In G. Sessions (Ed.), *Deep ecology for the twenty-first century* (pp. 225–239). Boston, MA: Shambhala.

Naess, A. (2005a). The basics of deep ecology. *The Trumpeter*, *21*(1), 1–11.

Naess, A. (2005b). The deep ecology 'eight points' revisited. In H. Glasser & A. Drengson (Eds.), *The selected works of Arne Naess, deep ecology of wisdom* (pp. 57–66). Dordrecht: Springer.

Öztekin, E., & Çolak, S. (2022). Eco city concept for sustainable and low carbon emission urban areas; suggestions for Zonguldak Province. *Ankara Üniversitesi Çevrebilimleri Dergisi*, *9*(2), 62–73.

Register, R. (1987). *Ecocity Berkeley: Building cities for a healthy future*. Berkeley, CA: North Atlantic Books.

Register, R. (1993). *Eco-city*. Berkeley: North Atlantic Books.

Register, R. (2012). *Report from China – Progress at Tianjin eco-city and promise at Nanjing*. Retrieved from https://ecocitybuilders.org/report-from-china-progress-at-tianjin-eco-city-and-promise-at-nanjing/. Accessed on January 04, 2023.

Ren, X. (2017). Green as urban spectacle in China. In F. Caprotti & L. Yu (Eds.), *Sustainable cities in Asia* (pp. 77–85). Abingdon: Routledge.

Roseland, M. (1997). Dimensions of the eco-city. *Cities*, *14*(4), 197–202.

Sengers, F. (2016). *Smart-eco cities in the Netherlands: Trends and city profiles 2016*. Exeter: University of Exeter (SMART-ECO Project).

Späth, P. (2017). *Smart-eco cities in Germany: Trends and city profiles*. Exeter: University of Exeter (SMART-ECO Project).

Tan-Mullins, M. (2017). Shanghai. In M. Tan-Mullins, A. Cheshmehzangi, S. S. Chien, & L. Xie (Eds.), *Smart-eco cities in China: Trends and city profiles 2016* (pp. 41–44). Exeter: University of Exeter (SMART-ECO Project).

T.C. The Ministry of Environment and Urbanisation. (2019). *2020–2023 national smart cities strategy and action plan*. Retrieved from https://www.akillisehirler.gov.tr/wp-content/uploads/EylemPlani.pdf. Accessed on July 12, 2022.

Tosun, K. E. (2013). The analysis of compact city model in the process of sustainable urban development. *Dokuz Eylül Üniversitesi Sosyal Bilimler Enstitüsü Dergisi*, *15*(1), 103–120.

UNEP. Sustainable Development Goals. Retrieved from https://www.unep.org/evaluation-office/our-evaluation-approach/sustainable-development-goals. Accessed on January 10, 2023.

Urban Ecology. (1990). *Eco-city conference 1990 report*. Retrieved from https://ecocitybuilders.org/ecocity-1-berkeley-usa-1990. Accessed on March 02, 2023.

Van Antwerp, J., & Heun, M. K. (2022). An introduction to a framework for sustainability thinking. *Numeracy: Advancing Education in Quantitative Literacy*, *15*(2), 1–8.

Welter, V. M., & Lawson, J. (2000). *The city after Patrick Geddes*. Oxford: Peter Lang.

Wiig, A. (2015). IBM's smart city as techno-utopian policy mobility. *City*, *19*(2–3), 258–273.

Wong, T. C., & Yuen, B. (2011). Understanding the origins and evolution of eco-city development: An introduction. In T. C. Wong & B. Yuen (Eds.), *Eco-city planning: Policies, practice and design* (pp. 1–14). Dordrecht: Springer.

Xie, L., Tan-Mullins, M., & Cheshmehzangi, A. (2019). Political ecology of Chinese smart eco-cities. In X. Zhang (Ed.), *Remaking sustainable urbanism: Space, scale and governance in the new urban era* (pp. 57–78). Singapore: Palgrave Macmillan.

Yigitcanlar, T., & Lee, S. H. (2014). Korean ubiquitous-eco-city: A smart-sustainable urban form or a branding hoax? *Technological Forecasting and Social Change*, *89*, 100–114.

Yu, L. (2016). Milton Keynes. In F. Caprotti, R. Cowley, A. Flynn, S. Joss, & L. Yu (Eds.), *Smart-eco cities in the UK: Trends and city profiles 2016* (pp. 48–56). Exeter: University of Exeter (SMART-ECO Project).

Zhou, N., He, G., Williams, C., & Fridley, D. (2015). ELITE cities: A low carbon eco-city evaluation tool for China. *Ecological Indicators*, *48*, 448–456.

URBAN TRANSPORT SOLUTIONS FOR A SUSTAINABLE AND SMART MOBILITY FUTURE: MACRO-ENVIRONMENTAL ANALYSIS

Marwa Ben Ali and Ghada Bouketaya

ABSTRACT

For decades, the fast population growth worldwide was interrelated with the adopted rapid lifestyle behavior that relies on the extensive use of fossil fuels. This primary energy source has caused various urban and environmental impacts, such as global warming, air pollution, and so forth. Consequently, the identified circumstance issues have caused many health, social, and economic hindering effects for global citizens. It poses an existential threat to humanity and the global earth's ecosystem. The alarming levels of urban pollution emissions are putting enormous challenges to the related stakeholders (governments, businesses, investors, automakers, and citizens) to admit the need to decarbonize the global economy and reach sustainable development goals (SDGs) for cleaner and smarter cities across borders. As such, a vital part of a smart city is the transport sector. The road transport sector, precisely, is one of the primary consumers of fossil fuels that contribute to high carbon dioxide emissions. Accordingly, it is essential to adopt novel and sustainable urban transport solutions and promote the achievement of the SDG's eleventh goal for sustainable cities and communities. This chapter provides insight into the present global energy situation with particular attention to the road transport sector. Indeed, it highlights different emerging technologies for a sustainable and smart urban mobility future that will mitigate the environmental situation thanks to the development of drive and internet telecommunication technologies. Furthermore, we aim in this chapter to study the international progress of the transition project using the Political, Economic, Social, Technological, Environmental, and Legal (PESTEL) analysis methodology. This study is to pinpoint opportunities for project development and the challenges that set back its evolution.

Smart Cities for Sustainability
Advanced Series in Management, Volume 32, 49–70

ISSN: 1877-6361/doi:10.1108/S1877-636120230000033005

Keywords: Smart cities; road transport sector; transition; ecosystem; smart mobility; emerged technologies; opportunities; challenges; PESTEL analysis

1. INTRODUCTION

More than half of the world's population now lives in urban areas, as highlighted in Fig. 4.1, which is expected to continue in the coming decades to attain possibly 70% (Ehrlich, Melchiorri, & Capitani, 2021; Haque, Bhushan, & Dhiman, 2021; Little, 2016).

Urban settings are a relatively new experience in human history where the population is very high and has the features of a built environment that provides basic facilities for human activity such as education, entertainment, economy, and health care (Huijsmans, Harteveld, van der Brug, & Lancee, 2021). However, rural is the geographical region located in the outer parts of the cities or towns with low population density. These communities often hold a reputation as being technologically backward and less advance. In addition, life is seen as moving at a slower pace, and opportunities are less abundant (Spicer, Goodman, & Olmstead, 2021).

Therefore, this transition has transformed the way of life, work, travel, and building networks, which are entirely dominated by the energy demand and rely on fossil fuel utilization that requires urgent solutions in different sectors: industry, residential, commerce, transport, and agriculture (Avtar, Tripathi, Aggarwal, & Kumar, 2019).

This entry presents an overview of urbanization across the world, extending from the distant past to the present, and projections of future trends and challenges for the development of smart, clean, and available urban mobility within smart cities that contribute to a more sustainable and intelligent ecosystem.

As cities' various opportunities and attractions draw in more and more countries, they have also created problems in fields such as the environment and transportation, especially the urban road transport sector (harmful emissions, congestion, accidents, and so forth) (Grincheva, 2022). The use of various technologies to solve such urban development problems comes under the banner of "smart cities." The world has been strongly

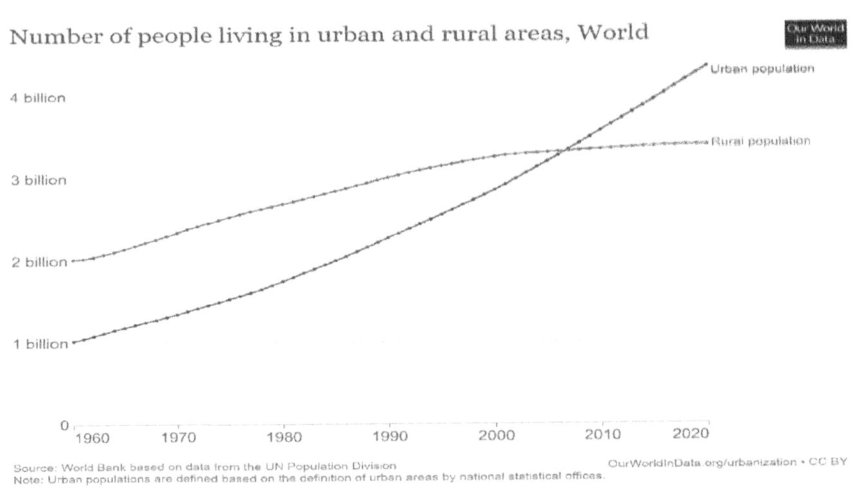

Fig. 4.1. Worldwide Urban and Rural Areas Development. *Source:* Our World in Data.

implicated in promoting better urban environments in the last decades with various advanced and emerging technologies diffusion. This special feature issue introduces the significance and features of smart city initiatives, along with particular attention to the urban sector, which entirely leans on fossil fuel consumption (Thondoo, Marquet, Márquez, & Nieuwenhuijsen, 2020).

The new emerging urban mobility technologies seem to be an efficient solution to deliver transport improvements and resolve road sector issues faster and smarter instead of extending physical infrastructure. They refer to the use of electronics, information, bidirectional communications process, and novel transportation technologies while saving money, and time, reducing environmental impact with lower fuel consumption, lower Carbon dioxide emissions, and better travel journey with minimum congestion and accidents (Pašalić, Ćukušić, & Jadrić, 2021).

Consequently, this chapter introduces a general insight into representative studies from the scientific literature surrounding the theme of smart cities and transportation systems innovations: (1) novel technologies to cut oil use in the automotive sector; (2) intelligent transportation systems and their positive impacts to improve the general transport system; and (3) the challenges and opportunities analysis affecting the discussed technologies.

The remainder of the present chapter is structured as follows. After the introductory part, the chapter remains with the presentation of the general concept of a Smart City. *This section* is important since it provides worthwhile details on the concept, features, and parties concerned with the related issues. The chapter then proceeds to a somewhat more detailed examination of the general energy present situation with particular attention to the transport sector worldwide in Section 3. Section 4 of the chapter presents and discusses the smart mobility actors and innovations. The challenges and opportunities of using macro-environmental analysis are discussed in Section 5. Finally, the conclusions of this study are extracted in Section 6.

2. CONCEPT OF SMART CITY

To deal with the Smart City concept, we considered comprehensive bibliometric data research from the online PubMed database between 1990 and 2022. Consequently, Fig. 4.2 highlights the dramatic growth of research in the studied field mainly after the worldwide COVID-19 pandemic. A deep insight into the Word cloud presentation in Fig. 4.3, the present research focuses on different solutions based on novel technologies such as internet of things (IoT), deep learning, wireless sensor network, artificial intelligence, big data, and machine learning not only to reduce the energy crisis but also the environmental issues.

The Smart City concept's appearance refers to various definitions in the research papers deepening with the intelligent, connected, green, and sustainable terms (Yuloskov, Bahrami, Mazzara, & Kotorov, 2021). Still, several examples of smart city definitions can be found in recent research papers concerning information and communications technology (ICT) development:

- *D1*: Smart Cities combine their objectives to enhance the quality of life, adopt ICTs in urban systems, implement new governance and focus on human capital and sustainability (Desdemoustier, Crutzen, & Giffinger, 2019).
- *D2:* A city that is supported by the pervasive presence and massive use of advanced ICT, which, in connection with various urban systems and domains and how these intricately interrelate and are coordinated respectively, enables the city to control available

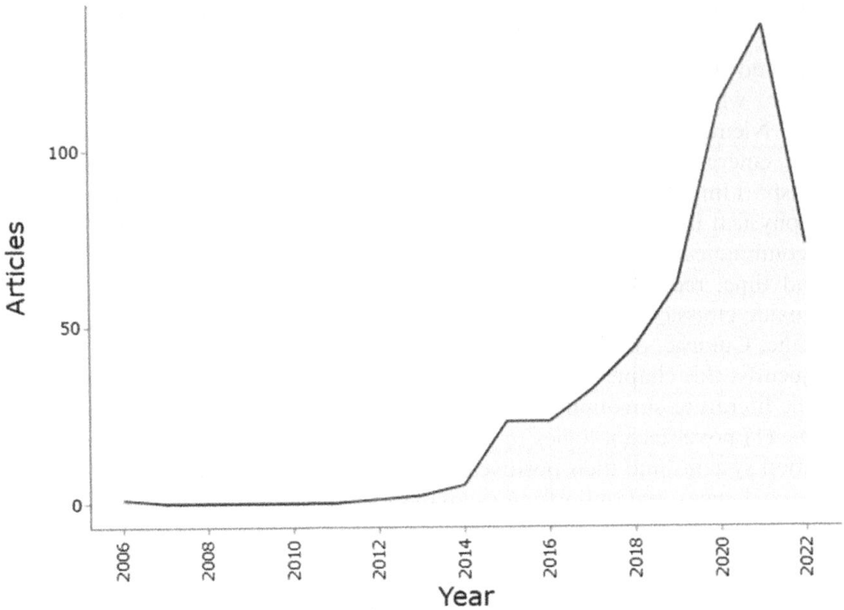

Fig. 4.2. Annual Scientific Production (1990–2022). *Source:* Authors' elaboration using biblioshiny.

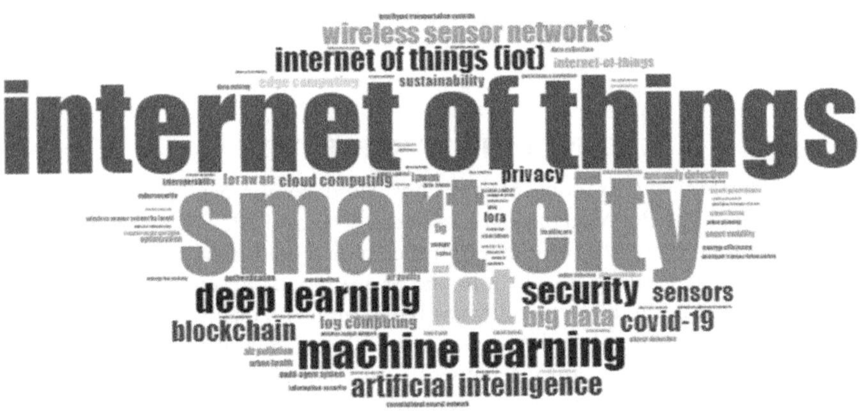

Fig. 4.3. Word Cloud for Smart City. *Source:* Authors' elaboration using biblioshiny.

resources safely, sustainably, and efficiently to improve economic and societal outcome (Bibri, 2018).

- *D3:* A smart city involves much more than ICT infrastructure; it contains other dimensions such as the social or ecological dimensions, which modify this technological scenario, adding social value through innovation and creativity (Barba-Sánchez, Arias-Antúnez, & Orozco-Barbosa, 2019).
- Meanwhile, it refers, in general, to the connected, interlinked, and innovative city-systems technologies with real-time management and organization toward better

citizens' life pace, better indoor environment, and sustainability of the global city systems based on digital shadow from various data collection using innovative technologies and emerging novel ITS generations.

The smart city concept, as presented in Fig. 4.4, is the combination of different urban systems using interconnected communication flow (Abu-Rayash & Dincer, 2021; Ribeiro, Dias, & Pereira, 2021) including:

- smart people,
- smart economy,
- smart mobility,
- smart environment,
- smart living,
- smart governance.

It is about the management and use of information more effectively between the different systems to develop adequate solutions for citizens based on the internet diffusion and bidirectional communication flow using advanced new technologies such as the IoT (Ang, Seng, Ngharamike, & Ijemaru, 2022). Hence, this technology base and backbone are made up of interconnected devices and sensors.

The IoT is a system of interrelated physical devices and objects with the ability to communicate and exchange information over a network. It is simply the relationship between person and machine (home loads, sensors, cameras, vehicles, robots) in a bidirectional data acquisition flow. It's a recent communication standard that aims at making the internet an even more universal tool all over the world (Echendu & Okafor, 2021; Eseosa & Temitope, 2019; Pradhan, Patankar, Shinde, Shivarkar, & Phadatare, 2017, pp. 2003–2006). As a result, internet diffusion is a crucial factor in this project.

In this vein, we can inspect the global access to the internet, which has been rising gradually as presented in Table 4.1. World internet usage and population statistics in the first quarter of 2022, 5.25 billion people, or 38% of the world's population, had Internet

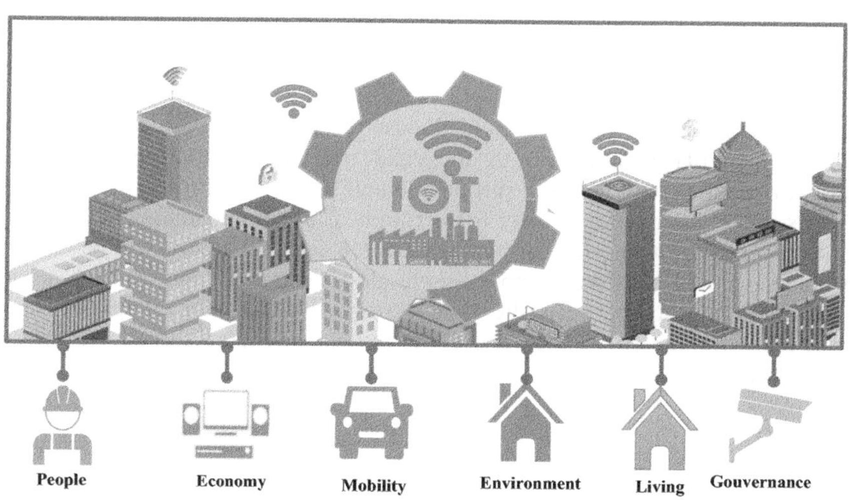

Fig. 4.4. Smart City Principal Systems.

Table 4.1. World Internet Usage and Population Statistics 2022 Year-Q1 Estimates.

World Internet Usage and Population Statistics 2022 Year-Q1 Estimates						
World Regions	Population (2022 Est.)	Population % of World	Internet Users Dec 31, 2021	Penetration Rate (% Pop.)	Growth 2000–2022	Internet World %
Africa	1,394,588,547	17.6%	601,327,461	43.1%	13,220%	11.5%
Asia	4,350,826,899	54.8%	2,790,150,527	64.1%	2,341%	53.1%
Europe	841,319,704	10.6%	743,602,636	88.4%	608%	14.2%
Latin America/ Carib.	663,520,324	8.4%	533,171,730	80.4%	2,851%	10.1%
North America	372,555,585	4.7%	347,916,694	93.4%	222%	6.6%
Middle East	268,302,801	3.4%	205,019,130	76.4%	6,141%	3.9%
Oceania/ Australia	43,602,955	0.5%	30,549,185	70.1%	301%	0.6%
WORLD TOTAL	7,934,716,815	100.0%	5,251,737,363	66.2%	1,355%	100.0%

Source: Copyright © 2022, Miniwatts Marketing Group.

access. Asia is the largest internet user with 53.1%, followed by Europe with 14.2%, and the Africa region with 11.5%, while Latin America/Carib, North America, Middle East, and Oceania/Australia are characterized by 10.1%, 6.6%, 3.9%, and 0.6% of internet usage respectively. These percentages will be duplicated in the recent future due to the new technology diffusion (smartphones) and the new connection generations, which are barely related to the global internet spread.

Meanwhile, different preceding studies addressed various topics and methods dealing with the sustainable smart cities concept and its connected systems. Table 4.2 summarizes some studied international and recent research papers and their adopted methodologies.

However, an essential part of a smart city is the transport sector. As highlighted before, the transport sector is one of the primary consumers of fossil energy and carbon dioxide emissions contributors, mainly the road sector. Consequently, we need to promote the transition to sustainable and eco-friendly urban mobility worldwide as a way to improve the quality of the environment through a decrease in fuel consumption and harmful emissions.

In this chapter, we will introduce and discuss the current state of the global energy situation with particular attention to the road transport sector, developments, and emerging advanced technologies for sustainable and smart mobility. In addition, we will focus on its ability to prepare society for a future of green and reliable cities. For this reason, diverse novel transportation: trends, technologies, and factors will be discussed in the upcoming sections that may help preserve global economic and ecological sustainability. The relationship between the road transport sector and energy consumption worldwide seems to be very strong. Therefore, the global energy context will be studied first in the upcoming section.

Table 4.2. Summary of Different Literature Studies.

Ref	Country	Year	Topic	Method/Device Used
Smart City Overall Concept				
Varró and Szalai (2021)	Central Eastern Europe: Hungary's big cities	2021	• Policies to improve the novel concept in Hungary	• Review based on two previous projects. • Qualitative methods.
Lim, Malek, Yussoff, and Yigitcanlar (2021)	Malaysia	2021	• Citizen's understanding and acceptance of smart city policies	• Fuzzy Delphi method • Survey
Yuloskov et al. (2021)	Russia	2021	• Current situation and insights for future development	• Analysis method of a case study.
Ang-Tan and Ang (2021)	Hong Kong and Singapore	2021	• Public sector innovation	• Comparative study
Echendu and Okafor (2021)	Africa	2021	• A smart city as a promising solution	• Literature review analysis
Syalianda and Kusumastuti (2021)	Indonesia	2021	• Case study of Jakarta smart City, Indonesia and an overview of its concept	• Qualitative research method literature review analysis
Pašalić et al. (2021)	Southeast Europe	2020	• Smart city innovation	• Review based on recent research papers • Plan-Do-Check-Act (PDCA)
Smart Economy				
Qian, Liu, Cheng, and Forrest (2021)	China	2021	• Urban economy	• Test and data analysis
Kumar and Dahiya (2017)	–	2017	• Smart economy	• Analysis
Smart Mobility				
Ruggieri, Ruggeri, Vinci, and Poponi (2021)	Europe	2021	• Electric mobility	• Review
Benevolo, Dameri, and D'auria (2016)	–	2016	• Smart mobility	• Analysis
Smart Governance				
Giuliodori, Berrone, and Ricart (2022)	–	2022	• Smart governance	• Data collection • An ordinary least squares (OLS) regression analysis with White's correction
Meijer and Bolívar (2016)	-	2016	• Smart governance	• Review
Smart People				
Ji, Chen, Wei, and Su (2021)	Taiwan	2021	• Citizens preferences and perspectives	• Questionnaire design • Data collection • Analysis methods
Capdevila and Zarlenga (2015)	Barcelona	2015	• Smart citizens	• Qualitative, based on a case study
Smart Environment				
Ullo and Sinha (2020)		2022	• Smart environment	• Review
Pradhan et al. (2017)	–	2017	• Smart environment	• IOT and communication system development
Aletà, Alonso, and Ruiz (2017)	Spain	2016	• Smart environment	• Qualitative, based on a case study

(Continued)

Table 4.2. (*Continued*)

Ref	Country	Year	Topic	Method/Device Used
Smart Building				
Apanaviciene, Vanagas, and Fokaides (2020)	–	2022	• Smart building	• Review
Han and Kim (2021)	–	2021	• Smart living	• Review
Chieng and Mustapa (2021)	Malaysia	2021	• Smart living concept in Malaysia	• Literature review analysis • Comparative study: Singapore, Finland, Thailand, and Ho Chi Minh
Eseosa and Temitope (2019)	Nigeria	2019	• The necessity to use the building management system	• Review

3. AN OVERVIEW

3.1 Global Energy Situation

According to the statistics of the British oil company (British Petroleum, 2019) illustrated in Fig. 4.5, the increase in global energy consumption is remarkable with the development of science, the growth of populations, and their different requirements (education, transportation, and health), and the rapid pace of life. Indeed, the lack of fossil sources with the increase in raw materials costs such as oil and gas requires an efficient, sustainable, green solution aiming to mitigate environmental issues attributed to human activities.

It is also noted that the evolution of the global consumption of renewable energy sources has experienced a particular growth over the last 20 years. Indeed, attention to the development of renewable sources for energy production makes it possible to meet the sustainable energy needs of future generations worldwide.

Nevertheless, the percentage of fossil fuel production in the world will decrease from 60 billion barrels to −5 billion barrels by the year 2050. Unfortunately, it is estimated that all

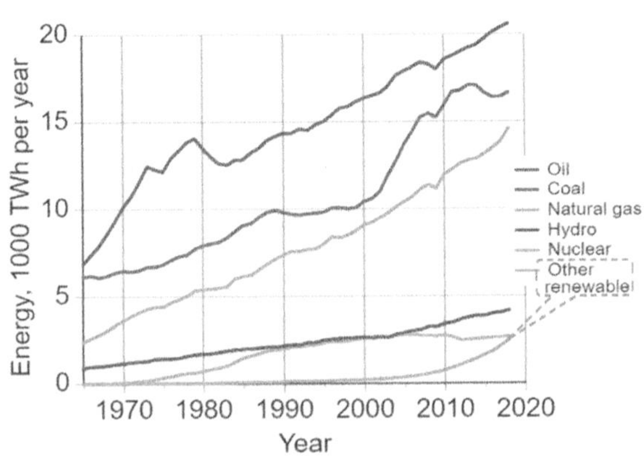

Fig. 4.5. World Consumption of Energy. *Source:* BP Statistical Review of world energy.

our fossil sources will be depleted by 2060 worldwide (Ibn-Mohammed, 2017). To the problem of energy resource depletion, including oil, coal, and gas, are added the environmental issues and climate change that have seriously threatened the world for the last 10 decades due to the intensive consumption of fossil fuels worldwide (Holechek, Geli, Sawalhah, & Valdez, 2022). Therefore, humanity has experienced the challenge of finding a compromise to ensure the best energy equilibrium and transit to a low-carbon economy while taking into account the ecological aspect concerning the vigorous consumption of fossil fuels in different regions and sectors.

In the present study, we will focus on the road transport sector. It has relied entirely on fossil fuels generating, therefore, an increase in greenhouse gas emissions in the environment, mainly the carbon dioxide emissions that will be examined in the following subsection.

3.2 Transport Sector General Situation

Upon closer inspection, it appears that in 2020 the transport sector was the second-largest carbon dioxide contributor, as presented in Fig. 4.6, accounting for 20.27% of the global energy-related emissions with passenger vehicles alone presenting 41% of the total, as highlighted in Fig. 4.7. Road transport sector emissions fell in 2020, but this was only due to the COVID-19 pandemic with high restrictions worldwide. The high level of carbon dioxide emissions is due to the reliance of the automobile sector on fossil fuel utilization. These harmful emissions cannot be reduced easily or eliminated unless switching to other technologies and promoting novel solutions to mitigate the environmental risks.

As a consequence, analyses of the effect of the energy crisis and climate change generate enormous public and political awareness with 96% concern growing over the past decade encouraging the different stakeholders to take an urgent action plan on this top international topic. Furthermore, the ambition of reducing oil demand should go beyond the emergency measures and assume a more structural nature to enhance energy security and climate change goals also in the longer term.

Given the predominant role of oil in the transportation sector, it is not a surprise that most of the problem potential is in the mobility sector as discussed. It is a worrying situation, which alerts us to think and plan intelligently about the quantity and type of energy

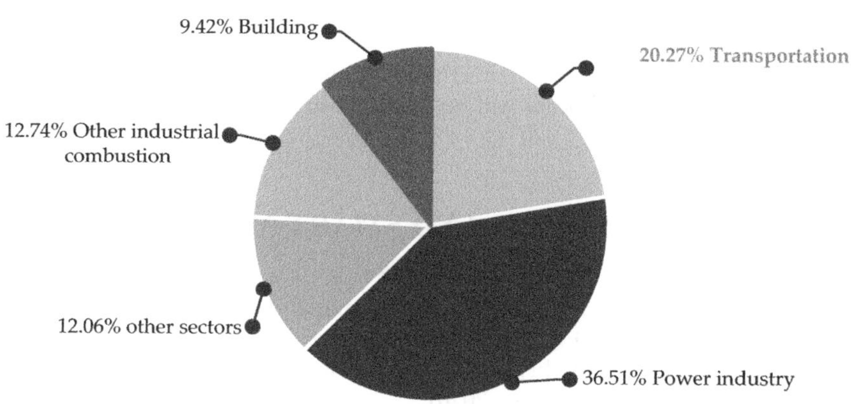

Fig. 4.6. Distribution of Global Carbon Dioxide (CO_2) Emissions in 2020, by Sector. *Source:* Statista.

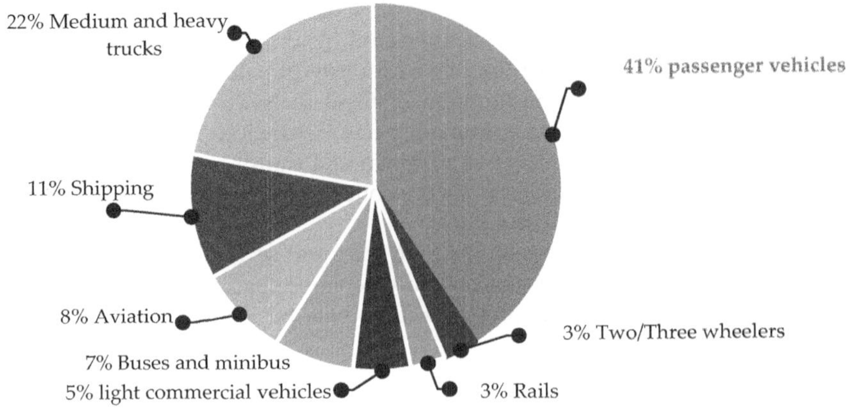

Fig. 4.7. Distribution of Carbon Dioxide Emissions Produced by the Transportation Sector Worldwide in 2020, by Subsector. *Source:* Statista.

resources to be used in the future. Indeed, the calls for governments to adopt an urgent action plan for a more sustainable and intelligent transition toward smart mobility sector development (Ben Ali & Boukettaya, 2022).

In this vein, various countries throughout the world have started their efforts in designing and implementing smart innovations, and technologies dedicated to the mobility sector to prepare society for the realization of future smart cities to manage different challenges (Ribeiro et al., 2021; Singh, Solanki, Sharma, Nayyar, & Paul, 2022) including:

- climate mitigation,
- reduce fuel consumption,
- reduce Carbone dioxide emissions,
- reduce congestion,
- facilitate the citizen's life pace,
- manage transport systems (users, machines, and so on) in real time,
- promote new ideas and solutions,
- change the citizen's negative perceptions toward the transition to cleaner solutions.

An examined study referred to the criteria of the sustainable and smart mobility elaboration for a smart city concept will be discussed in the upcoming section.

4. SUSTAINABLE AND SMART MOBILITY

4.1 Definition

Smart mobility is the integration and connection of numerous components of technologies and motilities as highlighted in Fig. 4.8. These innovative solutions are designed to increase the efficiency of the urban mobility sector for a sustainable future with economic and ecological objectives (Nagy & Csiszár, 2020).

The smart mobility concept surrounds in general the following criteria (Butler et al., 2020):

Fig. 4.8. Mobility Features. *Source:* Authors' elaboration using WordCloud.

- accessibility,
- availability of adequate and innovative infrastructure,
- sustainability,
- safety.

Mobility and innovation as a theme gained the moderate attention of researchers worldwide in different fields (electrical, management, civil, mechanical, and so on). Therefore, in this chapter, we will divide the literature studied themes into two major subjects:

(1) *Novel transportation technologies*: transition to electric and autonomous automobile technologies and its integration and connection mainly with alternative renewable energy sources (solar, wind).
(2) *Intelligent transportation systems*: intelligent solutions to help reach the target objectives with its connection with green automobile technologies. It includes different topics such as:
  - Parking solutions (Herrera-Quintero, Vega-Alfonso, Bermúdez, Marentes, & Banse, 2019; Mangiaracina, Tumino, Miragliotta, Salvadori, & Perego, 2017).
  - Traffic management system including public transport (Lingani, Rawat, & Garuba, 2019; Sumi & Ranga, 2018).
  - Digital mobility platforms and applications (Pflügler, Schreieck, Hernandez, Wiesche, & Krcmar, 2016).
  - Connected infrastructure with sensors, radars and cameras (Pandharipande & Thijssen, 2019), and so forth.

4.2 Novel Transportation Technologies

4.2.1 Electric Technology

More than 600 million vehicles are on roads around the world in 2012, to increase now to 1.2 billion, and this number is expected to increase significantly (Dahham, Wei, & Pan, 2022). It implies the significant impact of carbon dioxide emissions from the combustion of fuels during the propulsion the internal combustion engine vehicles (ICEVs), which has led to terrible global warming (Dahham et al., 2022). We argue that a concrete energy security plan in the road transport sector must reconcile strategies according to the transition and shift to sustainable and smart mobility. It became a necessity worldwide based on the evolution of the government's regulations and policy recommendations to overcome the global situation.

Focusing on road transport sector decarbonization, consequently, electrified automobile technology has become the most efficient alternative to conventional vehicles (CVs) all over the world, which offers zero emission, zero noise, low operational cost, and sustainable transportation. According to Fig. 4.9, we can investigate the electric vehicles (EVs) technologies market share growth from 0.2% to 8.3% between 2012 and 2021 respectively. However, this is not enough to mitigate mainly the environmental risks.

This significant attention is assigned to the many particular characteristics of the EV technology types: battery EV (BEV), hybrid EV (HEV), and plug-in HEV (PHEV) as presented in Table 4.3.

4.2.2 Autonomous Vehicle

It is also known as self-driving vehicles. It has recently emerged as an upcoming technology that will be a replacement for the vehicle driver. Several empirical studies (Acheampong, Cugurullo, Gueriau, & Dusparic, 2021; Chehri & Mouftah, 2019; Pakusch, Stevens, & Bossauer, 2018; Seuwou, Banissi, & Ubakanma, 2020) have focused on the necessity of implementing novel autonomous transportation modes to develop a more sustainable automobile future that aims to eliminate human driving error (Seuwou et al., 2020). It can be involved mainly as a taxi and car-sharing market to reduce travel journey

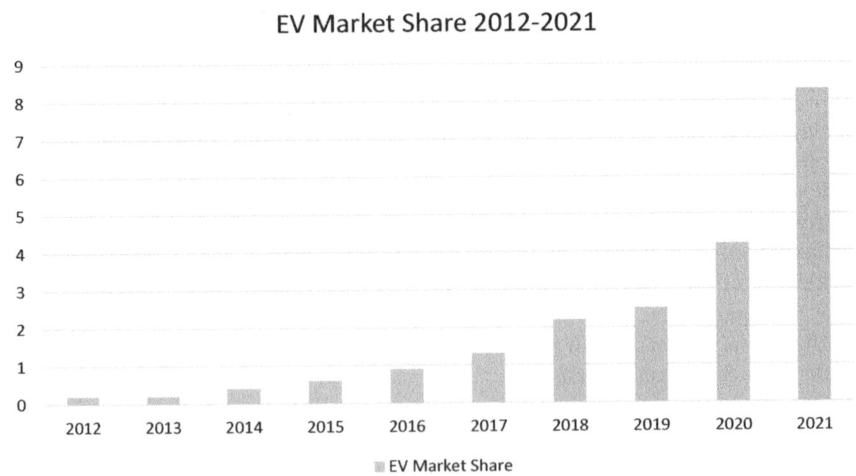

Fig. 4.9. Electric Vehicle Market Share. *Source:* Roland (2021).

Table 4.3. Characteristics of Electric Technology Types: Battery Electric Vehicle (BEV), Hybrid Electric Vehicle (HEV), and Plug-in Hybrid Electric Vehicle (PHEV) (Bagheri et al., 2021; Hasan, Mahmud, Habib, Motakabber, & Islam, 2021; Juda, 2002; Patyal, Kumar, & Kushwah, 2021).

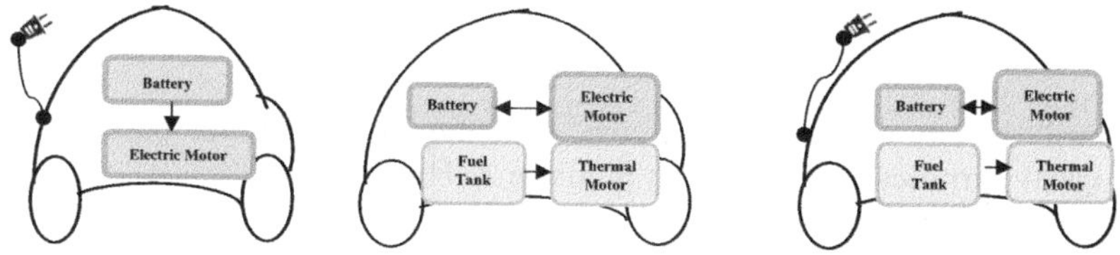

	BEV	HEV	PHEV
Constitution	• Electric motor • Battery pack.	• ICE. • Fuel tank. • Electric motor. • Battery pack.	• ICE. • Fuel tank. • Electric motor. • Battery pack.
Electrification degree	• 100% electric.	• Low.	• High.
Electric distance(km)	• 100–500.	• 10–25.	• 10–60.
Battery charging	• Public Charge stations. • Home.	• Regenerative braking. • Gasoline engine.	• Public Charge stations. • Home. • Regenerative braking.
Characteristics	• Plugged in. • 0% Carbon dioxide emission. • Sustainable Solution.	• Low Carbon dioxide emission. • Higher range of vehicle due to the extra fuel tank comparing to 100% EV. • Capable to displace petroleum energy for transportation with multi source electrical energy.	• Low Carbon dioxide emission. • Capable to displace petroleum energy for transportation with multi source electrical energy. • High range of vehicle due to the extra fuel tank. • Plugged in.

time and maximize the user's comfort. Meanwhile, the advantages of autonomous vehicle utilization are attractive and seem to be an intelligent solution for economic and environmental issues. It lead to a remarkable reduction in vehicles by 28%, parking places by 44%, and travel time by 30%; also it helped to reduce 66% of harmful emissions and 87% of accidents (Chehri & Mouftah, 2019).

Meanwhile, as discussed in previous subsections, the decrease in the potential of fossil energy and the degradation of the climatic conditions in the world ensure that the transition to the discussed technologies (electric and autonomous vehicles) and their integration into networks is unquestionably an irrevocable and necessary solution. Indeed, this connection requires reinforced effort and investment to incorporate an innovative and robust charge infrastructure network with the ITS implementation to facilitate the communication flow within an intelligent ecosystem. The ITS will be reviewed in the upcoming subsection.

4.3 Intelligent Transportation Systems

Intelligent transportation systems are considered one of the major elements of any smart city (Raza, Bukhari, Aadil, & Iqbal, 2021). Intelligent systems are real-time information technologies that can connect the different users, platforms, vehicles, applications, and governments for:

- better management with lower risks,
- efficient data acquisition,
- safe bidirectional information flow,
- healthy and rapid services with climate and time change.

It will directly affect the time, cost, and congestion of roads based on economic and ecological purposes (Mallik, 2014; Sumalee & Ho, 2018).

The ITS innovations include but are not limited to:

- sensors (Guerrero-Ibáñez, Zeadally, & Contreras-Castillo, 2018; Tasgaonkar, Garg, & Garg, 2020),
- cameras (Baran, Rusc, & Fornalski, 2016),
- bluetooth (Friesen & McLeod, 2015),
- drone (Akram et al., 2021; Sharma et al., 2018),
- Global Positioning System (GPS) (Aloi, Fay, Ronning, & Willer, 2003).

All these technologies are involved in different applications such as traffic management, driving assistance, automatic incident detection, vehicle-to-grid application (V2G) application, geographic and weather information, and automatic parking.... Thanks to the development of various novel communication and database technologies including big data, AI, IoT integrated with the new fifth-generation (5G) technology of wireless mobile communication (Gohar & Nencioni, 2021), which is capable to connect billions of devices mainly the vehicle to everything with high efficiency. Although, this amount of collected data must be controlled and managed, which can help in better outcomes, interpretations, and smarter decisions (Gharaibeh et al., 2017).

All discussed innovations have affected the transport sector in a smart city concept. A large number of recent studies have been carried out on this topic. They adopt different methods and strategies focused on the ITS integration and its positive effect on mobility change concerning time, security, emissions, and fossil fuel consumption.

For instance, Chen, Ardila-Gomez, and Frame (2017) developed a literature review about the role of ITS, for smart mobility in the smart city concept, with a deep focus on the energy-saving topic. Farid, Viswanath, Al-Junaibi, Allan, and Van der Wardt (2021) discussed the necessity of suitable ITS infrastructure integration with the diffusion of EV technology in the Abu Dhabi case. Indeed, Gohar and Nencioni (2021) briefly discussed the impact of the 5G methodology in the transportation sector in a smart city. They highlighted the capability of ITS with 5G to lead to a more sustainable and economic transport sector.

In the same context, Qoradi, Al-Harbi, and Aina (2021) studied the effect of the integration of ITS with geographic information system (GIS) using a wireless sensor network (WSN) at King Saud University city for female students in Riyadh, Saudi Arabia. They approved the efficiency of the implemented technology in the quality, security, and safety of the students with a lower time span using a bus as a transportation tool. Furthermore, Roy, Tariq, and Dey (2021) proved that the proposed socio-technical approach for cyber-attack detection is able to detect attack occurrences with reasonable accuracy and speed in connected transportation systems.

Hernafi, Ahmed, and Bouhorma (2016) proposed an innovative concept based on human driver behavior on ITS by proposing a new model of communication to prevent accidents from happening and to ensure smart mobility from a security aspect perspective. However, a combined Micro-Drones with vehicular ad hoc network (VANET) architecture, which successfully helps in reducing packet collisions and optimizing energy consumption has been proposed in (Raza et al., 2021). Furthermore, with the purpose to enhance the economic and ecological issues worldwide, Balasubramaniam, Paul, Hong, Seo, and Kim (2017) highlighted the advantages of integration of ITS with the autonomous vehicle, which seems to be promised solution for future generations.

Based on these research studies and beyond all the positive impacts of the emerged technologies, the transition process is still considerably low in developing countries such as Tunisia, Algeria, Egypt, Libya, and others. However, some developed countries are recording appreciable progress. Take the example of the United Kingdom, where the percentage of the EV market has increased in adoption by 291% just in the last two years due to adequate infrastructure implementation.

Consequently, in the next section, we will highlight, discuss, and analyze some of the open challenges that may set back advanced technologies as well as the opportunities that may scale up the transition process toward smart mobility in a smart city.

5. CHALLENGES AND OPPORTUNITIES: PESTEL ANALYSIS

As discussed in the previous sections and proved in different research studies, novel solutions for the road transport sector (vehicles, infrastructure, and digitalization) with changing habits (walking, cycling, sharing, and so forth) can improve the global situation and solve the energy crisis with a positive impact on the environment. Yet, various challenges need further investigation to overcome the general transport sector situation with particular attention to the road sector.

The direction of the road sector's future is affected by political, economic, social, technical, and environmental factors laying ahead for these technological nowadays countries' situations. Therefore, we will use the PESTEL analysis methodology to assess the macro-environment factors that affect the slow evolution of the sustainable transition project as presented in Fig. 4.10.

PESTEL is an analysis methodology that aims to identify and evaluate the external factors that play an essential role in shaping a project/technology environment and the chances of its success, which was invented by Francis Aguilar 50 years ago (Christodoulou & Cullinane, 2019).

5.1 Political and Legal Factors

Political and legal challenges are one of the major concerns for adopting and diffusion of the novel green, digital, and sustainable technology, as legislation, subsiding, investment, piloting, data protection law, and public–private stakeholders (governments, users, sponsors, policymakers, industries) cooperation and coordination and government's wiliness, as well as the political stability situation of the country. Those terms align with the promotion policy for new technologies and the attraction of the citizen's intention. Furthermore, the diffusion of novel technologies will give an impressive country image.

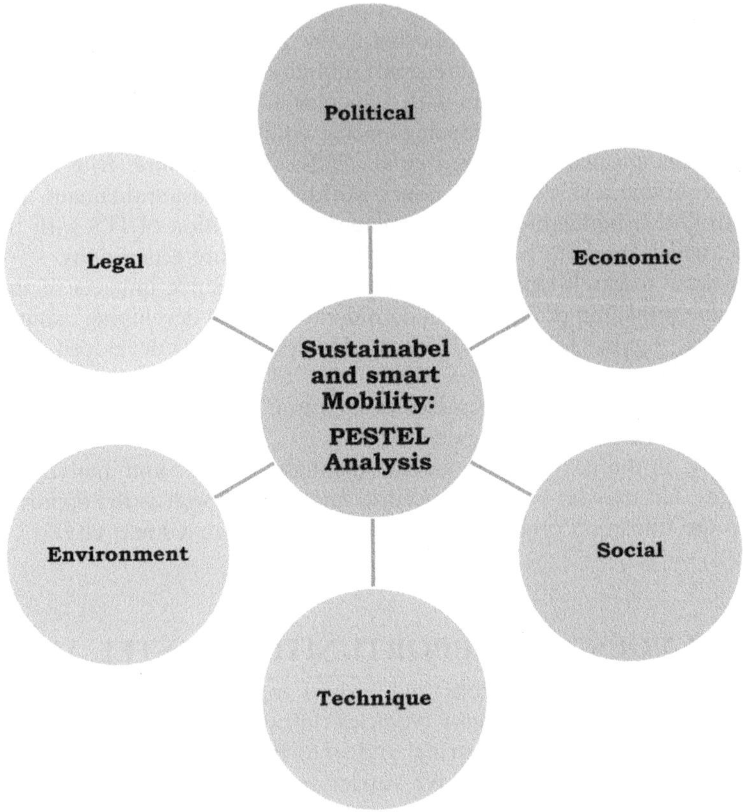

Fig. 4.10. Factors Influencing the Transition Process to a Sustainable and Smart Mobility in a Smart City Concept.

5.2 Economic Factors

The first cost of purchasing new technologies is costly, which must be supported firstly by the governments. Novel technology adoption is continuously growing in different countries such as Europe, China, the United States, and Canada. Hence, other countries are still in the planning phase. In addition, the global infrastructure also demands an immense investment for its development, which is very difficult for some countries. The transition project requires an urgent international collaboration to solve the challenges of the future "sustainability and climate change". It must be internationally not nationally diffusion.

However, knowledge and understanding of the new technologies' advantages: low operational cost, no maintenance cost, and no noise need to be improved through public education and awareness campaigns (internet and TV advertising...and so on) about the serious climate crisis. Marketing strategies are good tools to accelerate the whole process, which also can create a new business model and revenues for mobility as a service. The global objectives are better services for users and business revenues for the countries.

5.3 Social Factors

The biggest challenge for the electric and smart mobility transition is the consumers. Consumers do not have the same intentions to shift as the companies and government due to the lower prices in the second-hand vehicles market, which can be five times lower than the novel eco-friendly technologies. A deep vision of the decrease of the Gross Domestic Product (GDP) per capita in developing countries must be taken into account by the government's intervention, because of the noncapability of the users to afford the high first purchase, cost of the new technologies, and the sub price of the private charge at home. Fig. 4.11 shows the share of selected global regions in the GDP, adjusted for purchasing power, in 2021. We can remark the high difference between the purchasing power in a different region from 57.94% to 3.13% must be for sure reconsidered. Meanwhile, with the growth of population and the growth of education, young people should be the main group of vehicle buyers or buyers shortly. This opportunity might help in the acceleration and the adoption of green technology, while they had an open vision about the novel technology and its advantages through the education classes and new marketing technologies and strategies.

Indeed, they are very attractive and familiar with mobile internet telecommunication systems, applications, and information sharing. It can be a great factor to promote also the autonomous technology that will be diffused in the next few years. Furthermore, maximizing safety and reducing congestion is a positive point that may attract users to adopt novel technologies. However, the security and confidence of data sharing remain uncomfortable subjects for most citizens.

Furthermore, a key factor regarding mobility transition is the technical part, which says a lot about the economic luxury of a country that will be discussed in the upcoming subsection.

5.4 Technical Factors

The lack of sufficient and adequate infrastructure is one of the most challenging factors that set back the transition to smart, sustainable, green mobility in different countries, and needs to be prioritized. For this concept, a well-developed connected infrastructure (charging infrastructure, equipment, sensors, radars, traffics, and so on...) must be established and connected to everyone and everything. The plurality of global citizens is anxious about the charging infrastructure for electric technology, which reflects their anxiety about the vehicle range, security, and efficiency.

In addition, the implementation of smart technologies requires the realization of a robust digital infrastructure characterized by a high level of automation systems (radars,

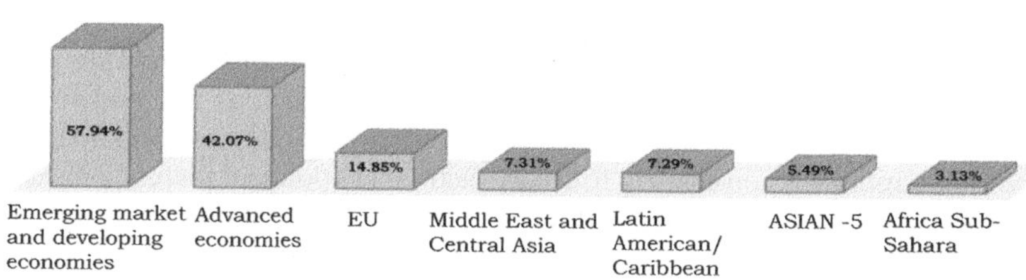

Fig. 4.11. Share of Global Regions in the Gross Domestic Product (Adjusted for Purchasing Power) in 2021. *Source:* Statista (2022).

sensors, and cameras) related to mobile applications. For that, different challenges must be considered: Coordinating the status of charging infrastructure, understanding the impact of charging on the power grid, considering the realization of a reasonable charging payment system and adequate business model respecting the consumer's economic conditions, data management, and data security.

In general, the high purchase price of novel technologies, a lack of sufficient infrastructure (charge, connection, cameras, radars, internet diffusion), in addition to competitive prices due to political disqualification are probably the most significant problems in various countries. Consequently, they directly limit the evolution of the mobility transition project that directly affects the global environmental situation. Thus, the environmental factors will be highlighted in the upcoming subsection.

5.5 Environmental Factors

The road transport sector relies mainly on petroleum products, which have affected the environmental situation reversely with high carbon dioxide emissions and global climate change with the extreme heat temperature. Consequently, this topic (climate change) is on top of the global agenda and putting enormous challenges toward a rapid transition to a more smart and sustainable road mobility sector. For that, electric technology as well as intelligent systems seem to be promising solutions to reduce harmful emissions and preserve environmental degradation guaranteeing a sustainable transportation future for all citizens by reducing congestion, fuel consumption, and harmful emissions. The transition to eco-friendly systems is a necessity and not only a worldwide trend.

Based on the arguments discussed above, we have to conclude that besides all the positive impacts on citizen's daily life that will be accomplished with sustainable and smart mobility, various challenges must be reconsidered to scale up the transition process. Indeed, the opportunities must be enhanced; however, the barriers must be mitigated.

6. CONCLUSION

In the present chapter, we have an insight into the novel technologies and methodologies innovations to create a sustainable and smart urban mobility sector worldwide. We have studied the global energy and transport sector's current situation. As a result, the need to transition from conventional to alternative sustainable and eco-friendly technologies has become a necessity in front of the current global economic and energy situation not only to reduce harmful emissions but also to create a new business model for the country taking into account the citizens' preferences for a better life pace.

Indeed, this transition needs also a smarter infrastructure based on the ITS implementation that facilitates the interaction between robots and users to reduce time, cost, and harmful emissions on the one hand and increase the comfort, security, and safety of users on the other thanks to the development of new technologies such as IoT and so on.

Meanwhile, the accomplishment of sustainable and smart mobility requires an integrated multidimensional approach considering four key factors: (1) political factors; (2) economic factors; (3) social factors; (4) technological factors; (5) environmental factors; and (6) legal factors to promote its diffusion. In particular, the approach should include the current vehicle technologies and digitization trends in transport systems together with citizens' perceptions, requirements, and expectations, and the support of governance transportation models that integrate all transition project stakeholders (governments, companies, sponsors. . .).

In short, it is imperative for governments to adopt policies that encourage the utilization of technology to develop sustainable cities. Smart mobility plays a vital role in achieving this goal, and it is equally important to address energy-related issues to ensure a better future for all humanity.

Here are different recommendations mainly for countries with slow progress in the studied project:

- Collaboration of the related stakeholders (embracing new project management approaches/techniques) nationally and internationally.
- Improve the consumer vision and intention (use a continuous survey approach to keep the feedback from users).
- Improve the infrastructure (charge stations, connected devices, with innovations, e.g., 5G and so forth).
- Proven its effectiveness and low maintenance and operational cost with marketing strategy.
- Promote the use of bicycles with convenient and safe infrastructure to cycle.
- Promote the eco-driving mode.

REFERENCES

Abu-Rayash, A., & Dincer, I. (2021). Development of integrated sustainability performance indicators for better management of smart cities. *Sustainable Cities and Society*, *67*, 102704. doi:10.1016/j.scs.2020.102704

Acheampong, R. A., Cugurullo, F., Gueriau, M., & Dusparic, I. (2021). Can autonomous vehicles enable sustainable mobility in future cities? Insights and policy challenges from user preferences over different urban transport options. *Cities*, *112*, 103134. doi:10.1016/j.cities.2021.103134

Akram, M. W., Bashir, A. K., Shamshad, S., Saleem, M. A., AlZubi, A. A., Chaudhry, S. A., & Zikria, Y. B. (2021). A secure and lightweight drones-Access protocol for smart city surveillance. *IEEE Transactions on Intelligent Transportation Systems*. doi:10.1109/TITS.2021.3129913

Aletà, N. B., Alonso, C. M., & Ruiz, R. M. A. (2017). Smart mobility and smart environment in the Spanish cities. *Transportation Research Procedia*, *24*, 163–170. doi:10.1016/j.trpro.2017.05.084

Aloi, D. N., Fay, L., Ronning, M., & Willer, M. (2003, June). The role of a bluetooth-equipped GPS receiver in intelligent transportation systems. In *Proceedings of the 59th annual meeting of the institute of navigation and CIGTF 22nd guidance test symposium (2003)* (pp. 372–378).

Ang-Tan, R., & Ang, S. (2021). *Understanding the smart city race between Hong Kong and Singapore* (pp. 1–10). Public Money and Management. doi:10.1080/09540962.2021.1903752

Ang, K. L. M., Seng, J. K. P., Ngharamike, E., & Ijemaru, G. K. (2022). Emerging technologies for smart cities' transportation: Geo-information, data analytics and machine learning approaches. *ISPRS International Journal of Geo-Information*, *11*(2), 85. doi:10.3390/ijgi11020085

Apanaviciene, R., Vanagas, A., & Fokaides, P. A. (2020). Smart building integration into a smart city (SBISC): Development of a new evaluation framework. *Energies*, *13*(9), 2190. doi:10.3390/en13092190

Avtar, R., Tripathi, S., Aggarwal, A. K., & Kumar, P. (2019). Population–urbanization–energy nexus: A review. *Resources*, *8*(3), 136. doi:10.3390/resources8030136

Bagheri, S., Huang, Y., & Walker, P. D. (2021). Strategies for improving the emission performance of hybrid electric vehicles. *Science of the Total Environment*, 144901. doi:10.1016/j.scitotenv.2020.144901

Balasubramaniam, A., Paul, A., Hong, W. H., Seo, H., & Kim, J. H. (2017). Comparative analysis of intelligent transportation systems for sustainable environment in smart cities. *Sustainability*, *9*(7), 1120. doi:10.3390/su9071120

Baran, R., Rusc, T., & Fornalski, P. (2016). A smart camera for the surveillance of vehicles in intelligent transportation systems. *Multimedia Tools and Applications*, *75*(17), 10471–10493. doi:10.1007/s11042-015-3151-y

Barba-Sánchez, V., Arias-Antúnez, E., & Orozco-Barbosa, L. (2019). Smart cities as a source for entrepreneurial opportunities: Evidence for Spain. *Technological Forecasting and Social Change*, *148*, 119713. doi:10.1016/j.techfore.2019.119713

Ben Ali, M., & Boukettaya, G. (2022). Integrating PESTEL and fuzzy logic methods for risk management process—Case of electric mobility transition in Tunisia. In *International conference on eco-innovation «ECONOV 2022»*. 07-09 Mai 2022. Tunisia.

Benevolo, C., Dameri, R. P., & D'auria, B. (2016). Smart mobility in smart city. In *Empowering organizations* (pp. 13–28). Cham: Springer.

Bibri, S. E. (2018). A foundational framework for smart sustainable city development: Theoretical, disciplinary, and discursive dimensions and their synergies. *Sustainable Cities and Society*, *38*, 758–794. doi:10.1016/j.scs.2017.12.032

British Petroleum. (2019). *BP statistical review of world energy*. London: British Petroleum.

Butler, L., Yigitcanlar, T., & Paz, A. (2020). Smart urban mobility innovations: A comprehensive review and evaluation. *IEEE Access*, *8*, 196034–196049. doi:10.1109/ACCESS.2020.3034596

Capdevila, I., & Zarlenga, M. I. (2015). Smart city or smart citizens? The Barcelona case. *Journal of Strategy and Management*. doi:10.1108/JSMA-03-2015-0030

Chehri, A., & Mouftah, H. T. (2019). Autonomous vehicles in the sustainable cities, the beginning of a green adventure. *Sustainable Cities and Society*, *51*, 101751. doi:10.1016/j.scs.2019.101751

Chen, Y., Ardila-Gomez, A., & Frame, G. (2017). Achieving energy savings by intelligent transportation systems investments in the context of smart cities. *Transportation Research Part D: Transport and Environment*, *54*, 381–396. doi:10.1016/j.trd.2017.06.008

Chieng, K. F., & Mustapa, F. D. (2021, December). Smart living implementation in Malaysia: A preliminary overview. In *Conference on the Human Environment* (Vol. 6, No. 24, pp. 79–89).

Christodoulou, A., & Cullinane, K. (2019). Identifying the main opportunities and challenges from the implementation of a port energy management system: A SWOT/PESTLE analysis. *Sustainability*, *11*(21), 6046. doi:10.3390/su11216046

Dahham, R. Y., Wei, H., & Pan, J. (2022). Improving thermal efficiency of internal combustion engines: Recent progress and remaining challenges. *Energies*, *15*(17), 6222. doi:10.3390/en15176222

Desdemoustier, J., Crutzen, N., & Giffinger, R. (2019). Municipalities' understanding of the smart city concept: An exploratory analysis in Belgium. *Technological Forecasting and Social Change*, *142*, 129–141. doi:10.1016/j.techfore.2018.10.029

Echendu, A. J., & Okafor, P. C. C. (2021). Smart city technology: A potential solution to Africa's growing population and rapid urbanization? *Development Studies Research*, *8*(1), 82–93. doi:10.1080/21665095.2021.1894963

Ehrlich, D., Melchiorri, M., & Capitani, C. (2021). Population trends and urbanisation in mountain ranges of the world. *Land*, *10*(3), 255. doi:10.3390/land10030255

Eseosa, O., & Temitope, F. (2019). Review of smart based building management system. *World Journal of Innovative Research*, *7*(2), 14–23.

Farid, A. M., Viswanath, A., Al-Junaibi, R., Allan, D., & Van der Wardt, T. J. (2021). Electric vehicle integration into road transportation, intelligent transportation, and electric power systems: An Abu Dhabi case study. *Smart Cities*, *4*(3), 1039–1057. doi:10.3390/smartcities4030055

Friesen, M. R., & McLeod, R. D. (2015). Bluetooth in intelligent transportation systems: A survey. *International Journal of Intelligent Transportation Systems Research*, *13*(3), 143–153. doi:10.1007/s13177-014-0092-1

Gharaibeh, A., Salahuddin, M. A., Hussini, S. J., Khreishah, A., Khalil, I., Guizani, M., & Al-Fuqaha, A. (2017). Smart cities: A survey on data management, security, and enabling technologies. *IEEE Communications Surveys & Tutorials*, *19*(4), 2456–2501. doi:10.1109/COMST.2017.2736886

Giuliodori, A., Berrone, P., & Ricart, J. E. (2022). Where smart meets sustainability: The role of smart governance in achieving the sustainable development goals in cities. *BRQ Business Research Quarterly*, 23409444221091281. doi:10.1177%2F23409444221091281

Gohar, A., & Nencioni, G. (2021). The role of 5G technologies in a smart city: The case for intelligent transportation system. *Sustainability*, *13*(9), 5188. doi:10.3390/su13095188

Grincheva, N. (2022). City museums in the age of datafication: Could museums be meaningful sites of data practice in smart cities? *Museum Management and Curatorship*, 1–27. doi:10.1080/09647775.2021.2023904

Guerrero-Ibáñez, J., Zeadally, S., & Contreras-Castillo, J. (2018). Sensor technologies for intelligent transportation systems. *Sensors*, *18*(4), 1212. doi:10.3390/s18041212

Han, M. J. N., & Kim, M. J. (2021). A critical review of the smart city in relation to citizen adoption towards sustainable smart living. *Habitat International*, *108*, 102312. doi:10.1016/j.habitatint.2021.102312

Haque, A. B., Bhushan, B., & Dhiman, G. (2021). Conceptualizing smart city applications: Requirements, architecture, security issues, and emerging trends. *Expert Systems*. doi:10.1111/exsy.12753

Hasan, M. K., Mahmud, M., Habib, A. A., Motakabber, S. M. A., & Islam, S. (2021). Review of electric vehicle energy storage and management system: Standards, issues, and challenges. *Journal of Energy Storage*, *41*, 102940. doi:10.1016/j.est.2021.102940

Hernafi, Y., Ahmed, M. B., & Bouhorma, M. (2016, October). An approaches' based on intelligent transportation systems to dissect driver behavior and smart mobility in smart city. In *2016 4th IEEE international colloquium on information science and technology (CiSt)*. IEEE. doi:10.1109/CIST.2016.7805013

Herrera-Quintero, L. F., Vega-Alfonso, J., Bermúdez, D., Marentes, L. A., & Banse, K. (2019, May). ITS for smart parking systems, towards the creation of smart city services using IoT and cloud approaches. In *2019 smart city symposium Prague (SCSP)* (pp. 1–7). IEEE. doi:10.1109/SCSP.2019.8805705

Holechek, J. L., Geli, H. M., Sawalhah, M. N., & Valdez, R. (2022). A global assessment: Can renewable energy replace fossil fuels by 2050? *Sustainability*, *14*(8), 4792. doi:10.3390/su14084792

Huijsmans, T., Harteveld, E., van der Brug, W., & Lancee, B. (2021). Are cities ever more cosmopolitan? Studying trends in urban-rural divergence of cultural attitudes. *Political Geography*, *86*, 102353. doi:10.1016/j.polgeo.2021.102353

Ibn-Mohammed, T. (2017). Application of mixed-mode research paradigms to the building sector: A review and case study towards decarbonising the built and natural environment. *Sustainable Cities and Society*, *35*, 692–714.

Ji, T., Chen, J. H., Wei, H. H., & Su, Y. C. (2021). Towards people-centric smart city development: Investigating the citizens' preferences and perceptions about smart-city services in Taiwan. *Sustainable Cities and Society*, *67*, 102691. doi:10.1016/j.scs.2020.102691

Juda, Z. (2002). Advanced propulsion systems for hybrid electric vehicles. In *28th international scientific conference on combustion engines–KONES*.

Kumar, V., & Dahiya, B. (2017). Smart economy in smart cities. In *Smart economy in smart cities* (pp. 3–76). Singapore: Springer.

Lim, S. B., Malek, J. A., Yussoff, M. F. Y. M., & Yigitcanlar, T. (2021). Understanding and acceptance of smart city policies: Practitioners' perspectives on the Malaysian smart city framework. *Sustainability*, *13*(17), 9559. doi:10.3390/su13179559

Lingani, G. M., Rawat, D. B., & Garuba, M. (2019, January). Smart traffic management system using deep learning for smart city applications. In *2019 IEEE 9th annual computing and communication workshop and conference (CCWC)* (pp. 0101–0106). IEEE. doi:10.1109/CCWC.2019.8666539

Little, W. (2016). *Population, urbanization, and the environment. Introduction to sociology* (2nd Canadian ed.).

Mallik, S. (2014). Intelligent transportation system. *International Journal of Civil Engineering Research*, *5*(4), 367–372.

Mangiaracina, R., Tumino, A., Miragliotta, G., Salvadori, G., & Perego, A. (2017, September). Smart parking management in a smart city: Costs and benefits. In *2017 IEEE international conference on service operations and logistics, and informatics (SOLI)* (pp. 27–32). IEEE. doi:10.1109/SOLI.2017.8120964

Meijer, A., & Bolívar, M. P. R. (2016). Governing the smart city: A review of the literature on smart urban governance. *International Review of Administrative Sciences*, *82*(2), 392–408. doi:10.1177%2F0020852314564308

Nagy, S., & Csiszár, C. (2020). The quality of smart mobility: A systematic review. *Zeszyty Naukowe. Transport/Politechnika Śląska*. doi:10.20858/sjsutst.2020.109.11

Our World in Data. (2022). Retrieved from https://ourworldindata.org/urbanization

Pakusch, C., Stevens, G., & Bossauer, P. (2018, May). Shared autonomous vehicles: Potentials for a sustainable mobility and risks of unintended effects. In *ICT4S* (pp. 258–269).

Pandharipande, A., & Thijssen, P. (2019). Connected street lighting infrastructure for smart city applications. *IEEE Internet of Things Magazine*, *2*(2), 32–36.

Pašalić, I. N., Ćukušić, M., & Jadrić, M. (2021). Smart city research advances in Southeast Europe. *International Journal of Information Management*, *58*, 102127. doi:10.1016/j.ijinfomgt.2020.102127

Patyal, V. S., Kumar, R., & Kushwah, S. (2021). Modeling barriers to the adoption of electric vehicles: An Indian perspective. *Energy*, *237*, 121554. doi:10.1016/j.energy.2021.121554

Pflügler, C., Schreieck, M., Hernandez, G., Wiesche, M., & Krcmar, H. (2016). A concept for the architecture of an open platform for modular mobility services in the smart city. *Transportation Research Procedia*, *19*, 199–206. doi:10.1016/j.trpro.2016.12.080

Pradhan, M. A., Patankar, S., Shinde, A., Shivarkar, V., & Phadatare, P. (2017, August). IoT for smart city: Improvising smart environment. In *2017 international conference on energy, communication, data analytics and soft computing (ICECDS)*. IEEE. doi:10.1109/ICECDS.2017.8389800

Qian, Y., Liu, J., Cheng, Z., & Forrest, J. Y. L. (2021). Does the smart city policy promote the green growth of the urban economy? Evidence from China. *Environmental Science and Pollution Research*, *28*(47), 66709–66723. doi:10.1007/s11356-021-15120-w

Qoradi, M. D., Al-Harbi, M. S., & Aina, Y. A. (2021). Using GIS-based intelligent transportation systems in the enhancement of university campus commuting in a smart city context. *Arabian Journal of Geosciences*, *14*(9), 1–13. doi:10.1007/s12517-021-07098-z

Raza, A., Bukhari, S. H. R., Aadil, F., & Iqbal, Z. (2021). An UAV-assisted VANET architecture for intelligent transportation system in smart cities. *International Journal of Distributed Sensor Networks*, *17*(7), 15501477211031750. doi:10.1177%2F15501477211031750

Ribeiro, P., Dias, G., & Pereira, P. (2021). Transport systems and mobility for smart cities. *Applied System Innovation*, *4*(3), 61. doi:10.3390/asi4030061

Roland, I. (2021). *Global EV sales for 2021*. Retrieved from https://www.ev-volumes.com/news/ev-sales-for-2021/

Roy, T., Tariq, A., & Dey, S. (2021). A socio-technical approach for resilient connected transportation systems in smart cities. *IEEE Transactions on Intelligent Transportation Systems*. doi:10.1109/TITS.2020.3045854

Ruggieri, R., Ruggeri, M., Vinci, G., & Poponi, S. (2021). Electric mobility in a smart city: European overview. *Energies*, *14*(2), 315. doi:10.3390/en14020315

Seuwou, P., Banissi, E., & Ubakanma, G. (2020). The future of mobility with connected and autonomous vehicles in smart cities. In *Digital twin technologies and smart cities* (pp. 37–52). Cham: Springer.

Sharma, V., You, I., Pau, G., Collotta, M., Lim, J. D., & Kim, J. N. (2018). LoRaWAN-based energy-efficient surveillance by drones for intelligent transportation systems. *Energies*, *11*(3), 573. doi:10.3390/en11030573

Singh, T., Solanki, A., Sharma, S. K., Nayyar, A., & Paul, A. (2022). *A decade review on smart cities: Paradigms, challenges and opportunities*. IEEE Access. doi:10.1109/ACCESS.2022.3184710

Spicer, Z., Goodman, N., & Olmstead, N. (2021). The frontier of digital opportunity: Smart city implementation in small, rural and remote communities in Canada. *Urban Studies*, *58*(3), 535–558. doi:10.1177/0042098019863666

Sumalee, A., & Ho, H. W. (2018). Smarter and more connected: Future intelligent transportation system. *IATSS Research*, *42*(2), 67–71. doi:10.1016/j.iatssr.2018.05.005

Sumi, L., & Ranga, V. (2018). Intelligent traffic management system for prioritizing emergency vehicles in a smart city. *International Journal of Engineering*, *31*(2), 278–283.

Syalianda, S. I., & Kusumastuti, R. D. (2021, March). Implementation of smart city concept: A case of Jakarta smart city, Indonesia. In *IOP conference series: Earth and environmental science* (Vol. 716, No. 1, p. 012128). IOP Publishing.

Tasgaonkar, P. P., Garg, R. D., & Garg, P. K. (2020). Vehicle detection and traffic estimation with sensors technologies for intelligent transportation systems. *Sensing and Imaging*, *21*(1), 1–28. doi:10.1007/s11220-020-00295-2

Thondoo, M., Marquet, O., Márquez, S., & Nieuwenhuijsen, M. J. (2020). Small cities, big needs: Urban transport planning in cities of developing countries. *Journal of Transport and Health*, *19*, 100944. doi:10.1016/j.jth.2020.100944

Ullo, S. L., & Sinha, G. R. (2020). Advances in smart environment monitoring systems using IoT and sensors. *Sensors*, *20*(11), 3113.

Varró, K., & Szalai, Á. (2021). *Discourses and practices of the smart city in Central Eastern Europe: insights from Hungary's 'big' cities* (pp. 1–25). Urban Research and Practice. doi:10.1080/17535069.2021.1904276

Yuloskov, A., Bahrami, M. R., Mazzara, M., & Kotorov, I. (2021). Smart cities in Russia: Current situation and insights for future development. *Future Internet*, *13*(10), 252. doi:10.3390/fi13100252

EVALUATION OF SMART CITY PROJECTS IN ECO-GASTRONOMY DIMENSION: THE EXAMPLE OF TURKEY

Fatih Varol, Merve Oksuz and Eren Yalcin

ABSTRACT

Cities were regions ruled by local governments, where people were supposed to live together and provide equal access to sociocultural opportunities. In the 21st century, global warming and overuse of scarce limited resources has made sustainability more examined about for our entire environment, particularly cities. With the rapid increase in the population in the cities, humankind has faced a lot of pollution, destruction, and social inequality. Many regions and countries have started to build new smart cities using technology to overcome crowded life, traffic, and air pollution, improve food production, and use scarce natural resources sustainable. Smart cities also provide residents to improve their quality of life and their health; therefore, eco-gastronomy is related to organic farming and cooking method that minimize the damage to the environment with organic ingredients. For a healthy and quality life, meals made with healthy ingredients are required. Smart cities have also started to implement eco-gastronomy projects by using technology. In this chapter, cities which can be associated with the eco-gastronomy dimension of smart cities such as Gaziantep, Izmir, and Konya from Turkey and Copenhagen were analyzed.

Keywords: Sustainability; smart cities; eco-gastronomy; eco-cities; sustainable tourism

1. INTRODUCTION

Cities, which play an important role in the global economy and are the starting point of social changes, are the center of consumption and production (Cohen, 2006; Neirotti, De Marco, Cagliano, Mangano, & Scorrano, 2014; Ramaprasad, Sánchez-Ortiz, & Syn, 2017). Especially with the industrial revolution that emerged in England in the 18th century, the increase in production and consumption and the fact that a large part of the population began to live in cities cause multifarious problems (Ramaprasad et al., 2017).

Smart Cities for Sustainability
Advanced Series in Management, Volume 32, 71–81

ISSN: 1877-6361/doi:10.1108/S1877-636120230000033006

Overuse of natural resources in cities causes various environmental and social deteriorations (Yin et al., 2015).

In 1950, only 30% of the world's population lived in cities, while this rate reached 54% in 2014. It is predicted by the United Nations that this rate will be 60% close by 2050 (Glasmeier & Christopherson, 2015; United Nations, 2022; Yin et al., 2015).

With the increase in population, technological innovations are needed in many areas, especially in security, the use of energy, traffic control, and communication (He, Stojmenovic, Liu, & Gu, 2014). Technology products emerge with smart systems as a crucial element for the management of cities. It is stated that smart cities are more efficient than existing systems (Glasmeier & Christopherson, 2015).

Cities, which are complex systems, are a combination of air and environmental pollution and social inequality (Neirotti et al., 2014). In addition to this, in cities where the management system is effective, people benefit from social opportunities and living conditions better (Ramaprasad et al., 2017). Since the beginning of the 21st century, digitalization and the prominence of technology have caused changes in people's lives. Many cities have started their own smart city project in order to eliminate the negativities of air pollution, produce better food production, and use scarce natural resources sustainably (Yin et al., 2015). In this context, close to 30 global cities would be smart cities by 2025; more than 50% of these smart cities are projected to be in Europe or North America (Koh, Gunawansa, & Bhullar, 2010).

Although there are different definitions of the concept of "smart city":

- Ensures social and cultural development by increasing economic development.
- It emerges within the scope of environmental and social sustainability as a strategic component (Albino, Berardi, & Dangelico, 2015; Caragliu, Del Bo, & Nijkamp, 2009; Ramaprasad et al., 2017).

The concept of smart city, in addition it also combines different disciplines, is a concept that expresses the increase of people's quality of life and the efficient use of limited resources (Ramaprasad et al., 2017).

In this study "Evaluation of Smart City Projects in Eco-Gastronomy Dimension: The Example of Turkey," "Sustainability," "What is a smart city? Why is it important?," "Sustainable cities-eco-cities and eco-gastronomy," and "Examination of smart cities in the world and in Turkey" are examined. Cities which can be associated with the eco-gastronomy dimension of smart cities, Gaziantep, Izmir, and Konya from Turkey and Copenhagen can be given as examples from the world.

1.1 Sustainability

The origin of Turkish culinary culture dates back to the Ottoman Empire, Seljuk, Central Asia, and even the prehistoric period, starting from the Republican period. The multinationality of the Ottoman Empire led to the diversification of the culinary culture and enabled very different tastes from the past to the present day with the gastronomic heritage. Gastronomy is a branch of science that includes not only the types of food and beverage, but also the relationship between food and culture, where many different disciplines work together, and where science and art are intertwined based on good eating habits (Kasli, Cankul, Koz, & Ekici, 2015). Gastronomy covers the stages of cooking hygienic foods, making them ready to eat and presenting them in a way that appeals to the eye.

Due to crowded population and the deterioration balance between the production and consumption, sustainability has become philosophical concept in society in order to overcome these negative effects.

Sustainability is the long-term use of scarce natural resources on earth without destroying them, taking into account future generations. The concept of sustainability is based on three main principles: economic dimension, sociocultural dimension, and environmental dimension. The economic dimension refers to regional economic planning and employment. The sociocultural dimension is that the people living in the region participate in the decision-making processes on every issue and interact with the visitors. The environmental dimension is the long-term use of limited natural resources without disturbing the ecological balance in the world (Kurnaz & Arman, 2020).

At present, food and gastronomy are increasingly recognized as potentially determining factors for the sustainable development of regions. Due to gain in competitive advantage to increase attractiveness and competitiveness of regions, the theory that they can contribute to their economic, social, and environmental sustainability is widely accepted in many research fields (Rinaldi, 2017). Many regions make marketing campaigns or branding their products in order to feature with their gastronomic tastes and gain comparative advantage. Because if a product becomes a brand and grab a market share, it supports the regional food industry, protects local foods, and improves region's identity (Oztopcu & Akar, 2019; Rinaldi, 2017).

When taken into consideration of production and consumption stage of the food, it is obvious that environmental aspects have huge impact on this process. With the globalization, natural resources and energy are used for food supply largely, and it is of great importance that it be sustainable within this system. In the light of this information, it is stated that all kinds of issues that may occur from the beginning of the production process of the food, including its quality, until it becomes the final product, and even after it, are within the scope of sustainable gastronomy (Gun & Kılıc, 2022).

Sustainable gastronomy encourages people to buy local products for environmental and sociocultural reasons. In the regional context, localized gastronomic tastes and products support environmental sustainability, economic development, and local markets even improving local brand. Furthermore, establishing regional markets for local tastes, gastronomic markets, organic and eco-farming and cuisine system, wine routes or eco-museums supports local development. Sustainable gastronomy supports communities to enrich them both socially and economically while maintaining an eco-nutrition commitment to environmental sustainability and the optimal health of community members (Yurtseven & Kaya, 2010).

City and business life, which has accelerated with globalization, has caused a change in eating and drinking habits. As a result of people tending to fast and quick consumption in order to save time, the culinary culture from the past to the present is slowly disappearing. Sustainable gastronomy studies carried out in the field of gastronomy aim to explain the importance of quality and conscious nutrition to the society in order to protect public health and to ensure its continuity in the long term by transferring the food and beverage culture to future generations. The basis of sustainable gastronomy is based on food safety. Food safety does not only mean the absence of chemicals and pesticides in agriculture. Food safety also refers to a long process involving the processing, trade, and supply of food (Scarpato, 2002a). In addition to this, sustainable gastronomy is a phenomenon that protects the gastronomic heritage, aims to spread biodiversity and local farmers, and aims to protect the cultural texture in the region (Durlu-Ozkaya, Sunnetcioglu, & Can, 2013).

It is an undeniable fact that sustainable gastronomy contributes to the region and gastronomic heritage, but it is limited to eating and drinking only in terms of sustainability. However, eco-gastronomy is based on all three dimensions of sustainability and adopts the "from farm to table" principle. When eco-gastronomy is examined in terms of the economic dimension of sustainability, it supports the participation of local producers in production, tries to keep the local markets and greengrocers in the region alive, and helps to create employment by supporting small-scale enterprises into the economy in rural areas and ensuring the development of the region in the long term, supporting local agricultural systems by encouraging small farms instead of large industries. Encouraging employment in the local area and supporting the promotion of local products also contributes to the region economically. Eco-gastronomy is not just about food and beverage production process.

It also argues that the region will develop socially and culturally, which is another principle of sustainability (Oztopcu & Akar, 2019). Eco-gastronomy is a process that protects the gastronomic heritage that reflects the local cuisine by preserving home-made foods, and the education path to children includes the transfer of pot food to future generations as a cultural heritage (Yurtseven & Kaya, 2010). In this way, while the gastronomic flavors specific to the region are transferred to new generations, it also contributes to the visitors to get to know the culture of the region because gastronomic tastes also reflect the cultural heritage of the region. According to sustainability's environmental dimension, the food needs to be prepared and produced in an environmentally friendly manner. Thus, it aims to feed people's minds by providing healthy ingredients (Scarpato, 2002b).

In the light of this information, eco-gastronomy cannot be considered separately from sustainability (Kurnaz & Arman, 2020). In summary, eco-gastronomy brings much more than just eating and drinking behavior, which is a physical need, to the local region in terms of both economic, environmental, and sociocultural sustainability.

The basic principles of eco gastronomy can be summarized as follows:

- Supporting the local economy and the flow of local farmers by focusing on production in the region and provides environmental friendly infrastructure.
- Supporting job creation by encouraging food at local restaurants and staying in local hostels.
- Helps to highlight and promote local products and brands.
- To pass on cooking techniques from generation to generation while keeping home cooking alive and to educate children and young people.
- To increase the orientation toward healthy and natural foods for human health.
- Paying attention to the fact that a good meal contains a real taste.
- Giving importance to cooking in a healthy environment and establishing good working standards for staff.
- Increasing visitor arrivals by bringing tourism to gastronomic authenticity.
- Eco gastronomy not only enhances the image of the destination, but also increases the value of local identity and encourages local people to feel proud of themselves.
- Creates unique internalized experiences by offering visitors the special flavors of the region (Oztopcu & Akar, 2019; Yurtseven & Kaya, 2010).

1.2 Definition, Importance, and Features of Smart City

The Republic of Turkey's Ministry of Environment, Urbanization, and Climate Change defines Smart City as "the effective integration of physical, digital, and human systems with a structured environment to deliver a sustainable, prosperous, and inclusive future to ecosystem assets" in the 2020–2023 National Smart Cities Strategy and Action Plan and the Smart Cities Capacity Building and Guidance Project. Furthermore, a "smart and sustainable city, while taking into account the economic, social, environmental, and cultural needs of current and future generations, is an innovative city that uses information and communication technologies and other tools to improve the quality of life, efficiency, and competitiveness of urban service delivery" (Akıllı Sehirler, 2022a).

The concept of the smart city is an important element for the economic and social development of a region (Eremia, Toma, & Sanduleac, 2017). At the same time, it is a concept that aims to reduce energy consumption and greenhouse gas levels by conserving resources in order to reach the sustainability of a certain region (Castro, Jara, & Skarmeta, 2013). Smart cities are an important part of the development of digital and smart technologies for a healthy community (Roche, Nabian, Kloeckl, & Ratti, 2012). The idea of a "smart city," which is connected to the Internet of Things (IoT) and technology, is an environmentally friendly strategy that protects natural resources, boosts productivity, and promotes responsible consumption (Castro et al., 2013).

When we examine "smart" as a word, it is derived from the initials of "secure and safe society," "mobility in green communication," "attain excellent education and health," "real estate and building," and "technology and transportation amplification" (Tomar & Kaur, 2019). Smart city refers to the digitization of the city by obtaining economic, social, and environmental data as well as the population. Information and communication technologies are part of a digital city, and it has a framework for its infrastructure that may help cities significantly (Kourtit & Nijkamp, 2012; Yin et al., 2015). Therefore, although every smart city is a digital city, a digital city is not a smart city. In addition, while technology is at the forefront in a digital city, sustainable qualities are taken into consideration with technology in a smart city (Yin et al., 2015).

With the use of information technology in smart cities:

- energy and water consumption can be reduced,
- gas emissions can be reduced and
- quality of life can be improved (Harrison & Donnelly, 2011).

Citizens living in cities face various difficulties such as traffic jams, air pollution, and decrease damage to historical places (Oliveira & Campolargo, 2015).

In this context, with the developments in technology, efficiency has become significantly applicable. The use of sensors, digital systems, traffic sensors, and wireless connectivity to obtain historical traffic data is becoming widespread (Harrison & Donnelly, 2011).

According to Ramaprasad et al. (2017), the structure, functions, focus, semiotics, stakeholders, and consequences of the concept of smart city are included in Table 5.1. While the structure of smart city creates architecture, infrastructure, systems, services, policies, processes and personnel, sustainability, quality of life, equality, livability and durability constitute its results.

The technologies used in smart cities are analyzed by obtaining data from the current situation and people. On the other hand, it can also be used to protect the security of public areas and to prevent negativity (Tomar & Kaur, 2019).

Table 5.1. The Structure, Functions, Focus, Semiotics, Stakeholders, and Consequences of the Smart City Concept (Ramaprasad et al., 2017).

Structure		Functions		Focus		Semiotics		Stakeholders		Outcomes
Architecture	[to]	Sense	[+]	Cultural	[+]	Data	[by/from/to]	Citizens	[for]	Sustainability
İnfrastructure		Monitor		Economic		İnformation		Professional		Quality of life (QoL)
Systems		Process		Demographic		Knowledge		Communities		Equity
Services		Translate		Environmental				İnstitutions		Livability
Policies		Communicate		Political				Businesses		Resilience
Processes				Social				Governments		
Personnel				Technological						
				İnfrastructural						

The dimension, subdimension, and description of the smart city concept are included in Table 5.2. Accordingly, the dimensions are divided into four different areas as "state," "citizen," "business," and "environment" (Yin et al., 2015).

1.3 Sustainable Cities-Eco Cities and Eco-Gastronomy

Eco-gastronomy concept is related to the cooking method in which the selection of ingredients, taking into account environmental impacts, the preparation of food, and the concern for the environment are expressed. It has emerged to meet the needs of consumers who adopt sustainable quality, demand healthy food, and adopt a different life habit to minimize the damage to the environment. According to the World Tourism Organization, it is possible to protect social and cultural values as well as economic development with the concept of eco-gastronomy where a sustainable understanding is adopted (Gunes, 2013).

The emergence of the negative environmental effects of urban development has led to the inclusion of cities within the framework of sustainability. In addition, the increase in industrialization and energy consumption has caused climate change and global warming. According to Richard Register's definition, an eco-city defines a city where people are in harmony with nature and the ecological footprint is reduced (Koh et al., 2010).

1.4 Examples of Smart Cities in the World and in Turkey

Examples of smart cities from across the world and in Turkey are provided under this section. Copenhagen, which is related to the eco-gastronomy aspect of smart cities, as well as Gaziantep, Izmir, and Konya in Turkey, are some examples of smart cities throughout the world.

Table 5.2. Dimension, Subdimension, and Description of the Concept of Smart City (Yin et al., 2015).

Dimension	Subdimension	Description
Government	Public safety Civil service Monitoring of the city	Ensuring the efficient operation of public services.
Citizen	Public transport Social solidarity Education and health Entertainment Tourism	Enabling individuals to live more efficiently.
Business	Entrepreneurship Novelty Advertising Agriculture Corporate governance	Increasing the level of prosperity by increasing the volume of production trade in a city.
Environment	Renewable energy Water management Waste management Environmental management	Using of renewable energy sources to ensure sustainability and reduction of urban pollution.

1.5 Smart City Examples in the World

1.5.1 Copenhagen Waste Management Program

Around 60% of urban waste is recycled, and only 2% of the total is sent to landfills. At the same time, the resulting waste is used in the production of heat for the city's district heating network. This is quite different from the city's waste management in 1988. In 1988, about 40% of waste was sent to landfills or incinerated (Akıllı Sehirler, 2022b).

1.6 Smart City Examples in Turkey

According to the Ministry of Environment, Urbanization and Climate Change, General Directorate of Geographic Information Systems, Department of Smart Cities, there are a total of 415 smart city projects in 81 provinces of Turkey. When the number of smart city projects is examined, smart city projects are 132 in Konya, 40 in Kocaeli, 36 in Gaziantep, and 1 each in Elâzığ, Isparta, Karaman, Kastamonu, Kırklaretli, and Rize (Akıllı Sehirler, 2022c).

1.7 Gaziantep – Gaziantep Ecological Village Project

In Gaziantep, it is planned to implement a project that respects the cycle of nature and will use energy resources effectively. Project green production will be implemented with the use of alternative energy and zero waste understanding. The village to be created will raise the awareness of individuals. In addition, it is foreseen that it will be an important destination in tourism (Akıllı Sehirler, 2022d).

1.8 Gaziantep – Another Gluten Free Cafe Online Sale

Various products are sold through the website created for celiac patients who are sensitive to gluten in barley, wheat, oats, and rye. On the website, orders are received only for Gaziantep (Akıllı Sehirler, 2022e).

1.9 Gaziantep – Smart Plant Factory in the City

Within the scope of the project, it is planned to create a production area where the soil is hygienic, healthy and with the least amount of computer control, and climate conditions are provided without the use of any pesticides. In the project, food safety is ensured and hygienic and health products are produced and high-quality products are grown with the knowledge of Agriculture 4.0. The project, which is limited to Gaziantep, aims to ensure continuous production for 24 hours a day, 365 days a week, to reduce water consumption by 95%, and to eliminate harmful effects caused by environmental effects (Akıllı Sehirler, 2022f).

1.10 Izmir – Izmir Agriculture Mobile Application

Producers who carry out agricultural production in Izmir can request professional support by requesting the analysis of satellite images. At the same time, the production area can be monitored remotely, and the decision on spraying, fertilization, and irrigation can be decided through the data obtained. This project also accelerates the transition to Industrial 4.0. The project covers farmers and producers producing in Izmir (Akıllı Sehirler, 2022g).

1.11 www.konyapedia.com

It is an internet application that aims to make Konya Encyclopedia accessible electronically. In the project, historical and natural beauties of Konya, daily living, education, culture, art and tourism as well as local delicacies are introduced (Akıllı Sehirler, 2022h).

1.12 E-Pattern

Within the scope of the project, an analysis of the agricultural and economic situation of Konya was carried out and shared on the website. It is aimed to share the most frequently grown products in villages and neighborhoods through the website. It is a website created to determine many values of economic value in the rural area of the city from agricultural production capacity to tourism (Akıllı Sehirler, 2022i).

1.13 Farmer Information System

It is an information system established to enable farmers who produce to reach expert personnel more easily against the problems they encounter. In other words, it is a support to the farmer who carries out the production. The project aims to create an agricultural inventory. With the Farmer Information System, the producer who is engaged in beekeeping, cattle/ovine breeding, and viticulture, etc., in Konya can find a solution to the problem he encounters by logging into the system 24 hours a day, 7 days a week (Akıllı Sehirler, 2022j).

1.14 Konya – Smart Tourism Guide Application

It is a mobile application where tourists who will visit Konya can easily and quickly get information, experience different cultural elements, create a travel route, and get information about touristic places. It is aimed at meeting the needs of individuals who will visit Konya for the first time. Thanks to the application, the number of days that the visitor will stay will increase, and the maximum efficiency will be obtained from the visit (Akıllı Sehirler, 2022k).

2. CONCLUSION AND EVALUATION

Cities, which are the starting point of social and economic changes, are centers of consumption. The increase in the population living in cities with the industrial revolution caused various environmental, social, and economic problems from the past to the present. The concept of smart city has emerged in order to ensure economic, sociocultural development, and sustainability.

In this chapter, smart city projects in Turkey and around the world are evaluated. Copenhagen from Denmark where eco-gastronomy can be associated with a smart city; in Turkey, Gaziantep, Izmir and Konya are explained. "Waste Management Program" in Copenhagen, "Gaziantep Ecological Village Project," "Another Gluten-Free Cafe Online Sales," "Smart Plant Factory in the City," "Izmir Agriculture Mobile Application," "www.konyapedia.com," "E-Pattern," "Farmer Information System," and "Smart Tourism Guide Application" in Gaziantep have been announced to express the relationship between smart city and eco-gastronomy.

While a total of 415 smart city projects belonging to 81 in Turkey were being realized or planned to be realized, eight of these projects were encountered only in Gaziantep, Izmir, and Konya where the relationship between smart city and eco-gastronomy could be expressed. With this study, it is foreseen that gastronomy tourism, which is an important element in ensuring sustainability within the scope of smart city concept, will contribute to the sustainability of the destination and will be effective in raising awareness of individuals.

This chapter examines the significance of sustainability and eco-gastronomy for a healthy, balanced and quality of life. Due to growing healthy generations, societies and cities need to be more attractive to improve eco-gastronomic projects for their destinations. Because, eco-gastronomy not only contributes to the region's economic and cultural values, but also helps to raise healthy generations for our future. Thus, societies need to develop their awareness and train people about the necessity of eco-gastronomy and smart city projects and sustainable life rather than unlimited consumption. However, it is clear to see that in Turkey, there are very few smart city projects about eco-gastronomy. In conclusion, for a better and healthy future, people need to be trained, and restaurants, entrepreneurs, farmers and local governments need to act together.

REFERENCES

Albino, V., Berardi, U., & Dangelico, R. M. (2015). Smart cities: Definitions, dimensions, performance, and initiatives. *Journal of Urban Technology*, *22*(1), 3–21.

Caragliu, A., Del Bo, C., & Nijkamp, P. (2009). Smart cities in Europe. In *Proceedings of the 3rd central European conference in regional science*. Košice, Slovak Republic (pp. 7–9).

Castro, M., Jara, A. J., & Skarmeta, A. F. (2013). Smart lighting solutions for smart cities. In *2013 27th international conference on advanced information networking and applications workshops* (pp. 1374–1379). IEEE.

Cohen, B. (2006). Urbanization in developing countries: Current trends, future projections, and key challenges for sustainability. *Technology in Society*, *28*(1–2), 63–80.

Durlu-Ozkaya, F., Sunnetcioglu, S., & Can, A. (2013). The role of geographical indication in sustainable gastronomy tourism mobility. *Journal of Tourism and Gastronomy Studies*, *1*(1), 13–20.

Eremia, M., Toma, L., & Sanduleac, M. (2017). The smart city concept in the 21st century. *Procedia Engineering*, *181*, 12–19.

Glasmeier, A., & Christopherson, S. (2015). Thinking about smart cities. *Cambridge Journal of Regions, Economy and Society*, *8*(1), 3–12.

Gunes, S. G. (2013). Eco-gastronomy, tourism and sustainability: The rise of sustainable restaurants in the world. *Organization*, *16*, 17.

Gun, S., & Kılıc, G. D. (2022). Eco gastronomy and sustainable gastronomy. *Journal of Tourism and Gastronomy Studies*, *10*(2), 999–1015.

Harrison, C., & Donnelly, I. A. (2011). A theory of smart cities. In *Proceedings of the 55th annual meeting of the ISSS-2011*, Hull, UK.

He, Y., Stojmenovic, I., Liu, Y., & Gu, Y. (2014). Smart city. *International Journal of Distributed Sensor Networks*, *2014*, 1–2.

Kasli, M., Cankul, D., Koz, E. N., & Ekici, A. (2015). Gastronomik miras ve sürdürülebilirlik: Eskişehir örneği. *Eko-Gastronomi Dergisi*, *1*(2), 27–46.

Koh, K. L., Gunawansa, A., & Bhullar, L. (2010). Eco-cities and Sustainable cities-whither? *Social Space*, 84–92.

Kourtit, K., & Nijkamp, P. (2012). Smart cities in the innovation age. *Innovation: The European Journal of Social Science Research*, *25*(2), 93–95.

Kurnaz, A., & Arman, A. (2020). In S. Gastronomi, A. Akbaba, & N. Çetinkaya (Eds.), *Gastronomide Yiyecek Tarihi* (pp. 161–178). Ankara: Detay Publishing.

Neirotti, P., De Marco, A., Cagliano, A. C., Mangano, G., & Scorrano, F. (2014). Current trends in smart city initiatives: Some stylised facts. *Cities*, *38*, 25–36.

Oliveira, Á., & Campolargo, M. (2015). From smart cities to human smart cities. In *2015 48th Hawaii international conference on system sciences* (pp. 2336–2344). IEEE.

Oztopcu, A., & Akar, M. (2019). Ecogastronomy and sustainable regional development. *International Social Mentality and Researcher Thinkers Journal*, *5*(24), 1432–1443.

Ramaprasad, A., Sánchez-Ortiz, A., & Syn, T. (2017). A unified definition of a smart city. In *International conference on electronic government* (pp. 13–24). Cham: Springer.

Rinaldi, C. (2017). Food and gastronomy for sustainable place development: A multidisciplinary analysis of different theoretical approaches. *Sustainability*, *9*(1748), 1–25.

Roche, S., Nabian, N., Kloeckl, K., & Ratti, C. (2012). Are 'smart cities' smart enough. In *Global geospatial conference* (pp. 215–235).

Scarpato, R. (2002a). Gastronomy as a tourist product: The perspective of gastronomy studies. In A.-M. Hjalager & G. Richards (Eds.), *Tourism and gastronomy* (pp. 51–70). Londra: Routledge.

Scarpato, R. (2002b). Sustainable gastronomy as a tourist product. In A.-M. Hjalager & G. Richards (Eds.), *Tourism and gastronomy* (pp. 132–152). Londra: Routledge.

Tomar, P., & Kaur, G. (Eds.). (2019). *Green and smart technologies for smart cities.* CRC Press.

United Nations. (2022). Retrieved from http://esa.un.org/unpd/wup/Highlights/WUP2014-Highlights.pdf. Accessed on December 06, 2022.

United Nations. (2022a). *T.R. Ministry of Environment, Urbanization and Climate Change General Directorate of Geographical Information Systems Smart Cities Department.* Retrieved from https://www.akillisehirler.gov.tr/wp-content/uploads/KapasiteGelistirme/Egitim_Pdf/Ornek_Uygulama_Incelemeleri.pdf. Accessed on December 06, 2022.

United Nations. (2022b). *T.R. Ministry of Environment, Urbanization and Climate Change General Directorate of Geographical Information Systems Smart Cities Department.* Retrieved from https://www.akillise-hirler.gov.tr/2018/05/15/kopenhag-atik-yonetimi-programi/. Accessed on December 06, 2022.

United Nations. (2022c). *T.R. Ministry of Environment, Urbanization and Climate Change General Directorate of Geographical Information Systems Smart Cities Department.* Retrieved from https://www.akillise-hirler.gov.tr/basarili-ornekler/. Accessed on December 06, 2022.

United Nations. (2022d). *T.R. Ministry of Environment, Urbanization and Climate Change General Directorate of Geographical Information Systems Smart Cities Department.* Retrieved from https://www.akillise-hirler.gov.tr/proje-envanteri/gaziantep-ekolojik-koy-projesi/. Accessed on December 06, 2022.

United Nations. (2022e). *T.R. Ministry of Environment, Urbanization and Climate Change General Directorate of Geographical Information Systems Smart Cities Department.* Retrieved from https://www.akillise-hirler.gov.tr/proje-envanteri/gaziantep-glutensiz-bir-baska-kafe-online-satis/. Accessed on December 06, 2022.

United Nations. (2022f). *T.R. Ministry of Environment, Urbanization and Climate Change General Directorate of Geographical Information Systems Smart Cities Department.* Retrieved from https://www.akillise-hirler.gov.tr/proje-envanteri/gaziantep-sehirde-akilli-bitki-fabrikasi/. Accessed on December 06, 2022.

United Nations. (2022g). *T.R. Ministry of Environment, Urbanization and Climate Change General Directorate of Geographical Information Systems Smart Cities Department.* Retrieved from https://www.akillise-hirler.gov.tr/proje-envanteri/izmir-tarimi-mobil-uygulamasi/. Accessed on December 06, 2022.

United Nations. (2022h). *T.R. Ministry of Environment, Urbanization and Climate Change General Directorate of Geographical Information Systems Smart Cities Department.* Retrieved from https://www.akillise-hirler.gov.tr/proje-envanteri/konya-www-konyapedia-com/. Accessed on December 06, 2022.

United Nations. (2022i). *T.R. Ministry of Environment, Urbanization and Climate Change General Directorate of Geographical Information Systems Smart Cities Department.* Retrieved from https://www.akillise-hirler.gov.tr/proje-envanteri/konya-e-desen/. Accessed on December 06, 2022.

United Nations. (2022j). *T.R. Ministry of Environment, Urbanization and Climate Change General Directorate of Geographical Information Systems Smart Cities Department.* Retrieved from https://www.akillise-hirler.gov.tr/proje-envanteri/konya-ciftci-bilgi-sistemi/. Accessed on December 06, 2022.

United Nations. (2022k). *T.R. Ministry of Environment, Urbanization and Climate Change General Directorate of Geographical Information Systems Smart Cities Department.* Retrieved from https://www.akillise-hirler.gov.tr/proje-envanteri/konya-akilli-turizm-rehberi-uygulamasi/. Accessed on December 06, 2022.

Yin, C., Xiong, Z., Chen, H., Wang, J., Cooper, D., & David, B. (2015). A literature survey on smart cities. *Science China Information Sciences*, *58*(10), 1–18.

Yurtseven, R. H. & Kaya, O. (2010). Eko-Gastronomi ve Sürdürülebilirlik. In O. E. Çolakoglu (Ed.), *Ulusal Turizm Kongresi Bildiri Kitabı* (pp. 57–65). Ankara: Detay Publishing.

BLUE-GREEN SMART U-CITIES VIA CLEAN TECHNOLOGIES: TOWARD HIGH SUSTAINABLE AND LOW GREENHOUSE GAS EMISSIONS URBAN AREAS

Hamid Doost Mohammadian

ABSTRACT

The technological revolution is fundamentally changing our lifestyle, work, and communication. There is still no accurate information about the future of this great transformation, but one thing is quite clear: the speed of the transformations in terms of scale and complexity, like a storm, will go through all aspects of a society's life and change all the current paradigms. One of the expectations from the fourth industrial revolution will be the expansion of smart cities in accordance with sustainable development criteria. In order for a city to be truly smart and innovative, city officials must pay attention to elements such as renewable and clean energy such as the internet of things, smart networks, smart parking, and smart transportation. In another sense, there must be a balance between economy, environment, and society in order to build a strong, sustainable, and flexible smart city that will survive the test of time. Of course, based on the 7PS model, the dimensions of culture and education are very important and fundamental for high sustainability. Researchers and those involved in the implementation of smart cities have explained the main indicators to identify these cities, which include some of the following items: For example, smart building, environmental protection, smart garbage collection, digitization of all government affairs and administrative work, widespread use of smartphones and electronic devices, complete, convenient, and universal access to the internet, car sharing service and online taxi. Other important indicators include the intelligentization of the traffic system and urban planning, citizen participation, economic ecosystem, optimization in electricity, water and energy consumption, development of electric and electric public transportation fleet, quality control and management, and reduction of air pollution. In this chapter, the theory of i-Sustainability Plus and i-Comprehensive Strategic Urban Plan is introduced as the beating heart of the ubiquitous blue-green smart city design. In such urban areas, it is all about the combination of real life, virtual reality, and in addition the future of

Smart Cities for Sustainability
Advanced Series in Management, Volume 32, 83–102

Published under exclusive licence by Emerald Publishing Limited
ISSN: 1877-6361/doi:10.1108/S1877-636120230000033007

clean technologies. In general, it can be said that sustainable smart cities are a solution to combat the challenges of urbanization.

Keywords: Blue-green ubiquitous city; sustainability; smartness; greenhouse gases; livability and quality of life; blue-green; U-cities

1. INTRODUCTION

The twenty-first century is characterized by increasing urbanization, population growth and huge social, economic, and environmental problems resulting from them, along with the growth and expansion of information and communication technology at the global level. The contradiction between economic growth and environmental preservation, the confrontation of the ultramodern citizen with the city of the industrial age (modern), and the inability to respond to the needs of the citizens of the present age led to the introduction of the sustainable smart city approach. This approach seeks to solve the problems with the help of clean technologies, including the internet of things. Therefore, sustainable innovative smart cities based on sustainability, advanced technologies, and digitalization can be a tool to create new urban areas with high livability quality and less global challenges, especially environmental problems, through a sustainable smart city master plan. The most important point in the development of smart cities is their sustainability. Environmental sustainability means the preservation of natural capital, which requires us humans to observe limits in the consumption of renewable materials and in the consumption of water and energy resources and not to consume more than what natural systems can provide. Sustainability also includes a wide range of different urban components that support citizens, depending on the level of development of the city, the inherent characteristics of the city, the desire to change the existing situation, the available resources, and the wishes of the citizens. In this section, ubiquitous blue-green cities are introduced as a major tool to combat the challenges of urbanization. The main goal of this study is to create a practical model to propose how to use inclusive blue-green cities as a tool for designing modern urban places in order to achieve high sustainability and sustainable development. The subgoals of livability and high quality of life achieve the main goal of the research. The first subgoal is to show what green-blue infrastructure is everywhere, and the last goal of the study is to find out how modern urban environments combat the challenges of urbanization by improving livability and reducing greenhouse gas emissions. To achieve the objectives of the research, several questions must be answered:

- How do ubiquitous blue-green cities deal with the challenges of urbanization?
- What is the role of inclusive concept in creating sustainable urban areas?
- How to achieve ubiquitous blue-green cities?
- How does rapid urbanization worsen the quality of life and create environmental challenges, especially the emission of greenhouse gases?
- What is the role of a modern and sustainable urban master plan in creating a high quality of livability and less emission of greenhouse gases?

In order to answer the research questions and achieve the objectives, it is important to follow a sound methodology. Systematic literature review (SLR), research-based survey, and review of case studies and their analysis are proposed as methodological tools in this research.

2. LITERATURE REVIEW

2.1 Sustainability

A sustainable smart city refers to an area where clean technologies are used to improve the quality of life, self-efficiency of urban services, and competitiveness. This indicates the importance of cities in social, economic, and environmental development and the balance of these three areas in the present era. In these cities, the economic, social, and environmental needs of future generations are taken into consideration. In the last few decades, the fear of population growth, consumption growth, lack of facilities, etc., has been the subject of many theorists of the world, and paying attention to environmental responsibilities has brought hope in this field. And the crisis caused by social anxiety is an important challenge that highlights the importance of social cohesion in the sustainability model. But the economic dimension supervises how to distribute and use the limited resources needed to improve people's lives. In order to survive, cities and communities in the age of intelligence, in addition to the traditional dimensions of sustainability (social, environmental, and economic), need more dimensions to be able to resist the extensive changes around them. In other words, planning future scenarios requires a new definition of sustainability in order to anticipate and prepare for shocks and crises. In the definition of high sustainability, cultural, technological, educational, and political dimensions are added to the traditional dimensions. Educational sustainability, with the sixth priority in line with other dimensions, should be developed relatively in universities, communities, and businesses to drive innovation and implement strategic plans for resilience against shocks. Fig. 6.1 shows these points in addition to peace and love (Doost Mohammadian & Rezaie, 2019).

2.2 Urbanization Challenges

Nowadays, people move to cities to benefit from higher livability and quality of life. That is, more than half of the world's population lives in cities, and it is predicted that by 2050, two-thirds of the world's people will live in urban areas. Despite the fact that urban residents have more comforts than rural residents, many problems threaten them. Basically, these challenges can be analyzed in two categories: environmental problems and challenges related to livability.

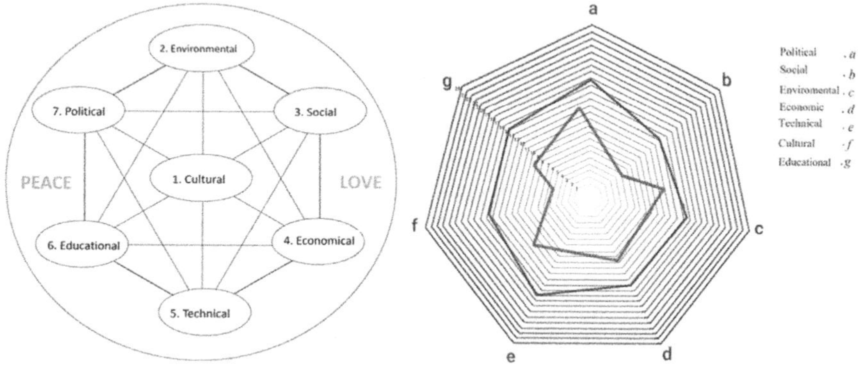

Fig. 6.1. 7PS Model With the Pillars' Priority, Connections, and PEACE/LOVE [Doost. H, 2017] (Author, 2017).

2.2.1 Environmental Problems

Ever since mankind stepped on the earth, it has had a continuous and stable interaction with its environment. But the growth of the population and the achievement of new scientific inventions and discoveries and the movement toward a modern and industrial society have caused today's man to live in a crisis-ridden world. In recent decades, in addition to social and economic crises that are considered serious threats, environmental problems and crises should also be added to the list of global crises. Global environmental problems are important threats to human sciences because the severity of the process of environmental destruction due to population growth, urbanization, forest destruction, noise pollution, endless consumption culture, energy crisis and limited durability of reserves, waste and destruction of natural resources, etc., have all caused extreme human intervention in nature. This approach has upset the balance of the natural environment and caused environmental pollution. In other words, the expansion of human societies and the development of urbanization have led to the emergence of phenomena such as destruction and pollution of the environment. Basically, the increase in greenhouse gases is the main cause of climate change.

CO_2 is the most significant contributor to climate change. CO_2 emission is related to the usage of fossil fuel and human activity. Since the 1970s, CO_2 emission has increased. Combustion of fossil fuels, land use changes, and population increasing could have an influence on CO_2 emission and make it increase (Cioca, Ivascu, Rada, Torretta, & Ionescu, 2015). In addition, land use and land cover change could make energy budget of the Earth be changed and make climate shifting (Desjardins, 2013). This is not only related to a country or a specific territory, but it is a problem for the whole world and includes various issues.

2.2.2 Challenges Threaten Livability and Quality of Life

Today, cities have become the main place of work and a major part of human life, and because attention to the quality of life, conditions, facilities, and needs of city residents is a vital issue, various approaches have been proposed for the development and improvement of living conditions in them. One of those conditions is livability. A city can be considered livable when the increasing growth of urbanization takes place without harming the needs of the future generation. On the other hand, citizens can receive more services and take a step toward urban comfort. The concept of livability has emerged alongside sustainability as a slogan in public discourse and planning. Planners and policymakers in relation to the creation or preservation of cities refer to the concept of livability as a guiding principle for investment and decision-making and shaping the social, economic, physical, and biological urban environment. In fact, livability is the standard of living and the overall desirability of people's lives in a settlement like a city. Due to the change of viewpoints from mere standardization to qualitative viewpoints and the introduction of sustainable development approach and theory instead of macroeconomic viewpoints and physical wisdom, the issues of quality of life and livability of the city have received serious attention. This issue is a multipurpose concept that is associated with many areas of the living environment in urban areas, including physical and sociocultural environments.

Not only environmental challenges but also the other urbanization problems such as slum, economic problems, health diseases, poverty, social instability, and insecurity would worsen livability and quality of life. In other words, urbanization challenges are significant menace to livability and quality of humans' life and fundamentally, humanities through influencing on humans' needs based on Maslow's Hierarchy and supplying needs. So, quality of livability could be threatened. Fundamentally, urbanization challenges are threats to humanities and human beings through reducing quality of livability and quality of life.

2.3 Restrictions

Fundamentally, restrictions are the other type of challenges, the word confront with. In addition, many innovations that changed life and the world were invented through restrictions. For instance, the Second World War is one of the most important proofs for humans to create new achievements through innovation. For instance, invention of the computer, nuclear bombs, aerospace, genetics, and others were invented after World War 2. In addition, Cold War between the United States and Soviets created the third wave, which Alvin Toffler indicated made new achievements through innovation such as internet. These innovations like computers, internet, and others created through restrictions of wars are altered humans' life. The most important restrictions are:

(1) wars,
(2) sanctions,
(3) challenges concerned on increasing population,
   - Increasing Demands
   - Depletion of Resources
(4) low Sustainability,
   - Low Quality of Livability+Low Quality of Life
   - Low environmental sustainability, low political sustainability, low educational sustainability, low cultural sustainability, low social sustainability, low technical sustainability, low economical sustainability
(5) risks.
   - Natural Disasters and Accidents
   - Project Risk Management (PRM)
   - Financial Risk Management (FRM)
   - Govern Mentality Risk Management (GRM)
   - Enterprise Risk Mangement (ERM).

2.3 Ubiquitous City

Information and communication technology has been proposed as the main axis of transformation and development of the third millennium. In this area, the city is also changing due to the evolution of the concept of urban space and the emergence of new information and communication technology infrastructure. In other words, today the role of information technology and communication infrastructures is extremely important; the reason for this is the emergence of environmental, social, and economic opportunities and the disappearance of time and place limitations.

U-City is defined as "a city in which ubiquitous based services are offered at any time and in any place in urban space by realizing high-tech IT, construction, traffic application technologies to enhance urban life quality and city competitiveness" [21].

In U-city, any citizen can get any service anywhere and anytime through any ICT device. ICT development has made the conventional city in terms of intelligence, innovation, and evolution to E-city and then to U-city. In addition, U-city is aligned with environmental sustainability and sustainable smart (or knowledge) city, which makes the ubiquitous computing available to the urban elements such as people, building, infrastructure and open space. High-speed networks, advanced communication services, digitalization, and smartness have an important role in U-cities. Digitalization such as ICT are used in several areas in U-cities like social science, geography, planning, and architecture (Lee, Cameron, & Hassall, 2019).

2.4 Blue-Green City

Blue-green cities are based on sustainable water management and green strategies (University of Newcastle). Basically, to get blue-green city, blue-green infrastructure is required. In recent decades, green infrastructure is not enough to adapt to the environment, and access to water is one of the important water infrastructures in smart cities. Following the growth of urban communities, water consumption is significantly affected. As a result, all urban residents have sustainable access to water with simple processes, and the use of efficient resource consumption solutions instead of the old ones and minimizing the wastage of this vital substance is of special importance among urban planners. Also, the lack of a precise concept for the smart city can be an important component in understanding the importance of providing a real plan for city managers. For all its advantages, digital information evokes in the minds of urban planners the fact that arguments about them are much more real than when managers want to discuss objective evidence and experience. More importantly, even if a precise definition of real digital information and smart cities is provided, it is still necessary to manage this information by experts, while the infinite understanding of smart management is also of particular importance. In other words, the blue-green urban environment can be a tool to fight the challenges of urbanization and create sustainable and livable urban areas.

3. GREENHOUSE GAS EMISSIONS AS SIGNIFICANT THREATS FOR SUSTAINABLE THREATS

In recent decades, environmental challenges are realized as one of the main threats of the world and humanities. So, environmental sustainability as a key to struggle with environmental challenges is required to maintain the nature and environment for humans' life. In addition, environmental sustainability is related to other aspects of sustainability. It means that it is needed to gain sustainable development. So, it is vital to achieve sustainable environment in order to maintain the planet for living.

Greenhouse gas emissions are the main reason for many environmental challenges, especially climate shifting named as the most significant threat for the preserving the planet and humanities. In other words, these gases not only make climate change but also the other environmental challenges such as migration of species, devastating of hurricanes, desertification of fertile areas, negative impacts on agriculture, and disturbing the nature and environment. All these challenges influence the planet and human beings for living. Therefore, reducing effects of these gases in order to preserve the planet and future life is suggested as a solution.

Carbon dioxide, methane, and nitrous oxide are the main greenhouse gases created through humans' activities and fossil fuels. Generally, carbon dioxide is made by fossil fuels and industrial processes, carbon dioxide created by forestry and other land use. Methane and Nitrous Dioxide made by different humans' activities create the main level of greenhouse gases. Further, it is necessary to make humans' activities align with reducing greenhouse gas emissions to develop environmental sustainability. It is required to use strategies to gain neutral greenhouse areas, especially carbon-neutral areas as a key to struggle with environmental problems through reducing greenhouse gas emissions. To gain such areas, it is needed to reduce usage of fossil fuels as main resources of energy in industry, transport, forestry through utilizing renewable and clean energy resources, improving public transport, clean transportation based on hybrid vehicles, nonpolluting vehicles like bicycle and electronic vehicles and walkable cities, utilizing other kinds of energy resources instead of fossil fuels like waste as a source of energy, especially in industry. In addition, promoting ecological awareness, environmental education among citizens, and consuming organic products could be used as tools to struggle with environmental challenges.

Fundamentally, growth in renewable energy, level of demands in electricity, improving vehicles efficiency and a shift in electricity generation from coal to neutral gas, improving ecological awareness, environmental education, and culture based on environmental friendly could be used as techniques to create neutral greenhouse gas areas in order to struggle with environmental challenges threaten humans' beings and the future of the world.

Based on greenhouse gases as the main reason of environmental challenges, it is vital to control emission of them to struggle with these problems in order to maintain the planet, environment, and the nature for future generation to live in this planet. Neutral greenhouse gas area is the best solution to struggle with environmental challenges, especially climate shifting created by greenhouse gas emission. In this research, blue-green smart city is suggested as a path to create neutral greenhouse areas.

4. TOWARD SUSTAINABLE CITIES WITH LIVABILITY AND HIGH QUALITY OF LIFE

Sustainable cities are cities that have high sustainability and sustainable development. Based on authors' researches, seven pillars of sustainability are required to gain sustainable development:

Educational Sustainability: The future of a country is concerned on their juveniles. In other words, youths of a country make its fate. Skillful, perseverance, and responsible youths are required to achieve sustainable society. So, sustainable education is needed to grow innovative and entrepreneur children who are able to create a sustainable and successful society. Generally, sustainable education is related to cultural, social, and technical sustainability and could improve sustainable economic. In addition, proper education would grow children who are responsible about their environment and endeavor to improve environmental problems. Fundamentally, sustainable education is required to achieve sustainability in a society.

Technical Sustainability: Technical sustainability could create new technologies and ways to struggle with environmental problems or improve economic systems. So, it would improve not only environmental and economic sustainability but also social sustainability.

Political Sustainability: Democracy, freedom and political stability are vital for a sustainable country. Generally, citizens of a society should be aligned with political policy to gain political sustainability. Democracy, freedom, equality, transparency and honestly are essential elements for political sustainability. Political problems might cause a revolution in a society that threatens a society. Fundamentally, political sustainability is required to gain social sustainability and improve livability and quality of life in a country.

Social Sustainability: Social responsibility and social sustainability could develop quality of life and livability. A balance between work and life, opportunities for job and education, welfare, social responsibility, high quality of life, certified future, and social security is necessary to achieve social sustainability. Generally, it is one of the most essential elements to preserve a society and a country.

Cultural Sustainability: In some categories, culture is a component of social sustainability, but based on the authors' concept, cultural sustainability is one of the aspects of sustainability. Cultural norms could be altered in a society by revolutions, immigrated citizens, etc. Cultural deviation is a significant threat for not only cultural sustainability but also social sustainability. In addition, cultural sustainability could influence political and economic sustainability.

Economic Sustainability: Economic policy makes human well-being, quality of life and equality without threatening the ecosystem's sustainability. Economic sustainability is needed to gain social sustainability and high quality of life. Proper management strategies, stable banking, economic stability, appropriate exports, proper strategies concerned on companies, especially SMEs, and stable communication in business are formed sustainable economic.

Environmental Sustainability: It is one of the most controversial issues in recent decades and generally concerned on global. Environmental sustainability is required to gain all aspects of sustainability. Sustainable environment means that humans' activities are aligned with the environment not to threaten the future of environment and resources. Climate change is the most significant threat for an environment. New and proper technologies should be utilized to deal with climate change and maintain the world.

These aspects of sustainability make sustainable cities with high quality of life and livability. So, there is a relation between sustainability, livability and quality of life. Generally, livability and quality of life are about humans' needs based on Maslow's hierarchy. Fundamentally, quality of life is about life satisfaction, including everything from family, physical health, education, wealth, employment, security and safety to freedom, the environment and religious beliefs, and livability is concerned on health, security, wealth, education, safety and job opportunities, transportation, public services, housing, recreation and environmental quality. Therefore, these two parameters will be gained by sustainability. Seven pillars of sustainability supply humans' needs in the appropriate path, and high quality of livability and quality of life will be achieved (Figs. 6.2–6.6).

To conclude, livability and quality of life could be named as indicators of sustainable cities. Although in recent decades, sustainable cities couldn't create appropriate livable urban areas, smartness is required to make sustainable cities kept up with humans' demands in order to supply humans' needs and create sustainable livable urban areas with

Fig. 6.2. Sustainability as a Mean Towards Sustainable Development (Author, 2017).

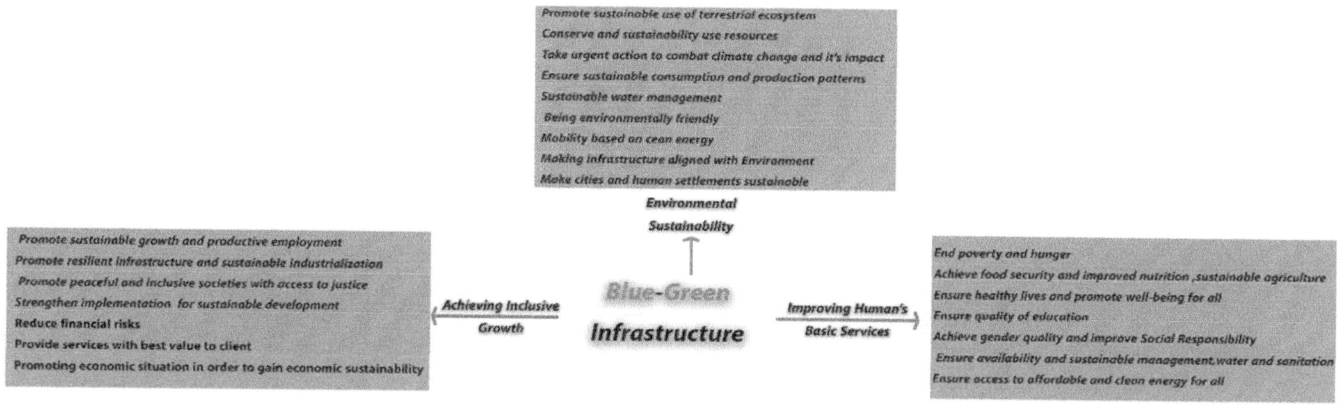

Fig. 6.3. Blue-Green Infrastructure (Author, 2017).

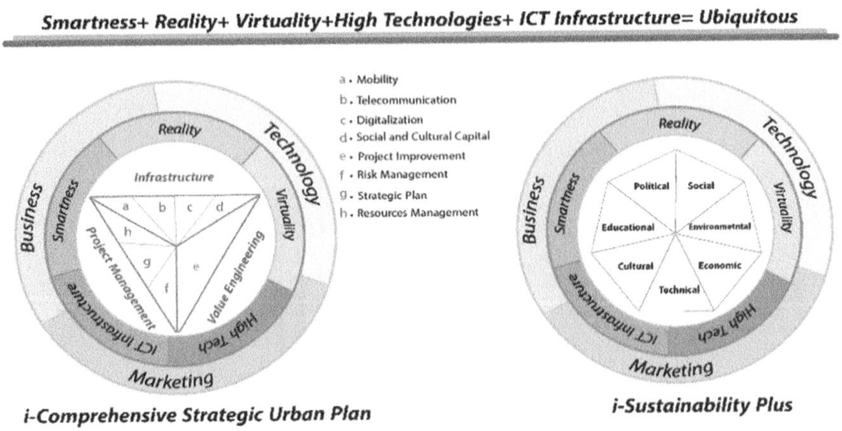

Fig. 6.4. Innovative Comprehensive Urban Plan and i-Sustainability Plus Theory Toward Blue-Green Smart Cities (Author, 2017).

high quality of livability. So, in this research, Blue-Green smart city is suggested as a path to create sustainable and livable urban areas. In addition, through this solution, neutral greenhouse gas areas will be gained because such urban areas are environmentally friendly, aligned with green strategies and sustainable water management.

5. BLUE-GREEN UBIQUITOUS CITIES AS A KEY TOWARD LIVABLE AND SUSTAINABLE URBAN SETTINGS WITH HIGH QUALITY OF LIFE AND LESS GREENHOUSE GAS EMISSIONS

We have been living in an era that needs to be changed in order to maintain the world for future generations and make high quality of livability and life for the present generation. In other words, global challenges created through rapid and unplanned urbanization such as poverty, slum, economic problems, environmental challenges, climate change, insecurity, social instability, and health diseases threaten the planet and humans' being. So, it is vital to struggle with these problems to maintain the world and humanities. Among these challenges, environmental problems and climate change made by greenhouse gas emissions are realized as important ones, which could directly and indirectly influence on other challenges and make them worse. Generally, environmental challenges make livability and quality of life worse through supplying humans' needs. As it was mentioned, growth in renewable energy, level of demands in electricity, improving vehicles efficiency and a shift in electricity generation from coal to neutral gas, improving ecological awareness, environmental education, and culture based on environmental friendly could be used as keys to reduce greenhouse gas emissions. To achieve these solutions, smartness, digitalization, high technologies such as IT, cloud computing like IoT and IoE, sustainability, innovation, and proper comprehensive strategic urban plan are required. So, in this research, Blue-Green Ubiquitous cities are introduced as a way to struggle with urbanization challenges and create sustainable, neutral greenhouse gas urban areas with high quality of livability and life for living. These cities could be kept up with rapid citizens' demands and aligned with environmentally friendly, sustainable water management and natural

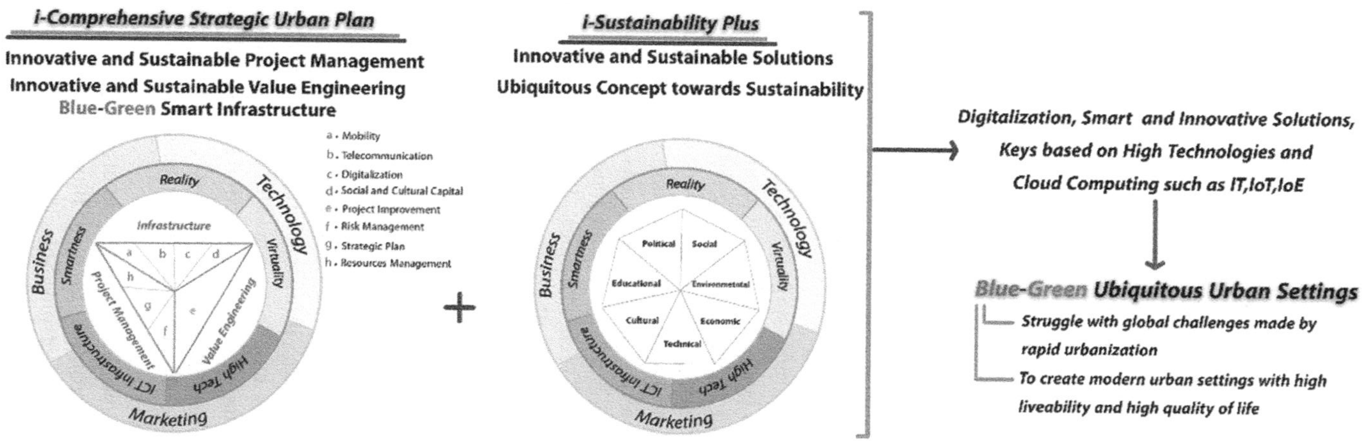

Fig. 6.5. Innovative Comprehensive Urban Plan and i-Sustainability Plus Theory Toward Blue-Green Ubiquitous Urban Setting (Author).

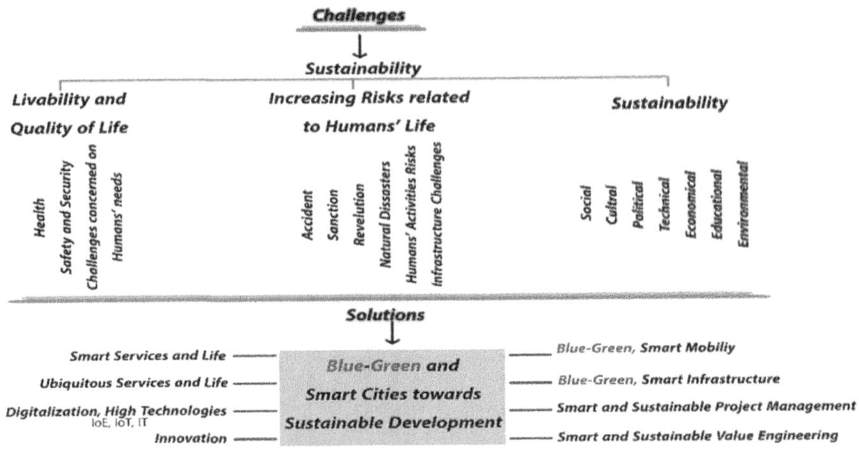

Fig. 6.6. Challenges and Solutions for Smart Cities (Author).

ecologic. So, such cities would make livable urban areas through sustainability and supplying humans' needs.

Nowadays, green strategies aren't sufficient enough to maintain the planet and deal with environmental challenges because many environmental problems are based on water management. So, besides green strategies and being environmentally friendly, proper water management as a response to struggle with important environmental challenges based on water problems like water shortage, flood, secretion, and others is needed. Therefore, to utilize Blue-Green strategies in designing urban areas, Blue-Green Comprehensive urban plan is required. In addition, digitalization, smartness, high technologies, and innovation could be used to achieve Blue-Green strategies. Therefore, Blue-Green Smart Comprehensive urban plan is vital to achieve Blue-Green Ubiquitous cities. Growth in renewable energy, level of demands in electricity, improving vehicles efficiency and a shift in electricity generation from coal to neutral gas, improving ecological awareness, environmental education and culture based on environmental friendly. Smartness, digitalization, high technologies, and innovation are techniques to achieve comprehensive urban plan aligned with Blue-Green strategies.

Blue-Green Ubiquitous cities are sustainable urban areas with high quality of livability and quality of life. To gain such urban areas, Smart Blue-Green comprehensive urban plan including Ubiquitous Blue-Green infrastructure, project management aligned with Blue-Green strategies and value engineering besides sustainability are needed.

Blue-Green infrastructure improves environmental sustainability, inclusive growth, and humans' basic services in urban areas. In other words, livability and quality of life will be increased through Blue-Green infrastructure. Smartness and ubiquitous concept are tools to achieve such infrastructure.

For instance, Blue-Green mobility through smartness, digitalization, and high technologies is one of the tools required to gain Blue-Green infrastructure. Nonpolluted vehicles like bicycle, hybrid and electronic vehicles, promoting public transport, walkable cities, utilizing clean and renewable energy and utilizing risk management to mitigate risks in mobility such as environmental, sustainability, health, standards and other risks could be tools to develop Smart Blue-Green mobility aligned with environmental friendly, sustainable water management and sustainable development.

Besides Smart Blue-Green infrastructure, sustainable and smart project management including risk management, resource management, and strategic planning are required as tools to achieve sustainable infrastructure based on Blue-Green strategies. In addition, value engineering is needed to improve challenges related to comprehensive urban plan through innovative and smart improvement and solution to design sustainable comprehensive strategic urban plan aligned with Blue-Green and smart strategies. Fundamentally, comprehensive strategic urban plan based on Blue-Green strategies, smartness strategies, and sustainable development is the best key to create modern urban areas to struggle with urbanization challenges and make better living areas with high quality of livability and quality of life through supplying and providing humans' needs appropriately.

Based on authors' experiences, i-sustainability plus and i-comprehensive urban plan are tools to achieve modern urban areas. These two theories are based on ubiquitous point of view, combination of reality and virtually, and innovative project management. In other words, ubiquitous concept and innovation are used as techniques to create a new version of sustainability and comprehensive urban plan named i-sustainability plus and i-comprehensive strategic urban plan. Fundamentally, these new models are required to be able to struggle with global challenges in recent decades. It means that in this era, because of significant challenges and growing demands of humans, sustainability and sustainable urban plan aren't sufficient enough to keep up with urbanization demands. So, innovation and ubiquitous are introduced as new paths to achieve sustainability in this era.

Fundamentally, according to authors' point of view, the main techniques and tools to design sustainable comprehensive strategic urban plan in order to create Blue-Green ubiquitous urban areas are: innovation as technique; seven pillars of sustainability including economic, cultural, educational, social, political, technical, and environmental aspects; digitalization and high technologies such as IT, IoE, IoT, and smartness as techniques; comprehensive strategic urban plan including value engineering as an improvement tool, Blue-Green, and smart mobility; and project management concerned on Blue-Green strategies and sustainability and Blue-Green and smart infrastructure.

Generally, illustrating these theories and the tools to achieve them isn't sufficient. Besides them, implementation of these theories to gain Blue-Green ubiquitous cities is required. In other words, how to utilize these theories in cities and make them practical are vital. Fundamentally, project management based on innovation, smartness, and sustainability is a tool to achieve i-comprehensive strategic urban plan. For instance, innovative and smartness in mobility such as nonpollution vehicles like hybrid and electronic cars or innovative risk management are proper examples of innovative and sustainable project management in order to create i-comprehensive strategic urban plan. In addition, innovative and sustainable value engineering needed knowledgeable and vocational social capitals, education, cultural and Blue-Green smart infrastructure are required to design proper urban plan introduced as i-comprehensive strategic urban plan in order to gain Blue-Green cities. Further, combination of reality and virtually as ubiquitous is required to be kept up growing urbanizations' demands and struggling with global challenges. In addition, i-sustainability plus as tools based on ubiquitous life and services and innovative solutions could be used to design cities aligned with sustainable development, environmentally friendly, sustainable water management and urbanizations' demands.

So, Blue-Green city based on smart solutions would improve safety, time, health, environment and nature, connectedness, jobs, cost of living, education, and situation for life. In addition, such urban areas could promote sustainable development, sustainability, livability and quality of life through struggling with global and environmental challenges. For instance, China is located as the first stage of emission of greenhouse gases. Therefore, utilizing Blue-Green and smart solutions are essential for China and high polluted counties to reduce their environmental impacts on global challenges. In addition, nonpolluted vehicles like bicycle, hybrid and electronic vehicles, improving public transportation and renewable energy could create Blue-Green mobility in order to make Blue-Green infrastructure. Blue-Green mobility is required to achieve sustainable environment through reducing emissions of greenhouse gases. Generally, mobility based on Blue-Green strategies is the best path to create neutral greenhouse gas areas. Because renewable energy and nonpolluted energy are solutions to reduce these gases and Blue-Green mobility are based on such solutions.

Fundamentally, Blue-Green

Generally, such urban areas could diminish global challenges such as climate change, emissions of greenhouse gases, low quality of livability, health diseases and other problems through creating urban settings aligned with environmentally friendly, sustainable water management and ecological maintenance.

6. CASE STUDY

6.1 Denmark

Denmark is known as a green state. Generally, Denmark follows strategies to gain high sustainability and smartness in not only big cities like Copenhagen but also in smaller cities like Glad Saxe. In other words, all cities in Denmark design a proper urban plan in order to achieve sustainability, high quality of livability, and high quality of life. Fundamentally, strategies utilized by governments, planners, decisions-makers, and citizens make this country one of the most sustainable and greenest country of the world. In other words, proper strategies create smart and sustainable infrastructure including sustainability, high quality of livability, and high quality of life. Generally, Blue-Green and smart mobility have an important role in creating a sustainable country. The most important strategies utilized to achieve Blue-Green and smart infrastructure are:

- bicycle as main transport vehicle,
- public Transport,
- utilizing Renewable and Clean Energy in Mobility Systems,
- independent of fossil fuels,
- high standards,
- transportation for everyone,
- Blue-Green Shipping,
- design different types of technology through innovation such as environmental science, pharmaceuticals, biomedicine, food and agriculture,
- high digitalization,
- becoming neutral CO_2 area in 2050,
- agriculture relied on renewable energy such as wind, solar, and biomass energy,
- utilizing renewable and clean energy as main resources in home, industry, agriculture; fundamentally, country based on clean energy, and
- High environmental sciences aligned with environmentally friendly, water control, and heating system to minimize environmental impact,

- sustainable, innovative, and smart solutions used by countries in order to be aligned with government policies toward creating sustainable country,
- architecture and construction aligned with sustainability,
- creating high livability and quality of life through policies concerned on citizens' life such as secure insurance, high quality of education, work, housing, etc.
- high education and teach high cultural norms, and
- high quality of life through supplying humans' needs.

Fundamentally, all strategies used by governments create high sustainability and quality of life in Denmark. In addition, policies utilized by companies make them aligned with sustainability. For instance, Maersk shipping countries is a Danish company aligned with environmentally sustainability and water management. Smart and innovative strategies, innovative risk management, and utilizing renewable energy and sustainable policies used by this company make it an important role in achieving not only Denmark sustainable but also dealing with global environmental challenges.

6.2 Finland

Finland is also one of the industrial countries that utilized strategies and policies to achieve sustainable urban areas aligned with green strategies, sustainable water management, and environmentally friendly and sustainable development. The main aim of Finland is to make cities and human settlements inclusive, safe, resilient, and sustainable. In order to gain such goals, the most important strategies and polices used by governments, planners, decision-makers, and citizens are:

- national strategies toward sustainable development,
- high education,
- education concerned on importance of sustainability especially environmental sustainability,
- education of norms and altering them as a kind of culture,
- conduct policy experiments,
- making high livability and quality of life through high quality of education, services, life, housing, etc.
- green economy,
- utilizing sustainable water management and environmentally friendly strategies as keys toward sustainable cities,
- invest in green economy,
- strategies aligned with environmental sustainability like improving resources efficiency, renewable energy, and solar power,
- utilizing strategies toward cleaner energy systems,
- clean transportation based on clean, inclusive, and renewable energies such as electric cars, hybrid cars, bicycle, and public transport worked with clean energy,
- policies to change transportation systems as a way to make public transport easier and safer,
- industry based on clean and renewable energy such as biofuels, waste as source of energy, etc., and banning of the use of coal in power generation by 2030,
- recycling as a solution toward sustainable choices,
- utilizing innovative solutions and strategies by companies and businesses to achieve sustainability,

- being independent of fossil fuels,
- creating cities' emission-free and carbon-neutral energy, and
- creating cities toward sustainable development with high quality of life.

Generally, governments, NGOs, businesses, companies, and citizens work together to achieve sustainability and create urban areas with high quality of livability and life through supplying humans' needs in the best way and making sustainable living areas. In addition, besides of mentioned strategies; education of importance of sustainability, courses related to sustainability, smartness and innovation could be used as techniques to help strategies be proceeded. Fundamentally, Finland is known as one of the most sustainable countries, further it utilized strategies, policies and solutions to gain more sustainable areas based on green strategies and water control.

7. DISCUSSION

Nowadays, the world confronts significant global challenges such as poverty, slum, economic problems, social instability, insecurity, health diseases, environmental challenges and climate change threaten preserving the planet and humans' life there. So, it is vital to deal with these challenges to not only maintain the world for future generation, but also to make better urban setting with high quality of livability and quality of life for present generation.

In this study, Smart Blue-Green city is introduced as a solution to struggle with global challenges especially environmental problems and climate change; in order to create neutral greenhouse gas areas with high livability and quality of life through developing sustainability. Fundamentally, smart Blue-Green urban areas are aligned with green strategies, sustainable water management, environmentally friendly, ecological sustainability and sustainable development.

To gain Smart Blue-Green cities; seven suggested pillars of sustainability by authors including economic, educational, technical, political, social, cultural and environmental sustainability, comprehensive strategic urban plan based on green strategies, sustainable and smart water management, ecological sustainability and high quality of livability, innovation, smartness, high technologies such as IT, IoT and IoE are required as tools.

Seven pillars of sustainability are needed to achieve sustainable development and livable urban areas. According to researches, high sustainability – cooperating of sustainability and ubiquitous concept – is the best key to develop sustainability nowadays. Basically, based on global challenges and rapid urbanization's demands, to achieve sustainability, sustainability through ubiquitous idea and innovation management named high sustainability by authors is required in recent decades. In other words, this kind of sustainability, is not only kept up with growing citizens "demands but also gain sustainable development and sustainability through reducing negative impacts of global and environmental challenges. A theory named i-Sustainability Plus is introduced by authors as a path to struggle with urbanization challenges and develop sustainability in this era confronted with significant challenges. In other words, in this era significant challenges needed to be solved through sustainability. Although sustainability is proper solution, it isn't" sufficient enough. Combination of sustainability, innovation, and ubiquitous is required as a way to achieve sustainability in urban settings.

In addition, comprehensive strategic urban plan is required to design sustainable and livable urban areas. Infrastructure, project management, and value engineering are components of comprehensive strategic urban plan. These dimensions should be concerned on Blue-Green strategies, sustainability, and smartness. Generally, Blue-Green strategies concerned on environmentally friendly, sustainable water management, sustainability, and green policies make infrastructure of cities sustainable. Further, sustainable infrastructure not only gains sustainable development but also improves livability and quality of life. Fundamentally, project management is utilized by urban planners and decision-makers as a tool to help them develop sustainable and smart infrastructure. For instance, risk management is an important tool in project management as a path to identify and mitigate risks in order to improve results through reducing risks. Risk management has an important role in creating sustainable and smart mobility. Public transport concerned on clean and renewable energy is about innovative risk management utilized in developed counties to reduce environmental risks, standards risks, and sustainability risks. In addition, high standards for transportation of dangerous goods is the other example of risk management to improve sustainable infrastructure. Therefore, project management's tools like risk management is needed to achieve proper infrastructure. Further, value engineering is a technique to increase infrastructure challenges. Fundamentally, these three dimensions could achieve comprehensive strategic urban plan aligned with sustainable development, sustainability, and high livability. A model named i-comprehensive strategic urban plan is introduced as an urban plan concerned on Blue-Green and sustainable strategic urban plan.

To conclude, comprehensive strategic urban plan and high sustainability are realized as tools to design modern and smart Blue-Green cities. Furthermore, innovation, high technologies like digitalization and smartness, IoE, IT, and IoT, could be techniques to develop high sustainability and Blue-Green comprehensive strategic urban plan.

8. CHALLENGES AND SOLUTIONS

Nowadays, the most important challenge that humanities confront with is sustainability. Sustainability directly and indirectly influences other global challenges. Generally, all global challenges threaten the planet and humans' life. So, it is vital to struggle with them.

Fig. 6.6 shows the most important global challenges and suggested solutions related to this research.

9. CONCLUSION

Nowadays, the world confronts a lot of global challenges such as economic problems, poverty, environmental challenges, slum, and climate change, social instability, health diseases, and unsafety. It is concluded that:

- These challenges threaten humans' life and the maintenance of the world. So, it is vital to struggle with these problems.
- Among all these challenges, environmental problems and climate shifting are the ones that directly and indirectly influence other challenges.
- Humans' activities and emission of greenhouse gases are the main reasons for environmental challenges. Therefore, neutral greenhouse gas areas are needed. Independency of fossil fuels, usage of renewable energies, improving public transportation, nonpolluted vehicles including hybrid, electronic vehicles, and bicycle, industry based on clean

energies, and clean energies as main energy resources are the best solutions toward such areas.

- Fundamentally, Blue-Green areas not only develop environmental sustainability but also create Blue-Green infrastructure made sustainable development, sustainable urban areas with high livability and quality of life.
- The IoT and its new applications have updated and enabled the ways of becoming a smart city on a global scale. This modern technology equips the city with the ability to remotely monitor, control, and manage urban facilities and devices as well as generate new insights and actionable information from quantitative streams of real-time data.
- Sustainability is one of the key parts of smart cities, and achieving it requires investing in efficient and environmentally friendly management. The rapid growth of urban clusters in recent decades has brought a sharp increase in the demand for energy resources, which is sometimes so high that it exceeds the available local resources. A long-term solution to meet the growing energy demand is the "smart energy chain", which relies entirely on renewable energy sources such as solar and wind energy. Through a smart digital-based system, this system provides the possibility of transferring clean and stable decentralized energy to any part of the urban area.
- Mobility is the heartbeat of a city, that's why smart mobility and transportation networks are prioritized in many smart city plans. Urban areas such as multimodal transportation, smart parking, and smart traffic lights are often included in this concept. This approach is based on rethinking the transport infrastructure used in everyday life, not only in terms of traditional cars, electric vehicles, and public transport but also in terms of completely new and innovative forms of transport such as on-demand, ride-sharing services and car-sharing programs. Considering such considerations can improve the accessibility and livability of a city.
- And finally, as we said above; One of the main features of the twenty-first century is urbanization and population growth; major social, economic, and environmental problems; and the expansion of information and communication technology in the world. The difference between economic growth and environmental protection, the confrontation between the ultramodern citizen and the modern industrial city, and the lack of response to the needs of the citizens have led to the idea of a sustainable smart city. This city seeks to solve the problems that arise in the new century with the help of information and communication technology such as the internet. The criticality of social, economic, and especially environmental problems in today's cities shows the importance of this issue. A sustainable smart city is formed based on the principle of equality and social inclusiveness, increasing the quality of life of citizens, creating productivity, creating flexible infrastructures, using information and communication technology, and preserving the environment. In a sustainable smart city, renewable energy, smart networks, smart parking, and smart transportation should be used and the social, economic, and environmental level of the city should be improved in an integrated manner.

REFERENCES

Akhtar, C., & Bahru, J. (2015). Can intellectual capital of SMEs help in their sustainability efforts? *Journal of Management Research, 7*.

Aotearoal, T. (2017). *Better urban planning. Chapter 10: Urban planning and infrastructure*. The New Zealand Productivity Commission.

Bahena-Álvarez, Cordón-Pozo, I., & Delgado-Cruz, E. (2019). A social entrepreneurship in the conduct of responsible innovation: Analysis cluster in Mexican SMEs. *Sustainability Journal*.

Bolden, R. (2007). Leadership development in SMEs: Designing a customized solution. *GITAM Journal of Management, 5.*

Camerer, C. (2005). Behavioural economics. In *The proceeding of world congress of the econometric society.* London 18–24 August 2005.

Cioca, L., Ivascu, L., Rada, L., Torretta, V., & Ionescu, G. (2015). Sustainable development and technological impact on CO_2 reducing conditions in Romania. *Sustainability Journal, 7.*

Criveanu, M., & Iacob, C. (2011). Role of management control in small and medium enterprises performance assurance. In *Annals of University of Craiova – Economic sciences series, University of Craiova, faculty of economics and business administration 3.* Craiova, Romania.

Department of Environment, Land, Water and Planning. (2017). *Planning a green-blue city. Produced with assistance from E2Designlab.*

Desjardins, R. (2013). Climate change—A long-term global environmental challenge. *Procedia – Social and Behavioral Sciences, 77.*

Doost Mohammadian, H. (2018). Internet of energy: A solution for improving the efficiency of reversible energy. In *Proceedings of the IEEE global engineering education conference (EDUCON),* Tenerife, Spain, 17–20 April 2018.

Doost Mohammadian, H. (2019). IoT – A solution for educational management challenges. In *The proceeding of 2019 IEEE global engineering education conference (EDUCON).* 8–11 April. Dubai, United Arab Emirates.

Doost Mohammadian, H. (2020, March 26). *I-sustainability plus theory as an innovative path towards sustainable world founded on blue-green ubiquitous cities (case studies: Denmark and South Korea).* Inventions, MDPI. Retrieved from www.mdpi.com/journal/inventions

Duran, D., Gogan, L., Artene, A., & Duran, V. (2015). The components of sustainable development – A possible approach. In *4th World conference on business, economics and management, WCBEM.*

Engineering Education Conference. (2020). (EDUCON). 27–30 April. Portugal: Porto.

Fink, M., & Hattak, I. (2010). *Current research on entrepreneurship and SME Management.* Turku: European Council for Small Business and Entrepreneurship.

Franco, M., & Matos, P. (2013). Leadership styles in SMEs: A mixed method approach. *The International Entrepreneurship and Management Journal.*

Fukuyama, M. (2018). *Society 5.0: Aiming for a new human-centered society.* Japan Spotlight.

Gehrels, H., Meulen, S., & Schasfoort, F. (2016). *Designing green and blue infrastructure to support healthy urban living.* TO2 Federatie.

GhofraniSposito, Z. V., & Faggian, R. (2017). A comprehensive review of blue-green infrastructure concepts. *International Journal of Environment and Sustainability, 6.*

GiapThye, T. W., & Grace, A. (2014). A new approach to measuring the livability of cities: The global livable cities index. *World Review of Science, Technology and Development, 11*(No. 2).

Hayatul, Y. (2019). *Society 5.0 for supper smart country.* Retrieved from https://www.researchgate.net/publication/331521659

Hecklau, F., Galeitzke, M., Flachs, S., & Kohlb, H. (2016). Holistic approach for human resource management in Industry 4.0. In *Proceeding of 6th CIRP conference on learning factories.* 29–30 June. Norway: Gjøvik.

Heynitz, H., & Bremicker, M. (2016). *The factory of the future: Industry 4.0-the challenges of the tomorrow.* KPMG International Cooperative.

Ilhan, A., Möhlmann, R, & Stock, G. (2015). *Citizens' acceptance of u-life services in the Ubiquitous City Songdo.* Citizen's Right to the Digital City. Springer Science+Business Media Singapore.

Jang, M., & Suh, S. (2010). *U-city: New trends of urban planning in Korea based on pervasive and Ubiquitous geotechnology and geoinformation.* Springer-Verlag Berlin Heidelberg.

Kim, Y. M., Kim, H. S., Moon, S. Y., & Bae, S. Y. (2009). *Ubiquitous eco-city planning in Korea. A project for the realization of ecological city planning and ubiquitous network society* (pp. 925–930). na.

Kuhlman, T., & Farrington, J. (2010). What is sustainability? *Sustainability Journal, 2.*

Lee, J., Cameron, I., & Hassall, M. (2019). Improving process safety: What roles for digitalization and Industry 4.0? *Process Safety and Environmental Protection, 132,* 325–339.

LWMOOCS VII. Learning with MOOCS 2020 in Antigua Guatemala from 30 September – 02 October 2020.

Mokaya, S., Namusonge, M., & Sikalieh, D. (2012). The concept of entrepreneurship; in pursuit of a universally acceptable definition. *International Journal of Arts and Commerce, 1.*

Momtaz, R., & Elsemary, Y. (2015). Qualitative conceptions of livability between theory and applications in Egypt. In *International conference on IT, architecture and mechanical engineering.* Dubai.

Monocha, S. (2012). Innovation and entrepreneurship in today's scenario. *International Journal of Marketing, Financial Services and Management Research, 1.*

Munasinghe, M. (2003). *Climate change and sustainable development linkages: Points of departure from the IPCC TAR.*

Nagy, J., Oláh, J., Erdei, E., Máté, D., & Popp, J. (2018). The role and impact of industry 4.0 and the Internet of Things on the business strategy of the value chain—The case of Hungary. *Sustainability*, *10*(10), 3491.

Negev, M., Garb, Y., Biller, R., Sagy, G., & Tal, A. (2010). Environmental problems, causes, and solutions: An open question. *The Journal of Environmental Education, 41.*

Nwankwoala, H. N. L. (2015). Causes of climate and environmental changes: The need for environmental-friendly education policy in Nigeria. *Journal of Education and Practice*, *6*, 30.

Rojko, A. (2017). Industry 4.0 concept: Background and overview. *International Journal of Innovation Management (iJIM), 11.*

Ruen, K. (2011). The quality of life and the environment. In *Asia Pacific international conference on environment-behavior studies,* North Cyprus.

Saif Obaid Alhefeit, F. (2018). *Society 5.0 A human-centered society that balances economic advancement with the resolution of social problems by a system that highly integrates cyberspace and physical space.* MSc Informatics Thesis, The British University in Dubai.

Sarma, A. (2016). *Smart city and urban sustainability.* Institution of Engineers (India). Tech Vol cum Souveni.

Seele, P., & Lock, I. (2017). The game-changing potential of digitalization for sustainability: Possibilities, perils, and pathways. *Sustainability Science Journal, 12.*

Skill Development for Industry 4.0. (2016). *BRICS skill development working group.*

Szczuka, M. (2015). Social dimension of sustainability in CSR standards. In *6th international conference on applied human factors and ergonomics (AHFE 2015) and the affiliated conferences, AHFE 2015.*

Trindade, E., Hinnig, M., Costa, E., Marques, J., Bastos, R., & Yigitcanlar, T. (2017). Sustainable development of smart cities: A systematic review of the literature. *Journal of Open Innovation: Technology, Market, and Complexity*, *3*(1), 11–24.

DRMM AND COMPREHENSIVE GLOBAL BLUE-GREEN CLEAN SUSTAINABLE URBAN MOBILITY RISK MITIGATION PLAN FOR MAPPING FUTURE SMART CITIES THROUGH THE 5TH WAVE THEORY

Hamid Doost Mohammadian

ABSTRACT

Based on the 5th wave/tomorrow age theory, we are living in the world that is in necessity to change. Rapid urbanization causes global challenges such as economic problems and recessions, environmental challenges, climate change, social instability, health diseases, biological attached, and crisis caused by technological dominations. These challenges threaten the world, humanity, and human beings. Therefore, it is vital to tackle and struggle with them in order to maintain the world and improve quality of livability and quality of life to achieve sustainability. Generally, modern Blue-Green urban areas and smart cities with high quality of livability and life are proposed to deal with urbanization challenges to maintain the world and improve quality of human life. Based on Prof. Doost's 5th wave theory, related theories, concepts and models like Doost Risk Mitigation Method (DRMM), and also his experience on sustainability as best practice such as cooperating with Danish Sustainable Platforms Company, working as an academic leader at IoE/EQ EU Erasmus Plus project in Germany during 2017–2020, cooperating with former mayor of Copenhagen, consulting the German MV State Minister of Energy, Digitalization, and Infrastructure to cooperate with Iran in 2016, more than 15 years holding lecture and research internationally about risk and risk management on mobility in different universities like (TU Berlin) Technical University of Berlin (EUREF Campus, Sustainable Mobility Management and Sustainability Building) and also achieving a honorary doctorate in sustainable development management, a practical model concerned on risk management in mobility to provide comprehensive global Blue-Green clean sustainable urban mobility risk

Smart Cities for Sustainability
Advanced Series in Management, Volume 32, 103–136

Published under exclusive licence by Emerald Publishing Limited
ISSN: 1877-6361/doi:10.1108/S1877-636120230000033008

mitigation strategic plan is given. Therefore, in this chapter, impact of risk management on mobility to provide sustainable global urban mobility plan in order to create modern Blue-Green sustainable urban area and future smart cities through the 5th wave theory are explored. Fundamentally, the main goal of the research is to have an applied study about mobility risk mitigation and utilize it as a key to create comprehensive global urban mobility risk mitigation plan toward Blue-Green sustainable clean mobility technologies to create modern sustainable smart cities through the tomorrow age theory in order to create livable urban area with high quality of livability and life. In addition, the risks in mobility through the DRMM are measured to analyze the risk and to do risk mitigation and mobility project improvement to move to sustainable mobility and high sustainability in future smart cities.

Keywords: Risk management; blue-green comprehensive sustainable strategic urban plan; sustainable mobility; sustainability; livable urban setting; risk measurement; the 5th wave/tomorrow age theory; DRMM method

1. INTRODUCTION

Nowadays, urban settings struggle with rapid urbanization that creates global challenges such as poverty, environmental problems, climate change, economic challenges, social instability, health diseases, etc. Although cities are becoming the predominant settings for living and working, it is vital to find solutions to deal with global challenges. Fundamentally, sustainable and green comprehensive strategic plan is required to develop sustainable mobility in order to gain modern green urban settings with high livability and quality of life. Tools and techniques such as risk management concerned on mobility are needed to achieve green and sustainable strategic plan. Generally, risk management could be used as an important tool to improve mobility in order to make livable urban areas.

The main aim/objective of this research is to find out practical model concerned on how risk management on mobility could influence urban plan to make modern livable urban settings. To gain the main aim of the research, two sub goals are proposed. The first sub goal is to realize the impact of comprehensive strategic urban plan on sustainable mobility and the second sub goal is to find out the importance of sustainable and smart mobility for creating sustainable and green urban setting with high livability and quality of life. In order to achieve the aims of the research, some questions should be answered:

- How would green and more livable urban areas be achieved?
- Why is it vital to create modern, sustainable, and green urban settings?
- How would risk management improve mobility in order to make livable urban settings?

In order to answer the research questions and gain aims, it is important to pursue a sound methodology. Systematic literature review (SLR), survey based on research, exploring case studies and analyzing them are proposed as methodologies' tools in this research.

2. LITERATURE REVIEW

2.1 Risk Management

At the beginning of the twentieth century, the first significant studies in business risk management were developed by Willet (1901), Leitner (1915), Knight (1921), Oberparletier (1930), Stadler (1932), and Sassi (1940) (Borghesi & Gaudenzi, 2013). Although risks create failure, unsuccessful projects followed by loss in projects, companies are willing to take risks to generate income and profits (Author, 2018).

2.1.1 Risk Management Classification

- Project Risk Management (PRM): (Steinkamp, 2014).
- Financial Risk Management (FRM): (Steinkamp, 2014).
- Govern Mentality Risk Management (GRM): (Author, 2018).
- Enterprise Risk Management (ERM).

2.2 Risk Analysis and Mitigation

Generally, to manage risks, risk management is utilized in projects. It has four steps.

Establishing the context is the first step. Aims, the objectives, and stakeholders of the projects are identified. Based on them, the risk assessments to avoid being overwhelmed are used by details and data. The second step is identifying the risk (Risk Management Handbook, 2016). Risk analysis is the third step (Padayachee, 2014). Risk mitigation is the fourth steps that includes prioritizing, evaluating, implementing, and maintaining the appropriate risk to reduce controls (risk-reducing measures) recommended from the risk assessment process. Risk mitigation has seven steps: prioritize actions, evaluating recommended control options, conducting cost-benefit analysis, selecting control, assigning responsibility, developing an action plan, and implementing selected control (www.oa.mo.gov).

2.3 Risk Management in Mobility

Mobility and transport security are required to achieve society development. Issues of risk management in mobility are explored at two levels: operation and governance (Author, 2018; ŠIMÁK, 2007). Generally, risks, influencing transport business and transport processes, can be derived from risks currently affecting society, human activities, material values, living environment and also the human life and health (ŠIMÁK, 2007). Risk management on mobility tries to determine and identify risks and failure concerned on mobility, in order to control them and improve goals of a project through minimizing the effects and consequences of potential probable crises. In addition, *it could help managers to make a balance between operating costs and financial costs.* The main aim of risk management is to do executive reduction of risks related to mobility processes up to acceptable level to obtain senior management approval (Author, 2018; Pande & Kathmandu, 2013–2017). Risk management consists of processes such as:

- Risk Assessments: (Author, 2018).
- Risk Analysis.
- Risk Mitigation: It is a process in which controls and safeguards are done to preventing identified risks. IT, digitalization, innovation, and value engineering could be used as tools to mitigate risks (Author, 2018).

2.3.1 Risk Management in Mobility Sector

Transport networks could be harmed by environmental risks such as natural disasters, extreme weather and pandemic, geopolitical risks like conflict and political unrest, export and import restrictions, terrorism corruption, nuclear/biological/chemical weapons, illicit trade and organized crime and maritime piracy, economic risks such as sudden demand shocks, extreme volatility in commodity prices, broader delays, currency fluctuations, global energy shortages and ownership/investment restrictions and technological risks like information and communications disruptions and transport infrastructure failures (Appel, 2012; Soeanu et al., 2015). Generally, risk management in transportation sector could be explored in two axes:

(1) Risk management in infrastructure, networks, and transportation systems (the safety aspect).
(2) Risk management in transport companies (from the perspective of investment and business).

To prioritize risk management processes in infrastructure, networks, and transportation systems, the following steps can be taken:

- Identify key elements and influential transport elements.
- Identify critical transport infrastructure and determine risk facilities.
- Implementation and operation of risk assessment in the transportation network.
- Codification and development of risk management control strategies.
- Implementing and operating control strategies.
- Performance monitoring throughout the process (Author, 2018).

2.3.2 Global Developments Affecting the Transportation System

Three global phenomena could influence on transportation systems:

(1) Changes in economic and commercial system.
(2) Multi-dimensional transportation approach.
(3) Environmental considerations and sustainable transport (Author, 2018).

2.3.3 Challenges and Solutions Concerned on Risks of the Road Transport System of the Global Countries

Insufficient and nonflexible supply, irregularities in transportation services, unbalanced dispersion of economic activities, failure to adhere to time of achieving destination, nonintegrated transport network system, truck waiting times, the inefficiency of the road transport system, the cost of national transport fleet, incomplete information flow of transportation backup system, and hazardous goods transportation problems are the most important factors involved in road transport risks.

It is vital to struggle with these risks to control and manage the road traffic. Some solutions are proposed such as utilizing digital and modern methods instead of old methods, making balance in dispersion of economic activities, using IT in transport network system, making transportation systems sustainable and optimized, utilizing policies of developing countries in order to improve transportation networks, making flexibility in transportation services, long term contracts and reliance, and utilizing standards concerned with transport of dangerous goods (Author, 2018).

2.3.4 Global Transport of Dangerous Goods

Transport of dangerous goods is one of the most controversial issues in the world that should be controlled. For instance, in Italy, about 80% of road traffic is related to the transport of goods and 18% of this 80% is related to the transport of dangerous goods. The

United Nations established mechanisms in terms of hazard classification criteria and communication tools (GHS) as well as transport conditions in the different modes (TDG). The ADR (European Agreement concerning the International Carriage of Dangerous Goods by Road, IATA DGR (air), IMDG Code (maritime), ADN (waterways), and RID (rail) are the most famous regulations used for transportation of dangerous goods (Author, 2018; Kervella, 2011).

2.4 Risk Management Patterns

In this research, few patterns of risk management are explored:

- PMBOK: (Author, 2018).
- Canadian National Risk Management Standard, No. Q850: (Canadian Standards Association, 1997).
- The English standard: provided by the Institute of Standards of England (assets.cambridge.org).
- National Risk Management Standard in Australia/New Zealand, No. 4360 and its Implementation Guide 2004: (Joint Technical Committee, 2004).
- ISO 31000: (www.iso.org).

2.5 Doost Risk Mitigation Method (DRMM)

DRMM is based on developing the failure modes and effects analysis (FMEA) model. It is one of the most common risk assessments methodologies. FMEA process has five steps: to identify potential failures and effects, determine severity, gauge likelihood of occurrence, failure of detection, and calculating risk priority number (RPN). (www. polarion.plm.automation.siemens.com).

2.6 Mobility

Mobility is defined as a demand for activities, where the cost is an integral part of demand (Kayal, Singh, & Kumar, 2014). Generally, transport infrastructures contribute to landscape fragmentation and the destruction of habitats and ecosystems, and many public transport systems do not keep pace with urban and spatial development (Berger, Berger, Feindt, Holden, & Rubik, 2014).

2.6.1 Blue-Green Sustainable Mobility

Sustainable mobility is the ability to meet society's need to move freely, gain access, communicate, trade, and establish relationships without sacrificing other essential human or ecological values today or in the future (Sustainable urban mobility). Sustainable mobility is related to land-use planning and sectoral allocation of land for different activities (Nijkamp, Verhoef, Ubbels, & Rodenburg, 2001).

Sustainable mobility has three aspects which focus on achieving sustainable people, planet, and profit development and the balance among global environment, economic success, and quality of life. There are four main types of actions to improve sustainable mobility:

(1) Making the best use of technology, including investment in technology in transportation modes, information systems and in the transport system itself, and in giving industry directions on priorities (e.g., on hybrid and fuel-efficient vehicles and alternative fuels).
(2) Regulation and pricing mean that the external costs of transport should be reflected in the actual costs of travel through higher fuel prices or through some form of road user

charging. Demand management reduces congestion and improves environmental quality, but it does require public support to work effectively.
(3) Land-use development, including planning and regulations, needs to be integrated, so that physical restraint measures and development patterns are used to support shorter travel distances. Improved levels of proximity would help reduce distance traveled and contribute to trip reduction and modal split changes.
(4) Clearly targeted personal information, including social pressure, awareness raising, demonstration, persuasion, and individual marketing, is also crucial. Acceptability is an essential (yet often neglected) element of sustainable mobility (Benevolo et al., 2016).

2.6.2 Smart Blue-Green Mobility

Not only sustainable mobility is needed, but also smart mobility is required to make urban setting more livable (Jeekel, 2017). Smart mobility has four main dimensions:

(1) Smart mobility is about vehicle technology.
(2) Smart mobility is related to Intelligent Transport Systems (ITS).
(3) Smart mobility is about data.
(4) Smart mobility is about new mobility services.

Fundamentally, smart mobility is user-oriented, technology-oriented, mostly car-oriented, IT-oriented, developed world-oriented, and action-oriented. Sustainable mobility looks fundamental, smart mobility looks pragmatic. Smart mobility could be considered the clever and active younger brother of sustainable mobility (Narang, 2007).

2.7 Comprehensive Blue-Green Sustainable Strategic Urban Plan

Rapid and unplanned urbanization causes challenges such as poverty, depletion of resources, infrastructure problems, etc. Strategic planning could be used as a management tool to determine the direction in which an organization is moving, and how it will get there. Urban strategic planning is a key to respond to rapid urbanization to manage urban challenges and improve quality of life (Akkiah, 2017).

Based on authors' research, comprehensive strategic urban plan has three pillars: project management including strategic planning, resources management and risk management, infrastructure containing social and cultural capital, telecommunication, digitalization and mobility, and value engineering including project improvement (Sofeska, 2017).

Fundamentally, governance and corporation, mobilization of stakeholders, and consensus toward sustainable cities are introduced as significant advantages of strategic urban plan (Momtaz & Elsemary, 2015).

2.8 Livable Urban Setting

Todays, urban settings need to be changed to gain sustainability, resilience, and livability. Generally, rapidly urbanization causes challenges for urban settings and makes them less livable. So, it is vital to find solutions to make high quality of livability in cities. Fundamentally, livability is known from ancient time (Gehrels et al., 2016). Quality of livability is measured with some factors such as political stability, safety, health care, education, public services, transportation, recreation, housing, and environmental quality (Kuhlman & Farrington, 2010). To gain livable urban area, sustainability and sustainable urban development are required. Some factors such as green urban development through sustainable urban infrastructure, sustainable and smart mobility and digitalization, proper and sustainable project management through risk management and resources management, and value engineering through project improvement could improve sustainable urban development. In addition, sustainability could be promoted through social, economic, and environmental sustainability. Fundamentally, livable urban area should have some aspects like smart and sustainable mobility, sustainable infrastructure, sustainable urban plan, stable city government and social, economic, and environmental sustainability (Gehrels et al., 2016; Sofeska, 2017).

2.9 The 5th Wave/Tomorrow Age Theory or Theory of Comprehensive Everything

2.9.1 Main Points

(1) The 5th wave theory is to measure the readiness to change into the new age/wave we are just entering.
(2) The 5th wave theory is getting ready for tomorrow's concerns.
(3) Future crises/shock, tomorrow's readiness.
(4) Modern business' economy/future of business economy.
(5) Edge of tomorrow (the first step of this edge 2020–2030).
(6) Combination of future of I4.0 (I5.0) as west cultural symbol and future of Society 5.0 (Society 6.0) as non-west cultural symbol to avoid gender differences in west and non-west cultures.
(7) Hybrid SMEs/SME 5.0/Tomorrow's SMEs as a new business economic concept.
(8) Human rights, women, freedom, and equality.
(9) Balance for opportunities for leadership positions for men and women.
(10) Proceeding of future of Industry 4.0 (Industry 5.0) as a symbol for western culture:
- Industry 4.0 is a western concept which is similar to DIGITALIZATION
- is about intelligent and smart networking of products based on fourth technology areas:
  - embedded systems,
  - smart factories,
  - strong networks,
  - cloud computing, and
  - Information Technology (IT) security.

(11) Proceeding of future of Society 5.0 (Society 6.0) as a symbol for non-west culture:
- society 5.0 is non-western concept similar to UBIQUITOUS
- is about being or seeming to be everywhere at the same time
- in an intelligent society, it is based on:
- REAL LIFE + HIGH TECH. + VIRTUAL REALITY.

2.9.2 What

(1) Is invented, introduced, and improved to support all businesses especially SMEs to:
- FORECAST,
- PREVENT,
- FACE,

For the:

(1) TODAY'S CHALLENGES,
(2) TOMORROW'S CRISIS/SHOCKS or future concerns.

2.9.3 Where

TODAY'S CHALLENGES and TOMORROW'S CRISIS have impacted at level of KNOWLEDGE, TECHNOLOGY, and BUSINESS, which is called KTB model by Prof. Hamid Doost.

2.9.4 Why (Aims)

TO PROVIDE SOCIETIES:

(1) Blue-Green Sustainability (a new concept, Doost H. 2017).
(2) Innovative digital readiness.
(3) Recovery and CSR 1.0, CSR 2.0, CG, and CSE strategies.

2.9.5 Why

BY LEVERAGING ON: CSR 1.0, which is Corporate Social Responsibility and CSR 2.0 which is Corporate Sustainability Responsibility, Corporate Governance (CG) and Corporate Sustainability Governance (CSG) approaches.

2.9.6 How

BY USING INNOVATIVE DIGITAL INFRASTRUCTURES:

(1) Innovation,
(2) Implementation, Development, and Application of
(3) Future of the 4.0 technologies (5th technologies).

Which can influence, change, and improve business economy, education, and training on digital infrastructure as well as our lives.

2.9.7 Background

The 5th wave theory is based on some other theories, concepts, methods, and models which are invented, introduced, and developed by Prof. Dr. Hamid Doost Mohammadian (2010–19) e.g.,:

(1) i-Sustainability Plus Theory,
(2) DCT theory (Doost Cultural Theory),
(3) seven Pillars of Sustainability model (7PS),
(4) Knowledge, Technology, and Business (KTB) model,
(5) 3D Socio-Eco-Environment SMEs model,
(6) DRM methodology (Doost Research Methodology),
(7) D-SDIC model (Doost Sustainability Digitalization Impact Comprehensive Plan),
(8) DRMM (Doost Risk Mitigation Method),
(9) SME 5.0/Hybrid SMEs/Tomorrow's SMEs concept.

2.9.8 Impacts

"The 5th wave theory" – EXPECTED IMPACT:

(1) Education and Training with a focus on HR competencies, EQ and conversational intelligence,
(2) health-care systems reinforced (Welfare 5.0),
(3) new opportunities for affected citizens to be employed created,
(4) citizens' basic services enhanced (Urban 6.0/Utopia),
(5) world economies more resistant and sustainable,
(6) education engineering and vocational training with a focus on HR competencies, EQ, LQ, conversational intelligence, and women,
(7) welfare 5.0 as a very good, sustainable, and trustable social system,
(8) equal opportunities for both genders' citizens for the leader position,
(9) urban 6.0/Utopia for all gender community,
(10) human right, women, freedom, and equality,
(11) women sustainable innovative engineering Edu leadership and economies more resistant and sustainable,
(12) new opportunities for women leadership in small and medium enterprises (SMEs) to achieve cultural sustainability,
(13) based on Prof. Hofstede's six cultural dimensions, masculinity and femininity will make all these findings different in west because of being masculine culture and non-west culture being feminine culture.

2.9.9 Results

"The 5th wave theory" – EXPECTED RESULTS to create:

(1) communities (Urban 6.0/Utopia), Societies (Society 6.0),
(2) mapping women sustainable innovative business economy,
(3) tackling future concerns,
(4) make the world a better place for living,
(5) changing each person and each person's life,
(6) human right, women, freedom, and equality,
(7) equal opportunities for women and men to obtain leadership positions in organizations,
(8) cities (Tomorrow's Cities), Businesses (SME 5.0) capable of:
   - mapping a sustainable future,
   - tackling future concerns,
   - make the World a better place for living,
   - change each person's mindset for changing the world.

2.9.10 Conclusions in Business Economy, Especially SMEs

In "The 5th wave theory", we have a new concept for SMEs named *"HYBRID SMEs/SME 5.0/Tomorrow's SMEs', and based on the 3D Socio-Eco-Environment SMEs model whose* priorities are (*Based on the* Figure 13):

(1) environmental responsibility,
(2) social cohesion, and
(3) economic efficiency.

Based on the 5th wave/tomorrow age theory or theory of comprehensive everything, at the beginning of the 21st century, three revolutions (D3 Revolutions) which are:

(1) Decarbonization,
(2) Decentralization, and
(3) Digitalization

took place, shape, change, and influence our century and as a result of digital sustainable economic developments, governments also changed.

These revolutions and technologies can make the businesses especially SMEs in economic sector successful for making high livability and quality of life for mapping the future, sustainable smart cities, and make the world a better place for living. In today's fast-paced world, many factors, both internal and external, affect business and quality of life, and societies and organizations need a new definition of sustainability. To be able to understand the abilities and management of themselves and others for growth and development, to create processes different from the previous standard procedures in the organization and to be able to respond appropriately and promptly in the face of change. High sustainability adds cultural, educational, political, and technical dimensions to the traditional environmental, social, and economic dimensions that are required to address the anticipated issues, and all seven pillars must be developed equally. Figs. 7.1 and 7.2 present seven aspects of sustainability (7PS model) and its classification in addition PEACE and LOVE. This model is focusing on seven pillars of sustainability. Fundamentally, we are settling in edge of future that confronts future shocks including tomorrow's economic challenges, new businesses' barriers, modern training style risks, and human competencies challenges and so on. The figure below presents the point that we are perched on: As in previous section was declared, solutions are required to deal with ubiquitous and digitalization challenges to be able to apply them in different educational businesses and services toward creating modern areas. The 5th wave/tomorrow age theory or theory of comprehensive everything could be a comprehensive strategy to find out solutions to deal with such challenges, even to tackle future shocks for mapping the future business economy (Figs. 7.3–7.8).

In recent decades, besides sustainable development, digitalization (one of the D3 revolutions) is one of the most remarkable phenomena. When digitalization and smartness work together in a city, a smart city is created. Smart city is a combination of (1) information city, (2) digital city, and (3) sustainable city. According to the above theories, smartness in economy, mobility, education, environment, society, people living, and governance form the tomorrow's cities super-smart. These cities cope with the future demands of their citizen. Tomorrow's cities use new Internet of Things (IoT) facilitates, smart administration, and good governance using tools such as U-governance and U-democracy.

This theory mentions that it is required to get ready for the edge of tomorrow, today's challenges, and tomorrow's crises and shocks. In addition, this theory makes societies and

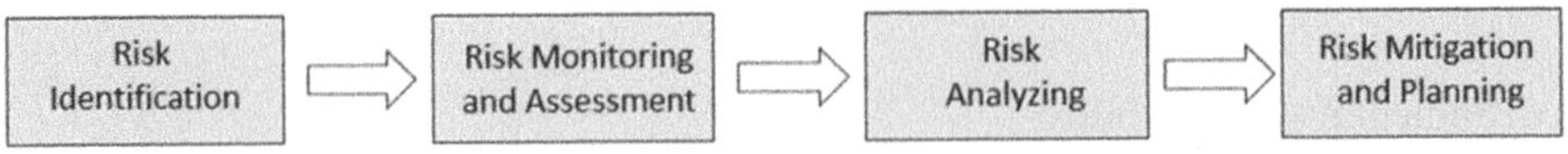

Fig. 7.1. Risk Management Steps (Author, 2017).

$$\text{FAR} = \frac{\text{Number of fatalities} \times 10^{8}}{\text{Totals hours worked by all employees during period covered}}$$

The FAR is:

- The number of casualties per 100 million working hour in the transport system, based on hours
- The number of operation hours of the transportation system leading to a casualty unit, based on hours

$$\text{AIR} = (\text{Loss/year}) \times (\text{number of trips per person}) / (\text{number of trips per year})$$

The AIR is:

- The probability that a person will be lost during a trip year.

The probability of this index is also a number between zero and one and the unit of calculation is the casualty or fatality / person

- The number of people involved in the transportation system throughout the year will result in the loss of one person

Fig. 7.2. Fatal Accident Rate (FAR) and Average Individual Risk (AIR) Formula of Water Transport (Author, 2017).

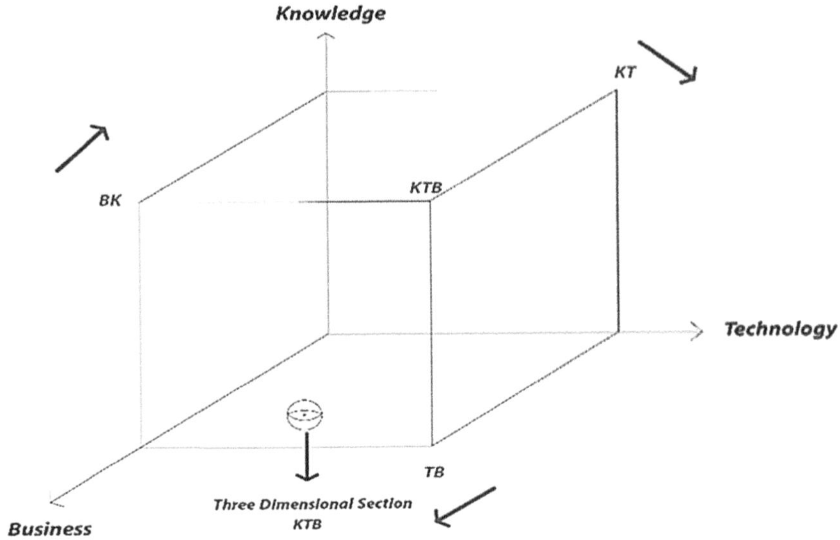

Fig. 7.3. KNOWLEDGE, TECHNOLOGY, BUSINESS (KTB) Model (Doost, H. 2010–17).

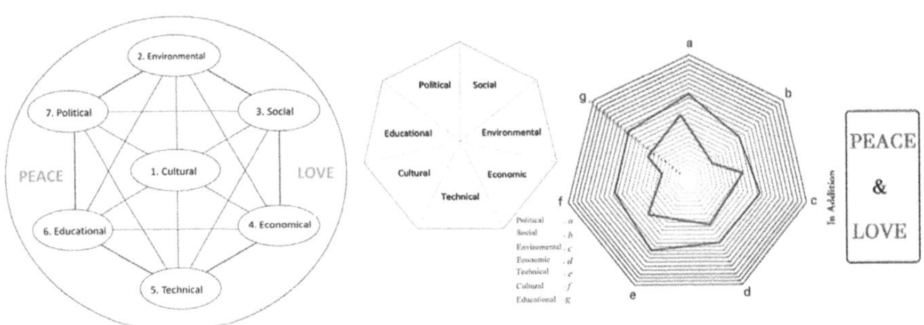

Fig. 7.4. Seven Pillars of Sustainability Model With Priority and Connections (7PS Model) (Doost, H. 2010–17).

urbans founded on high technologies, digitalization, appropriate HR, strategies concerned on sustainability which can create new concept of cities which tackled with future concerns and readiness for tomorrow shocks entitled: "Tomorrow's Society for Tomorrow's Cities".

Doost Mohammadian's concept of tomorrow's SMEs states that hybrid SMEs should entail the sustainable development criteria such as environmental responsibility, social cohesion, and economic efficiency. In addition, SMEs should create innovative projects which focus on future planning, clean economy, and succession planning. HR talents, competencies, qualifications, skills, and training (focus on EQ and conversational intelligence).

A utopia is an imagined community, society, or urban space that possesses highly desirable or nearly perfect qualities for its citizens.

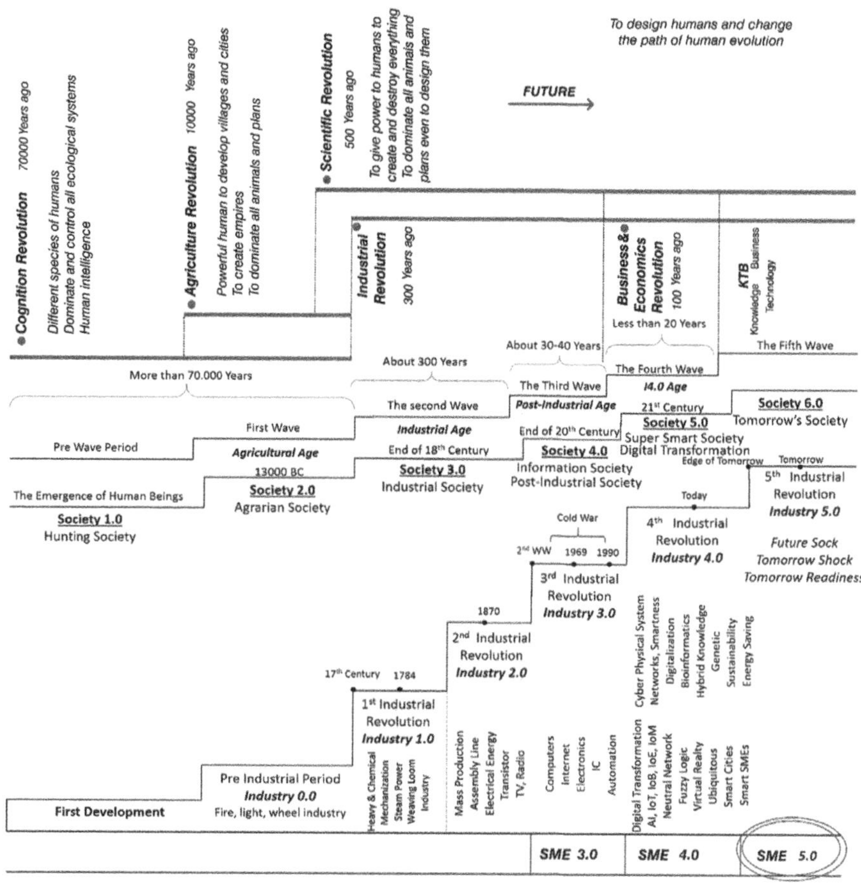

Fig. 7.5. The 5th Wave Theory, Ages, Society, Industries, and Technologies (Doost. H, 2010–17).

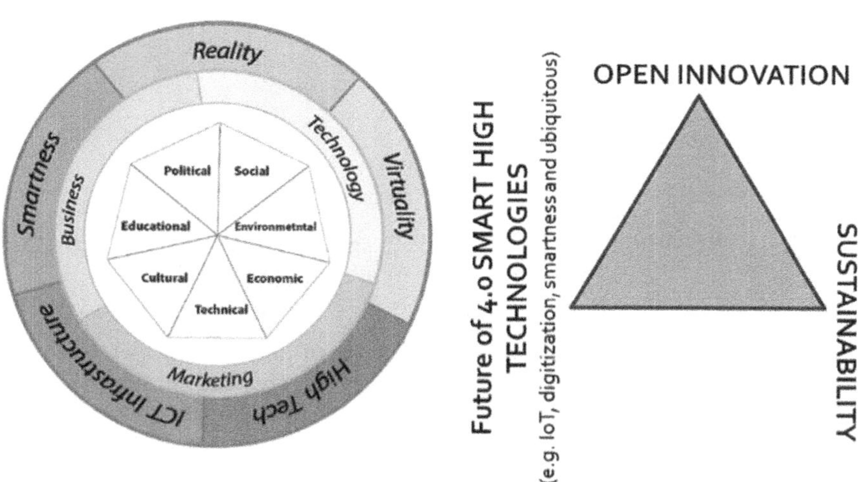

Fig. 7.6. i-Sustainability Plus Theory (Doost. H, 2017).

SME	Society	Industry	Waves/Ages	Revolutions			Year
	Hunting Society Society 1.0		Pre wave period	-)Cognition Revolution -)To Dominate and Control all ecological System -) Human Intelligence	The Emergence of Human Beings	-)First Development -)Different Spices of Human	70000 Years ago
	Society 2.0 Agrarian Society	-) Pre Industrial Period -) Industry 0.0 -) Fire, Light, Wheel Industry	-)The First Wave -)Agricultural Age	-)Agriculture Revolution -) Powerful Human to Develop Urban Areas -) To Create Emprises To Dominate All Animals, Plans and planets			13000 Years ago
				-) Scientific Revolution -)To Give Power to Humans to Create and Destroy Everything In the Planet -) Just one Human Specie			500 Years ago
	Society 3.0 Industrial Society	-)1st Industrial Revolution -) Industry 1.0	The Second Wave Industrial Age	Industrial Revolution			17th Centaury
		-) Heavy and Chemical Industry -) Mechanization -) Steam Power -) Wearing Loom					1784
		-)2nd Industrial Revolution -) Industry 2.0 -)Mass Production -)Assembly Line -) Electrical Energy -) Transistor, TV, Radio					1870
SME 3.0	-) Society 4.0 -) Information Society -) Post Industrial Society	Industry 3.0 • Computers • Internet • Electronics • IC • Automation	-) The 3rd Wave -) Post Industrial Age	-)Business and Economics Revolution1 Future	To Design Humans and Change the Path of Human Evolution	100 Years 40 Years The Cold War	2nd WW 1969 1990 2000
SME 4.0 Smart SME	-)Society 5.0 -) Smart Citizen	The 4th Industrial Wave • AI, IoT, IoB, IoE • IoM, Neutral Network • Fuzzy Logic • Ubiquitous • Networks	-) 4th Industrial Wave -) Digitalization Age -) Digital Transformation -) Virtual Reality -) Cyber Physical Systems -) Smartness -) Digitalization	-)Business and Economic Revolution 2 -)Hybrid Organization -)Cloud HR -)Greenhouse Gases Reduction -)Energy Saving -) CSR	Bioinformatics Hybrid Knowledge Genetics Sustainability	10 Years	2006 2011 Today
SME 5.0 SMEs for Tomorrows' Shocks	Society 6.0	Industry 5.0	-) The 5th Industrial Wave -) Tomorrow Age		KTM Model Future Shocks Tomorrow Shocks	Edge of Tomorrow	Tomorrow

Fig. 7.7. Histomap of 5th Wave/Tomorrow Age Theory (Doost. H, 2010–17).

The 5th wave theory forecasted a crisis at the edge of tomorrow, which has led to the creation of communities, societies, cities, and businesses founded on high technologies, IoT, and on appropriate business strategies concerned with sustainability which can create new concepts and new businesses, societies, and cities which are capable of tackling future

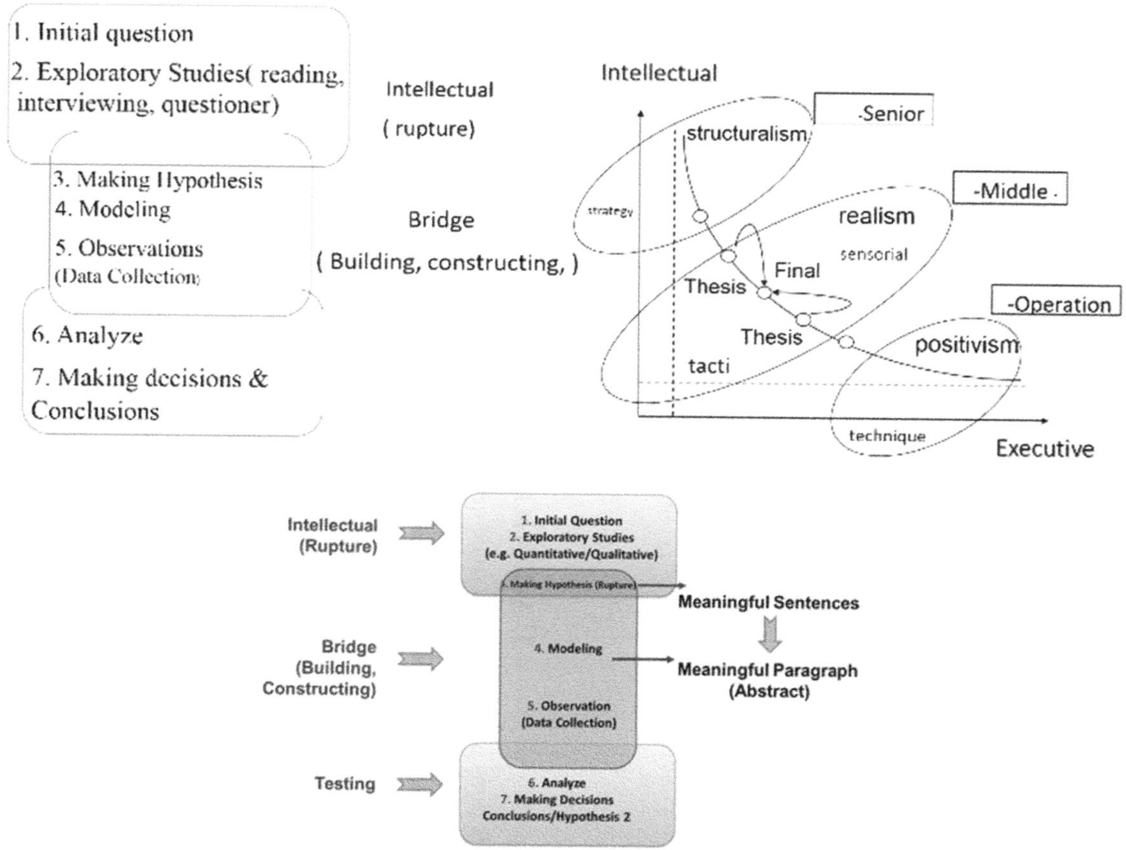

Fig. 7.8. Doost Research Methodology (DRM) and Its Levels (Doost. H, 2009–10).

concerns. These are commonly known as "Tomorrow's Society and Business" and "Super-Intelligent Society", and they function in a super-intelligent business environment.

Another theoretical basis is the "i-sustainability plus theory", which is made up of the trinity open innovation, sustainability, and 4.0 smart high technologies, e.g., digitization and ubiquitous and super-intelligent society (Society 6.0).

This construct, which includes the idea of a sustainable ubiquitous and smart society and city, is evaluated as a new idea of urban living in tomorrow's business, and societies entitled URBAN 6.0 (Prof. Dr. Hamid Doost Mohammadian, 2017–2019).

i-sustainability plus theory comprises three characteristics: open innovation, sustainability pillars, and future 4.0 smart high technologies. This construct explores ideas derived from the combination of real life, high technology, and virtual reality, remarkable sustainable development drivers, to reach a higher quality of life. The i-sustainability theory allows communities to avoid impacts of global crises on sustainability, today's sustainability challenges, and tomorrow's sustainability shocks. Generally, there has been an increased interest in the quality of life by researchers who have made it a key concept in the field of global and human issues. Thus, today, along with the ideas of sustainable development, the concept of quality of life has become a priority. In other words, the quality-of-life index in today's community indicates the degree of achieving welfare and providing greater citizen satisfaction. Any community with a high score in the quality-of-life index is a desirable community for living. Instead of accelerating and moving forward, societies want more attention to the viability, quality of life of citizens, and sustainable development.

3. METHODOLOGY

To address the aims of this chapter, some questions should be answered:

- How the 5th wave/tomorrow age theory could be a method for mapping the future Smart Cities?
- How tomorrow age theory could be as solutions for the mobility challenges in the edge of tomorrow?
- How the DRMM & Comprehensive Global Blue-Green Clean Sustainable Urban Mobility Risk Mitigation Plan can be a challenge for having a sustainable future?
- Why is it necessary for the future Smart Cities to utilize high mobility technology through 5th wave/tomorrow age theory?
- What are the privileges of the 5th wave theory for the future Smart Cities?
- What are the main challenges that the future Smart Cities confront to utilize future of Industry 4.0 and Society 5.0? And how DRMM could be a solution?
- Why do the future Smart Cities play significant role in future of not only countries, but also the world and quality of human life?
- What is 5th wave theory and how it could map the future sustainable on the future Smart Cities function, process, performance, and outcome?

- How through the 5th wave theory the future Smart Cities is changing from a traditional business economy to INNOVATIVE DATA and Blue-Green SUSTAINABLE mobility technology?

In this chapter, three stages were developed to implement the study:

Stage 1: The development of a research methodology.

This stage was developed based on the literature review in the area concerned on objectives of the 5th wave theory. The author has investigated most of the previous research in the context of future Smart Cities and Urban 6.0 concept, DRMM, Mobility Risk mitigation, and sustainability. In addition, the impact of the 5th wave theory on future Smart Cities function and performance was explored.

Stage 2: The development of the research tool.

The research tool was developed on the basis of the relevant literature, knowledge, questionnaire, and personal experience (Do-How/Know-How). The purpose of the surveys was to indicate policies, strategies, methods, and solutions in order to find out the roles of the theory of comprehensive everything in creating sustainable future Smart Cities. Impacts of proper future Smart Cities and DRMM & Comprehensive Global Blue-Green Clean Sustainable Urban Mobility Risk Mitigation Plan for Mapping Future Smart Cities through the 5th Wave Theory as a component of society infrastructure in altering the world as better place for living and sustainable future through the related theories, concepts, and models to the 5th wave theory. In addition, role of KTB model in achieving Blue-Green sustainable Smart Cities is realized. The methodology of DRM, which is based on the dialectical thinking and BESQol (built environment sustainability and quality of life) were used as a tool for analyzing. DRM methodology is used in social science, engineering, and management, DRM is based on three levels and seven questions.

Stage 3: Gathering the study results and analyzing.

Based on literature reviews, surveys, case studies and analyses, some results were obtained, and these results were analyzed and discussed by the author.

3.1 How Mobility Risk Management Provide Comprehensive Blue-Green Sustainable Urban Plan to Move to Blue-Green Sustainable Mobility

Sustainable mobility is required to gain proper and sustainable comprehensive strategic urban plan. Generally, sustainable mobility is needed to satisfy needs of people and businesses on mobility in cities and their surroundings for better quality of livability and life. In other words, mobility could directly influence on livability and quality of life in urban settings.

Generally, risk could threaten mobility and influence quality of humans' life. So, it is vital to utilize risk management on mobility to control and reduce impacts of risk on mobility in order to improve sustainable mobility. To control risks of mobility, the first step is to categorize transportation systems, the second step is to find out risks related to each category, and the last step is to utilize stages of risk management to manage and mitigate risks.

- Road Transport.
- Rail Transport.
- Air Transport.
- Sea Transport.
- Other Modes of Transportation.

The other types of transportation like pipeline transport, cable transport, and space transport form this category. Safety risks, financial risks, environmental and technical risks, environmental concerns, and transportation of dangerous goods are the most significant threats in this category.

In addition, in this step three questions should be answered:

(1) What will happen if each of these risks occurs?
(2) What is probability of happening of each risk?
(3) How much does it cost to control each risk, is it cost-effective or not?

Risk assessments will help organizations related to mobility to make the right decision for choosing the right solutions. For instance about flash flood, a natural event. It should be answered and estimated.

3.2 How to Measure the Risk With Doost Risk Mitigation Method (DRMM)

DRMM methodology could be utilized as proper tool to measure risks of mobility and to control them. The main goal is to identify risks, mitigate them, and improvement plan is proposed.

The first step is to prepare risk tree concerned on mobility risks.

In this research, one sub-criteria defined is to be measured without computer programs. More than two sub-criteria need to be calculated through computer programs like Software Failure Modes Effects Analysis (SFMEA), Ishikawa diagrams, the 5-Why method, etc.

The second step, after defining main and sub-criteria is weighing ratio of each main criteria are calculated. Weighting ratio should be between 0 and 0.1. Generally, r is estimated for normalization. Then probability of occurrence (P) and impact of occurrence (I) are estimated through questionnaire. Thirty people including senior managers for main factors, experts, and specialist for supporting factors answer the questions. Fundamentally, main factors are answered by senior managers in the organization and senior managers who cooperate with other mobilization organization and national transportation. In addition, weighting ratio of each sub-criteria is estimated for normalization among other sub-criteria in one group. Finally, total organizational effective risk is estimated to find out the situation of a company through risks measurement. Fig. 7.9 shows the formula of estimating total effective risk. In addition, different ranges of total risks are demonstrated.

3.3 How to Calculate Risk Through the DRMM Method

Based on the DRMM method, Figs. 7.9–7.12, it is possible to measure the risk according to the sustainability compass, which is presented in Table 7.1 (Doost. H, 2017).

Table 7.1 presents how to measure the risk made of the trinity of Impact (I), Probability (P), and ratio (r).

Table 7.1. DRMM Risk Measurement [Doost. H, 2017].

Index	Description	Row
Ri	Risk	1
Pi	Probability of Risk Criteria	2
Ii	Impact of Risk Criteria	3
ri Normal	Normalized ratio of Risk Criteria	4
Risk (Ri) = $\sum$ (Pi * Ii * ri Normal)		

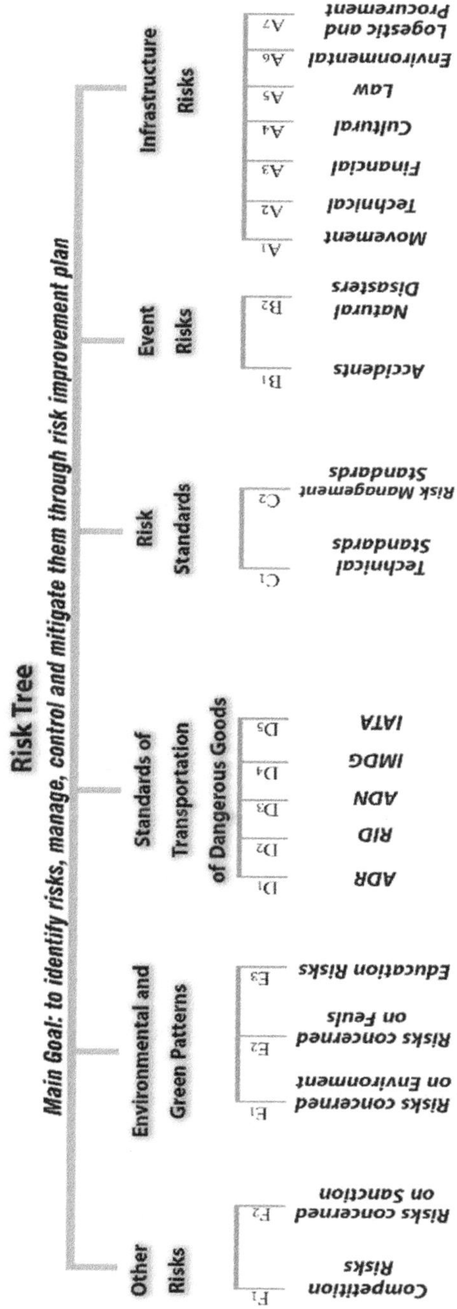

Fig. 7.9. Risk Tree Concerned on Mobility Based on DRMM Method (Author, 2017).

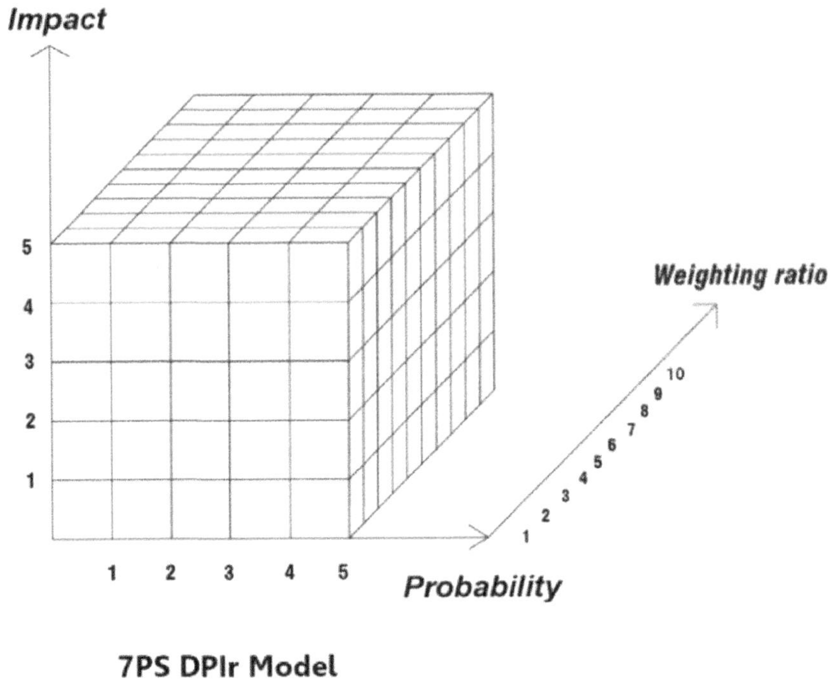

Fig. 7.10. DPIr Model Based on DRMM Method [Doost. H, 2017].

3.4 How to Do Improvement Projects like Share, Control, and Mitigation

Fundamentally, according to pattern introduced such as ISO 31000, PM book, and others, risk improvement could be done. In this research, a new pattern based on authors' experience is proposed. A comprehensive framework and plan are needed to enhance the execution of strategies and improve organizations' goals through low-risk probability and impact. This pattern has seven steps: Risk Identification, Risk Estimation, Risk Analysis, Risk Evaluation, Risk Treatment, Risk Management Implementation, and Risk Monitoring Review. Generally, four categories are introduced for Risk Treatment:

(1) *Mitigating risks:* If activities are high in likelihood of occurring and low in financial impact; the best technique is to utilize risk control system in order to reduce the risk of potential loss.
(2) *Avoiding risk:* If activities are high in likelihood of occurring and high in financial impact; the best technique is to avoid activities.
(3) *Managing risk:* If activities are low in probability of occurring and in financial impact, some supporting solutions like purchasing insurance, hedging, outsourcing, or entering into partnerships should be used. In other words, risks should be managed and controlled to gain best results.
(4) *Accepting risk*: If mitigation of risks' cost is estimated more than bearing the risks, accepting, and continually monitoring the risk are suggested as the best solution.

Effective Risk = $P_i * I_i * r_{i\ Normal}$

Total Organizational Effective Risk = *Effective Risk** $r_{i\ criteria}$

Total Organizational Effective Risk = *∑ Total Organizational Effective Risk i for each subcritria*

1 ≤ Total Effective Risks ≤ 7	⟹	Low risk and has little impact
8 ≤ Total Effective Risks ≤ 18	⟹	The risk within the limits of control, people need to find the factors, which have great impact and take the ordered strategy
18 < Total Effective Risks	⟹	The risk is so dangerous, should pay more attention and run project risk improvement

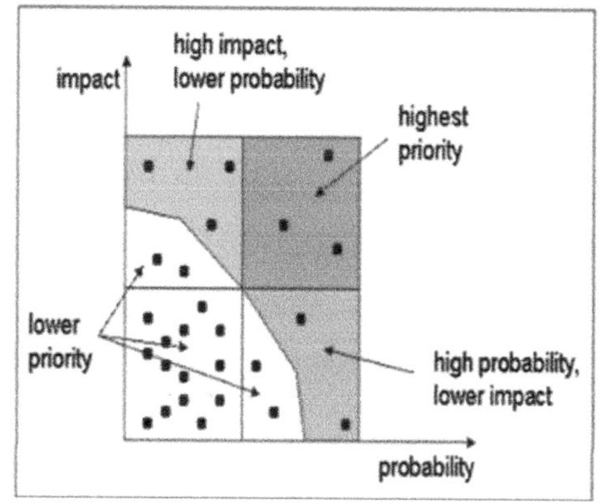

Fig. 7.11. Total Effective Risks Based on DRMM Method (Author, 2017).

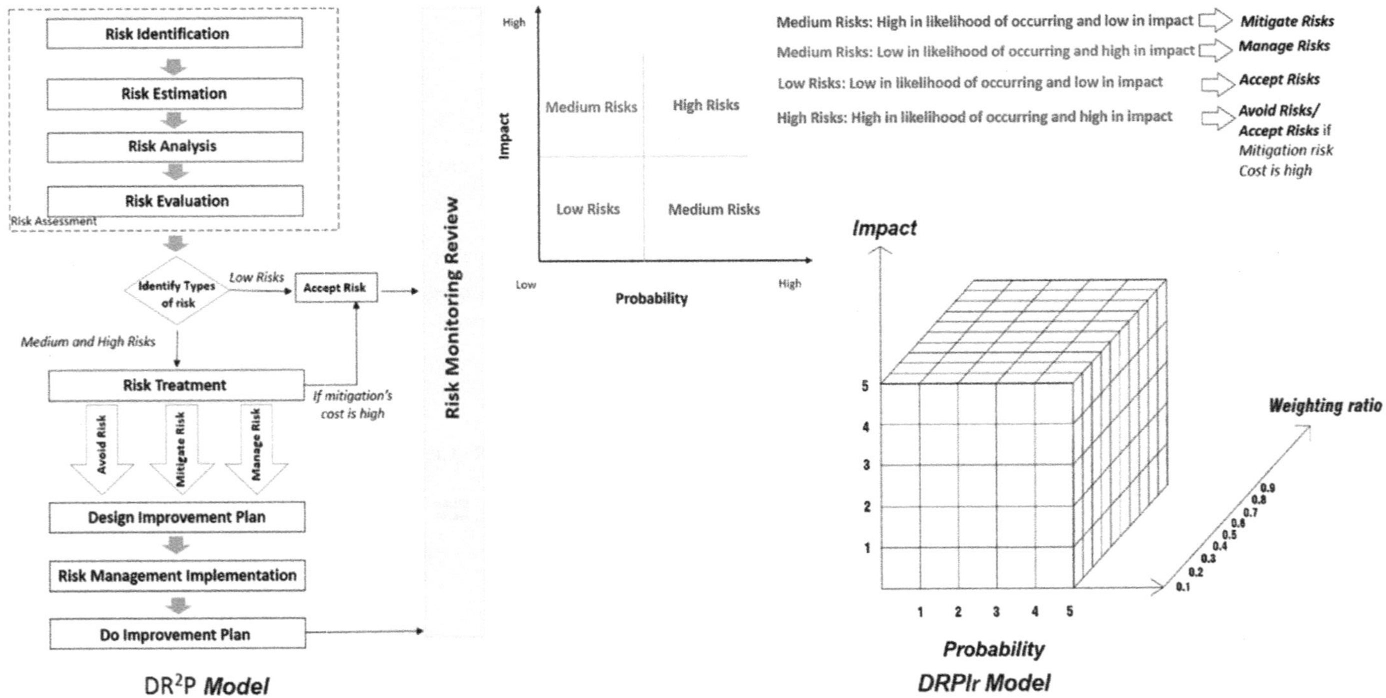

Fig. 7.12. DR²P and DRPIr Models Based on DRMM Method (Author, 2017).

4. CASE STUDY

In this study, Germany and Iran are chosen as case studies. The situation of mobility, their challenges, and their risks on mobility are explored.

4.1 Germany

Germany is one of the best country in mobility. Strategies utilized by mobility and transportation organization are aligned with green, smart, and sustainable mobility in Germany. So, these strategies make Germany as a country with high quality of mobility in order to provide high quality of livability and life for inhabitants.

Fundamentally, Germany is one of the countries with the best infrastructure in the world. Around 830,000 kilometers of road (including 13,000 kilometers of motorway), 38,500 kilometers of rail, and 24 major airports ensure excellent transport links throughout the country, making high quality of movement infrastructure in Germany (www.research-in-germany.org). In addition, proper technical infrastructure, financial infrastructure, cultural infrastructure, environmental infrastructure, logistical infrastructure, and law infrastructure make risks related to infrastructure in Germany be decreased. For instance, high quality of public transport makes citizens to use it as the main transport vehicles in Germany and it could help the government with struggles related to environmental concerns. In addition, high taxes for cars or considering benefits for hybrid cars are law infrastructure which help the country to create smart green sustainable mobility. Not only infrastructure risks are low in Germany, but also event risks are low. For example, traffic and accidents data statics by Federal Highway Research Institute shows that total accidents, total killed, and total injured are decreasing in Germany every year. High standards like tuff standards make vehicles with high technical aspects which could diminish rates of injured and killed statistics in accidents. Furthermore, such standards are profitable for environment and could decrease environmental challenges concerned with fuel pollutions, etc. Generally, risks of transportation of dangerous goods are very low. It is not only aligned with basis of international provisions, but also The German Federal Institute for Risk Assessment (BfR). It means that BfR helps to ensure that toxic and corrosive chemical products, biological agents, and materials are correctly assessed and classified in these provisions. Not only dangerous goods are controlled, but also all goods are controlled for taxes through governments like ZOL polices. Urban and street design in Germany makes sustainable mobility. It means that this country is livable for all people, the needs of handicaps and children with carriage are attended and the country mobility systems are aligned with their needs. Basically, cultural norms such as attending laws and regulations, ambulance and fire trucks could achieve sustainable mobility. In addition, strategies concerned on education of mobility, sustainability and their importance are techniques used in Germany to improve green, smart and sustainable mobility. For example, Technical University of Berlin holds courses such as Sustainable Mobility Management and Sustainability Building in EUREF Campus. These strategies could be used as tool to improve green mobility.

Although Germany has a high quality of mobility, all German mobility companies aren't successful. For example, Germania was declared as a bankrupt company in 2019. Competitive risks, financial risks are the most important reason for the failure of this company. On the contrary of Germania, Deutsche Bahn is one of the most successful transportation organizations. Strategies such as profitable quality leader, environmental strategies, considering social capital, and utilizing risk strategies make this company successful.

Generally, comprehensive plan, strategies, risk management processes such as tuff standards, high taxes for cars, promoting green transportation like riding bicycle and hybrid cars by German governments and mobility organizations make *Green Smart Sustainable Mobility* in Germany. Basically, it could be used as a tool in green comprehensive urban plan to create green livable urban areas in Germany.

4.2 Iran

Iran is a country with unstable and unsustainable mobility system confronted with a lot of mobility and transportation challenges. The main reason is about risk concerned on mobility. Infrastructure risks, environmental risks, event risks, risks of transportation of dangerous goods, and risks of standards are high in Iran. High risks on mobility infrastructure are one of the main challenges. For instance, Iran is one of the 10 countries with the highest number of road accidents according to the World Health Organization (WHO) and the University of Michigan Transportation Research Initiative. Based on research, the killed people by car accidents between 1998 and 2004 are more than eight years of war in Iran. Statistics of death and injured due to accidents is high. Low quality of Iranian's cars, not attending to traffic laws, and old infrastructure of roads (around the 2nd World War) are the main reasons for accidents with high injured and killed statistics. Sanction makes automobile companies to use low quality of Chinese pieces in cars. In addition, to gain more profit, the companies use materials and pieces of low quality. These make Iranian's car less effective and capable. High taxes of imported cars make citizens to buy Iranian's cars with low quality (Although, Denmark also establishes such regulation but its goal is different. Denmark encourages people to use public transport through high taxes of cars). In addition, cultural norms could influence on infrastructure risks. For example, not attending to laws related to cultural norms create fundamental risks such as accidents. Generally, air transport is one of the transportations with high risks in Iran. Since 2000, 18 aircraft accidents have happened in Iran. Air transport infrastructure has problems because they are old and follow old processes. In addition, sanction has significant impact on it. Based on sanction, old pieces and old airplanes are used. These make standards of air transport in Iran low and increase aircraft risks. Fundamentally, water transport struggles with important risks. Processes used are old ones with low effectiveness and it could increase risks of mobility in water transport. Basically, sanction makes problems in providing new pieces with high quality in this industry. These risks make hazardous risks such as high environmental pollution, water pollution, low profit, low speed of processes and high wasting. Further, there is no balance between financial cost and operation cost. One reason is existing Fraud Risk in air transport in Iran. This case would be discussed in discussion section. Fundamentally, air transport deals with high risks. Besides of all these transport systems, rail transport is related to high risks. Old infrastructure of railways related to World War 2 is the most important reason of risks on rail transport. For instance, in 2004 an explosion of train led to death of 320 people and 460 were injured. 20 kilometers was the incident ray. Such events show rail transport has risk too in Iran. Furthermore, this industry is controlled by road ministry. It seems that this ministry doesn't follow comprehensive mobility plan in order to reduce risks and struggle with environmental concerns. So, risks like accidents risk, health risks, environmental risks, etc. threaten rail transport of Iran. Generally, cultural norms and infrastructure could influence on mobility. For example, in traffic, fire trucks and ambulances confront a problem related to drive as fast as they should, because people don't allow them. Furthermore, traffic makes high risks in mobility, for instance, traffic affected on Plasco explosion in 2016, fire trucks could arrive at the time they had to, and a huge calamity happened.

Urban and street design in Iran, aren't aligned with handicaps and children with carriage. Fundamentally, confronting with risks and challenges isn't proper in Iran and it could increase the severity of risks. In other words, risk management and confronting with risks aren't appropriate in Iran and cultural norms could influence on them.

Basically, infrastructure risks, event risks, standards risk and environmental risks are high in Iran, and they make mobility in Iran worse. So, it is vital to have risk management in mobility and comprehensive urban plan concerned on risks of mobility to be utilized to gain green, smart, and sustainable mobility in order to improve quality of livability and life.

5. RESULTS AND DISCUSSION

In this section, importance of risk management in mobility to achieve Blue-Green comprehensive sustainable plan in order to create green livable urban setting is discussed.

5.1 Risk Management in Mobility

Generally, mobility and transportation systems confront dangerous risks which could make challenges such as health diseases, death challenges, environmental problems, social instability, destroying cities and rural settings, increasing social concerns and economic challenges. Mobility systems are one of the important aspects of urban plan in cities and these challenges would influence on livability and humans' life. In addition, mobility risks could make quality of human life worsen. So, it is vital to struggle with risks in mobility to improve mobility systems to make livable settings.

According to Chinese language, risk is a combination of danger and opportunity. Therefore, it is possible to change risks to opportunities and utilize risks as a new chance. Risk management could be used as a tool to alter risks to new opportunities. Basically, risk management is recognized as being concerned with both positive and negative aspects of risks. In other words, by utilizing risk management, negative impacts of risks could be decreased. So, risk management in mobility could manage and control risks concerned on mobility and transportation and create new opportunities through managing risk.

Risk management could be achieved through patterns of risk management like PM book, ISO 31000, Canadian National Risk management, etc., all these patterns have same basic and in details are different. Identifying risks, estimating, and analyzing risks, treating risks through evaluating them, implementation of risks and risk monitoring review are the main steps of risk management.

There are different risks that could threaten the profit and effectiveness of mobility systems. Movement infrastructure including road, waterway, airway, and railway infrastructure, cultural infrastructure, law infrastructure, regulation risks, education infrastructure, financial risks, pollution and environmental risks, logistic and procurement risks, accidents, natural disasters, technical and environmental standards, competitive risks, sustainability risks, risks concerned on sanction, fraud risks, and safety risks are the most significant risks in mobility and transportation. In addition, not balancing between financial cost and operational cost is an important risk in mobility. Based on Prof. Dr. Hamid Doost Mohammadian's experience about cooperation with P.O.G.C Company (Pars Oil and Gas Company, one of the biggest gas companies); when he was Logistics and Procurement Manager there, fraud risks and difference between financial cost and operational cost besides infrastructure risks, environmental risks, standards risks and sanction existed in water transport in Iran. For instance, ships traveled a destination in 60 hours instead of 18 hours and weather problems were used to be stated as the reason for delays. According to exploration, it was found that the weather wasn't the main problem. Instead of four engines of ships, just one engine was working, and it made the speed of ships slower. In addition, extra fuels were sold. It should be mentioned, fuel was bought with subsidies price and sold on higher price. It shows that fraud risks existed in mobility of Iran, and it is a dangerous threat. Not only in Iran, in many countries such risks existed that threaten the mobility systems of a country. Generally, there are two issues about risks in mobility: risk management and managing risks. Risk management is about preventing and managing risks with fewer disadvantages as it was mentioned. Risk management says that each individual should be changed in order to alter all aspects. Furthermore, culture has powerful impact on it. Although risk management is about preventing, managing risk is concerned on managing and controlling situation after the occurance of risks. For instance, about Iran's flash flood in 2019, main reason is destroying the river blanking out way of water due to road construction. However, improper risk management made natural disaster worse, inappropriate managing risk is the main cause of death. Iranians couldn't manage the hard situation and they faced the crisis. So, managing risk is concerned on efforts and actions should be taken after a disaster occurs in order to reduce negative impacts of it and control the situation to gain less damages. Fundamentally, risk management and managing risks are important tools in mobility in order to gain sustainable, smart, and green mobility.

5.2 Blue-Green Smart Sustainable Mobility

Risk management is one of the tools used to gain green, smart, and sustainable mobility, however it isn't sufficient.

Generally, green mobility is environmental friendly. In other words, green mobility is concerned on environmental risks. The most important environmental risks are climate change, natural disasters, environment concerns such as destroying forests, construction in nature, air pollution, water pollution, degradation of soil, risks concerned on waste, waste and graywater, environmental standards, education, and cultural infrastructure. These risks not only threaten the world and environment, but they are also significant threat for human life. So, it is vital to reduce such risks to maintain the nature for living. Green mobility is a solution to struggle with these risks. Risk management identifies, analyzes, treats, and manages risks to reduce their negative impacts. In addition, new technologies such as digitalization and IT create smart mobility aligned with green mobility. It means that smartness through technologies could reduce environmental risks and make mobility smart and green. For instance, there is an application for transportation that shows the time and destination of journey with low price through private cars, and it is possible to choose such transportation instead of public ones. It is a proper method to reduce cars with one passenger in order to reduce pollution and environmental risks concerned on fuels of vehicles. It is a proper case showed advantage of smartness for being environmentally friendly. Fundamentally, smartness is a concept aligned with reducing environmental concerns. Smartness through technologies, digitalization and IT create e-banking, e-learning, e-shopping, hybrid cars, smart energy, etc. All these could reduce transport in urban areas. Therefore, environmental risks will be reduced.

Sustainable mobility is mobility that provides social, economic, and environmental sustainability. For instance, in Germany, high quality of public transportation makes people to transport through them and it helps the economic situation of a country. In addition, in Denmark, government establishes high taxes of private cars to promote utilizing public transportation and green transportation like riding bicycle to reduce buying cars and make economic situation better. Furthermore, mobility with low risks could develop social sustainability. Generally, green, smart, and sustainable mobility is aligned with sustainable development. It is a key to gain green sustainable strategic urban plan in order to improve livability and quality of life in urban area. The best case of green smart sustainable mobility company is Maersk. Maersk is a Danish business conglomerate with activities in the transport, logistics, and energy sectors and it is the largest shipping company. It follows strategies to gain green, smart, and sustainable mobility such as: In 2011–12, Maersk Line cooperated with the US Navy on testing 7–100% algae biofuel on *Maersk Kalmar*, from 2007–2014, and mainly due to slow steaming, Maersk Line reduced its CO_2 emissions by 40% or 11 million tons, about the same reduction as the rest of Denmark, Maersk set a goal in December 2018 to be carbon neutral by 2050, In 2017 the company's ships emitted 35.5 million tons of CO_2, and it hopes to eliminate that by using biofuels to power its fleet, using especial category management including: stakeholder management, category strategy and sourcing, supplier relationship management, proactive category management and renegotiate agreements and special risk management in different categories. All these strategies make green, smart, and sustainable mobility.

Fundamentally, companies related to mobility like Maersk should use strategies and risk management to gain green, smart, and sustainable mobility.

New concept of Blue-Green mobility is followed by different company of mobility like Ramboll and Dress and Sommer Company. Based on Dress and Sommer strategies, digital blue mobility, energy management in mobility to reduce CO_2 emission, risk control and risk management in mobility are required to gain digital blue cities with high sustainability.

5.3 Comprehensive Blue-Green Sustainable Strategic Urban Plan

Comprehensive strategic urban plan is a technique used by governments to improve livability and quality of life in cities. Based on authors' article named: "An applied comparative study about the countries with low sustainability and high CO_2-emission and advanced sustainability knowledge in Europe and giving practical models to improve and implement sustainability solutions presented in 6th International Conference on Civil, Architectural and Environmental Sciences"; risk management and mobility are the components of comprehensive strategic urban plan. Generally, risk management could be used as a tool to improve mobility in cities. Blue-Green, smart, and sustainable mobility are required to gain green infrastructure to gain green sustainable strategic urban plan. Fig. 7.14 shows the importance of risk management and mobility to gain proper, green, and sustainable urban plan. In addition, Government Financial Organization Association (GFOA) indicates importance of risk management in comprehensive strategic urban plan and suggests governments to utilize risk management to achieve their goals. It declares that:

- Governments should develop a comprehensive risk management program that identifies, reduces, or minimizes risks to its property, interests, and employees.
- Cost and consequences of harmful and damaging incidents arising from those risks should be contained.

All show that green, smart, and sustainable mobility besides blue-green mobility is required to achieve comprehensive blue-green urban plan in order to gain cities with sustainability and high quality of livability. Blue-green mobility is about adaption of mobility to climate change, environmental concerns via highlighting importance of being friendly with environment.

5.4 Modern Livable Urban Setting

Generally, rapidly urbanization leads to challenges such as poverty, slum, increasing natural disasters, climate change and environmental problems, economic challenges, social instability, safety and security problems, and health diseases. These problems threaten the planet and humans' being. So, it is vital to struggle with them to maintain the world, nature for humanities. Creating modern urban area aligned with sustainability and environment is the best solution to deal with urbanization's challenges. Fundamentally, these modern urban settings are based on Green-blue concept. Blue-Green city is an urban area designed through green and environmentally friendly infrastructure and water management. In addition, these new settings are smart ones. In other words, new urban settings are concerned on smartness, sustainability, and Blue-Green ideas. Such urban settings not only struggle with urbanization's challenges, but also could provide high quality of livability and quality of life for their citizens. Basically, to gain such areas, different indicators are required. Comprehensive Blue-Green urban area is one of the indicators required to achieve modern livable urban settings and mobility has an important role in improving urban plan. In this research, impact of risk management in improving Blue-Green and smart mobility is explored. Fundamentally, risk management could enhance mobility in order to achieve proper urban plan. Based on authors' researches, a practical model concerned with creating livable urban settings through comprehensive Blue-Green smart strategic urban plan and risk management in mobility is proposed (Figs. 7.13 and 7.14).

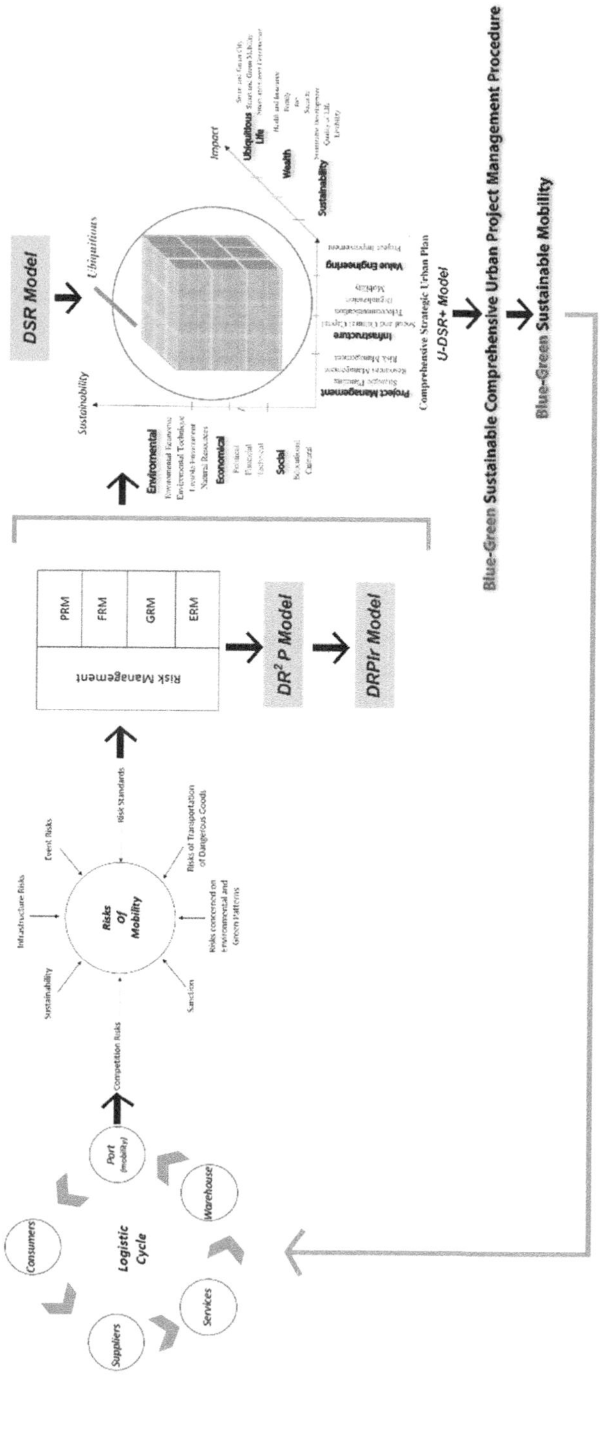

Fig. 7.13. D-BGSCUP Model Based on DRMM Method (Author, 2017).

5.5 Challenges and Solutions

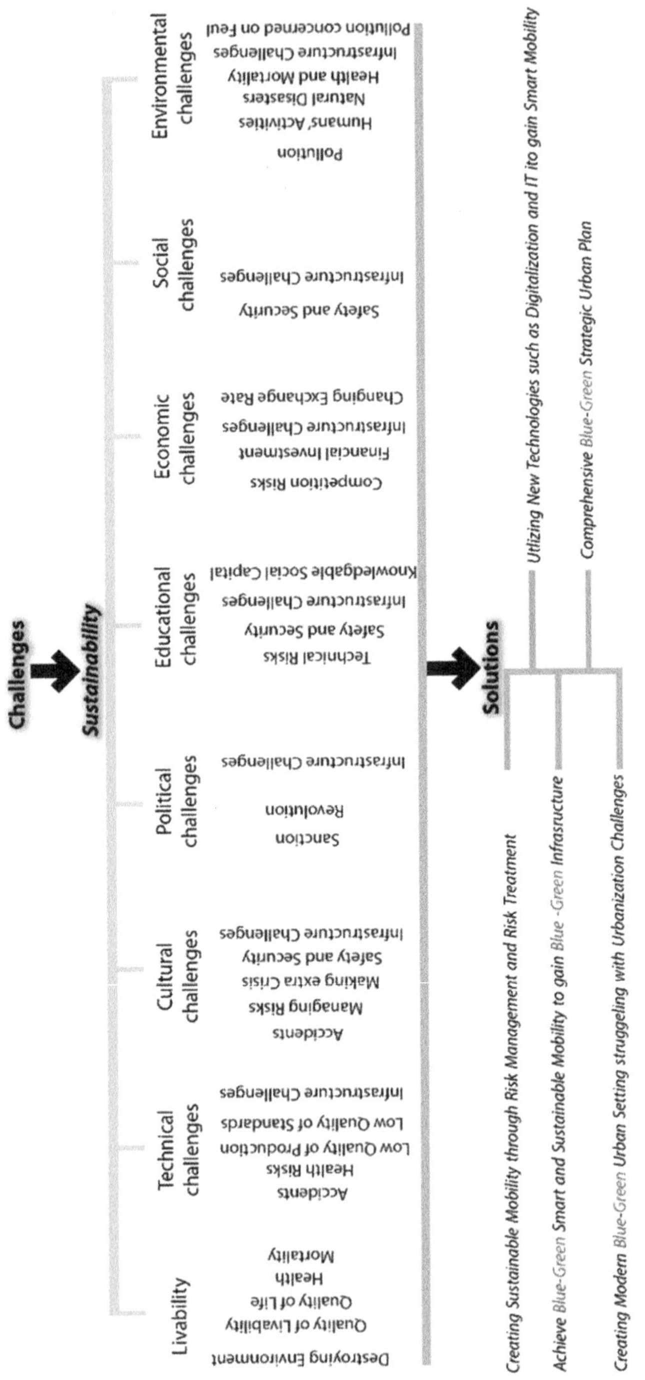

Fig. 7.14. D-CS Model Based on DRMM Method (Author, 2017).

6. CONCLUSION

Generally, rapid and unplanned urbanization causes significant challenges such as poverty, economic problems, environmental challenges, climate change, social instability, security and safety problems, health diseases, etc. These challenges make the urban areas unlivable. So, modern urban setting with high quality of livability and quality of life should be designed to struggle with these problems.

To gain such urban areas, Blue-Green comprehensive strategic urban plan is required. Fundamentally, to design such urban plan, mobility has an important role. Mobility could be used as a tool to design proper urban plan. Besides smartness, strategies concerned with environment and sustainability are needed to create Green-Blue smart mobility. In this research, impact of risk management in mobility is explored. Basically, risk management is a technique to achieve Blue-Green smart mobility through identifying risks in mobility, estimating, analyzing, treating, and mitigating them in order to reduce risk-threatened mobility systems. Further, risk management concerned on control risks and situation after any mobility disaster occurs is required as a kind of risk treatment. In this study, a practical model to gain Blue-Green mobility in comprehensive strategic urban plan through risk management is proposed.

REFERENCES

Akkiah, P. (2017). *Strategic urban planning perspective*. Ethekwini Municipality.

Albishi, S., Soh, B., Ullah, A., & Algarni, F. (2017). Challenges and solutions for applications and technologies in the Internet of Things. In *Proceedings of 4th information systems international conference*. 6–8 November 2017, Bali, Indonesia.

Alter, S. (2019). Making sense of smartness in the context of smart devices and smart systems. *Information Systems Frontiers*. doi:10.1007/s10796-019-09919-9

Aoun, C. (2014). *Smart cities cornerstone series: Urban mobility in the smart city age*. Council Advisor the Climate Group and ARUP.

Appel, F. (2012). *New models for addressing supply chain and transport risk*. World Economic Forum.

Aralu, U. (2015). Influence of information and communication technology on digital divide. *Global Journal of Computer Science and Technology: H Information & Technology, 15*.

Ardnt, W., Jain, A., Emberger, G., Lah, O., Sohr, A., & Tomaschek, J. (2014). *Future Megacities2: Mobility and transportation concepts for sustainable transportation*. Berlin: Jovis.

Aulbur, W., Arvind, C. J., & Bigghe, R. (2016). Skill development for Industry 4.0. Roland Berger, BRICS Skill Development Working Group, India Section (pp. 1–50).

Author. (2018). *Sustainable mobility management lecturing materials*. TU Berlin.

Benevolo, C., Dameri, R., & D'Auria, B. (2016). *Smart mobility in smart city: Action taxonomy, ICT intensity and public benefits*. Springer International Publishing Switzerland.

Berger, G., Feindt, P., Holden, E., & Rubik, F. (2014). Sustainable mobility—Challenges for a complex transition. *Journal of Environmental Policy and Planning, 16*(3).

Borghesi, A., & Gaudenzi, B. (2013). *Risk management: How to assess, transfer and communicate critical risks*. Springer.

Bullock, C., Brereton, F., O'Neill, J., Clinch, P., & Russell, P. (2008). *Quality of life and the environment*. Environmental Protection Agency.

Canadian Standards Association. (1997). *Risk management guideline for decision-makers CAN/CSA-Q850-97*. A National Standard of Canada.

Clemente, C., Ottorino, V., & Dario, A. (2018). A novel platform for the experimental training on internet of energy. In *Proceeding of IEEE global engineering education conference (EDUCON)*. 17–20 April. Tenerife, Spain.

(2012). *Danish smart cities: Sustainable living in an urban world*. Copenhagen Cleantech Cluster.

David Jiboye, A. (2011). Sustainable urbanization: Issues and challenges for effective urban governance in Nigeria. *Journal of Sustainable Development, 4*.

Department of Environment, Land, Water and Planning. (2017). *Planning a green-blue city*. Produced with assistance from E2Designlab.

Doost Mohammadian, H. (2020a). The 5th wave and i-sustainability plus theories as solutions for SocioEdu consequences of Covid-19. In *2020 IEEE conference LWMOOCS VII, learning with MOOCS 2020 in Antigua Guatemala,* from 30 September – 02 October 2020.

Doost Mohammadian, H. (2020b, July 2). *Blue-green smart mobility technologies as readiness for facing tomorrow's urban shock toward the world as a better place for living (case studies: Songdo and Copenhagen)*. Technologies. MDPI. Retrieved from https://www.mdpi.com/2227-7080/8/3/39

Doost Mohammadian, H., & Rezaie, F. (2019). Sustainable innovative project management: Response to improve quality of life and livability. *Inventions, 4.*

Doost Mohammadian, H., & Rezaie, F. (2020). The role of IoE-education in the 5th wave theory readiness and its effect on SME 4.0 HR competencies. In *Proceeding of 2020 IEEE global engineering education conference (EDUCON).* 27–30 April. Porto, Portugal.

Dutzler, H., Schmaus, B., Schrauf, S., Nitschke, A., & Hochrainer, P. (2016). *Industry 4.0: Opportunities and challenges for consumer product and retail companies.* Service Mark of PwC Strategy& LLC.

Feil, A., & Schreiber, D. (2017). Sustainability and sustainable development: Unraveling overlays and scope of their meanings. *Cadernos Ebape BR., 14.*

Finnveden, G., & Gunnarsson-Ostling, U. (2016). Sustainable development goals for cities. In *JPI urban Europe symposium "Shaping common ground in urban sustainability". Brussel.*

Fukuyama, M. (2018). *Society 5.0. Aiming for a new human-centered society.* Japan Spotlight.

Gehrels, H., Meulen, S., et al. (2016). *Designing green and blue infrastructure to support healthy urban living.* TO2 Federaties.

Ghofrani, Z., Sposito, V., & Faggian, R. (2017). A comprehensive review of blue-green infrastructure. *The International Journal of Environmental Sustainability, 6.*

Giap, T., Thye, W., & Grace, A. (2014). A new approach to measuring the livability of cities: The global livable cities index. *World Review of Science, Technology and Sustainable Development, 11*(2).

Gillis, D., Semanjski, I., & Lauwers, D. (2015). How to monitor sustainable mobility in cities? Literature review in the frame of creating a set of sustainable mobility indicators. *Sustainability Journal, 8.* No one.

Greco, I., & Cresta, A. (2015). A smart planning for smart city: The concept of smart city as an opportunity to re-think the planning models of the contemporary city. In *Proceeding of 15th international conference computational science and its application.* June 22–25. Banff, Canada.

Hayatul, Y. (2019). *Society 5.0 for supper smart country.* Retrieved from https://www.researchgate.net/publication/331521659

Hecklau, F., Galeitzke, M., Flachs, S., & Kohlb, H. (2016). Holistic approach for human resource management in Industry 4.0. In *Proceeding of 6th CIRP conference on learning factories.* 29–30 June. Gjøvik, Norway.

Heynitz, H., & Bremicker, M. (2016). *The factory of the future: Industry 4.0-the challenges of the tomorrow.* KPMG International Cooperative.

Ho Lee, S., Hoon Han, J., Taik Leem, T., & Yigitcanlar, T. (2008). *Towards ubiquitous city: Concept, planning, and experiences in the Republic of Korea.* Hershey: IGI Global, Information Science Reference.

Hui, S. (1996). *Sustainable architecture and building design (SABD).* Retrieved from www.arch.hku.hk

Hussein, A. (2019). Internet of Things (IoT): Research challenges and future applications. *International Journal of Advanced Computer Science and Applications, 10.*

Idowu, A., & Esere, M. (2013). ICT and higher educational system in Nigeria. *Academic Journals, 8.*

Ilhan, A., Möhlmann, R., & Stock, G. (2015). *Citizens' acceptance of U-life services in the Ubiquitous City Songdo.* Springer Science+Business Media Singapore.

(2003). *Information and communication technology development indices.* New York and Geneva: United Nation.

Jeekel, H. (2017). Social sustainability and smart mobility: Exploring the relationship. In *World conference on transport research – WCTR 2016 Shanghai.*

Joint Technical Committee. (2004). *Australian/New Zealand risk management standards.*

Kafle, Y. R., Mahmud, K., Morsalin, S., & Town, G. E. (2012). Towards an Internet of Energy. In *Proceeding of IEEE international conference on power system technology (POWERCON).* Oct 30, Nov 2. New Zealand.

Kayal, P., Singh, R., & Kumar, M. (2014). *Defining sustainable urban mobility.* The Energy and Resources Institute. TERI-NFA Working Paper No.11.

Kervella, O. (2011). *Dangerous goods and special cargoes section.* Retrieved from http://www.unece.org/trans

Klarin, T. (2018). The concept of sustainable development: From it is beginning to the contemporary issues. *Zagreb International Review of Economics and Business, 21*(1).

Köhler, J. (2013). Globalization and sustainable development: case study on international transport and sustainable development. *The Journal of Environment & Development, 23.*

Kranenburg, R., & Bassi, A. (2019). *IoT challenges.* Retrieved from http://www.comcjournal.com/content/1/1/9

Kuhlman, T., & Farrington, J. (2010). What is sustainability? *Sustainability Journal, 2.*

Lang, L. (2023). *What is smart city.* Retrieved from https://www.wien.gv.at

Lichá, A. (2015). *Songdo and Sejong: Master-planned cities in South Korea*. halshs-01216229f.

Mahmoud, R., Yousuf, T., Aloul, F., & Zualkernan, I. (2015). Internet of Things (IoT) security: Current status, challenges and prospective measures. In *Proceeding of 10th international conference for internet technology and secured transactions*. 14–16 December, London, United Kingdom.

Maksimovic, C., Mijic, A., Smith, K. M., & Suter, I. (2017). *Blue-green solutions. A systems approach to sustainable, resilient and cost-efficient urban development*. Climate-KIC.

Momtaz, R., & Elsemary, Y. (2015). Qualitative conceptions of livability between theory and applications in Egypt. In *International conference on IT, architecture and mechanical engineering*. Dubai.

Mühlhäuser, M., & Gurevych, I. (2010). Introduction to ubiquitous computing. In *Ubiquitous and pervasive computing: Concepts, methodologies, tools, and applications* (pp. 1–19). IGI Global.

Nagy, J., Oláh, J., Erdei, E., Máté, D., & Popp, J. (2018). The role and impact of industry 4.0 and the Internet of Things on the business strategy of the value chain—The case of Hungary. *Sustainability Journal*. MDPI.

Narang, S. (2007). Inclusive and sustainable urban planning: A guide for municipalities. In *An introduction to urban strategic planning* (Vol 1). UNON.

Nijkamp, P., Verhoef, E., Ubbels, B., & Rodenburg, C. (2001). *Sustainable mobility*. Retrieved from https://www.researchgate.net/publication/4795140

Nureni, Y. (2014). *Information communication technology (ICT): Concepts and applications* (pp. 13–15). Nigeria: Hasfem Publication Center.

Padayachee, K. (2014). *Techniques used in risk analysis of software development*. Retrieved from https://www.researchgate.net

Pande, K., & Kathmandu, N. (2013–2017). Risk assessment and risk management plan: Transport sector. Retrieved from http//www.adb.org

Papa, E., & Lauwers, D. (2015). Smart mobility: Opportunity or threat to innovate places and cities? In *Proceeding of 20th international conference on urban planning and regional development in the information society, Proceedings*. Gent, Belgium.

Patel, K., Patel, S., Scholar, P., & Salazar, C. (2016). Internet of Things-IoT: Definition, characteristics, architecture, enabling technologies, application and future challenge. *International Journal of Engineering Science and Computing (IJESC), 6*.

Quality of city life multiple criteria analysis – Scientific figure on ResearchGate. Retrieved from https://www.researchgate.net/figure/Indicators-most-widely-applied-for-analysing-sustainable-city

Ratheeswar, K. (2018). Information communication technology in education. *Journal of Applied Science and Research, 3*.

Rojko, A. (2017). Industry 4.0 concept: Background and overview. *International Journal of Innovation Management, 11*.

Ruen, K. (2011). The quality of life and the environment. In *Asia Pacific international conference on environment-behaviour studies, North Cyprus*.

Saif Obaid Alhefeit, F. (2018). *Society 5.0 a human-centered society that balances economic advancement with the resolution of social problems by a system that highly integrates cyberspace and physical space*. MSc Informatics Thesis, The British University in Dubai.

Sarhan, Q. (2018). Internet of Things: A survey of challenges and issues. *International Journal of Internet of Things and Cyber-Assurance, 1*.

Sassi, P. (2016). Built environment sustainability and quality of life (BESQoL) assessment methodology. In W. Leal Filho & L. Brandli (Eds.), *Engaging stakeholders in education for sustainable development at University level. World sustainability series*. Cham: Springer.

Seele, P., & Lock, I. (2017). The game-changing potential of digitalization for sustainability: Possibilities, perils, and pathways. *Sustainability Science Journal, 12*.

Sen, J. (2012). *Embedded systems and wireless technology. Chapter: 1*. CRC Press, Taylor & Francis Group. USA.

Šimák, L. (2007). Risks in operation of transport processes. *Academic Journal of Mechanics Transport Communications, 3*.

Smart Transportation Systems. (2014). Copenhagen Capacity. Retrieved from https://issuu.com/.../smart_transportation_system

Soeanu, A., Debbabi, M., Alhadidi, D., Makkawi, M., Allouche, M., Bélanger, M., & Léchevin, N. (2015). Transportation risk analysis using probabilistic model checking. *Expert Systems with Applications, 42*(9), 4410–4421.

Sofeska, E. (2017). Understanding the livability in a city through smart solutions and urban planning towards developing sustainable livable future of the city of Skopje. *Procedia Environmental Sciences, 37*, 442–453.

Steg, L., & Gifford, R. (2008). *Sustainable transport and quality of life. Chapter 11 of building blocks for sustainable transport policy: Obstacles, trends, solutions*. Bingley.

Steinkamp, R. (2014). *Risk management for government*. Brown Smith Wallace LLC.

Sustainable urban mobility (Methodology and indicator calculation method for World business council for sustainable development). (2015). Sustainable Mobility Project 2.0 (SMP2.0) Indicators Work Stream – 2nd Edition. ISBN: 978-2-940521-26-5.

Tregua, M., D'Auria, A., & Bifulco, F. (2014). Digital city towards smart city: A theoretical overview. In *Proceeding of 2nd international virtual scientific conference.* Slovakia. 9–13 June 2014.

Vuksanović, D., Ugarak, J., & Korčok, D. (2016). Industry 4.0: The future concepts and new visions of factory of the future development. In *Proceeding of international scientific conference on ICT and E-business related research 21 April. Belgrade, Serbia.*

Woods, M., & Dowd, K. (2008). *Financial risk management for management accountants.* The Society of Management Accountants of Canada, the American Institute of Certified Public Accountants and The Chartered Institute of Management Accountants.

PART 2

DIGITALIZATION AND SUSTAINABILITY

ECO-CITY TOURISM IN SMART CITIES FOR SUSTAINABILITY

Burcu Kıvılcım Zorba

ABSTRACT

Cities and the tourism industry are closely related. Cities are destinations that serve for the welfare of the people and the tourists visiting the city. The tourism sector, on the other hand, uses the environmental and sociocultural resources of the cities, ensures that these resources are transferred to the future by protecting them, and contributes to the economic development of the cities by creating employment. Several urban models have been created within this framework to make urban tourism viable. Eco-city tourism is one of these urban design concepts. Eco-city tourism aims to protect cities from sociocultural, economic, and environmental factors while promoting their growth. Eco-city tourism, which includes a number of practices such as prevention of pollution, protection of biological diversity, consumption of renewable energy, production of local products, employment, and protection of cultural values, is easily implement in cities in an integrated manner with smart technological systems. Smart building and smart energy systems for energy saving, mobile applications, Wi-Fi, big data, Internet of Things (IoT), and cloud computing are among the smart technological systems used in the tourism sector.

Keywords: Sustainable tourism; environmental cleaning; social equality; local welfare; physical integrity; biodiversity; employment; cultural preservation

1. INTRODUCTION

Cities have been important settlements for people from the past to the present. Along with the development of cities in the modernizing world, cities have begun to host not only their residents but also tourists. Cities are essential tourism destinations with their natural, historical, and cultural attractions. The intact resources from the past to the present maintain the attractiveness of the cities. Nonetheless, ensuring the sustainability of these resources for the benefit of future generations has revealed sustainable urban tourism. It contributes to the welfare of both local residents and city visitors by ensuring the sustainability of the city's tourism resources (Van der Borg, 1992). The sustainability of urban

Smart Cities for Sustainability
Advanced Series in Management, Volume 32, 139–150

ISSN: 1877-6361/doi:10.1108/S1877-636120230000033009

tourism is possible by providing environmental, social, cultural, and economic development (Yin, 2016, p. 5). Certain urban models have been developed for the sustainability of cities. The eco-city model is one of these models. In eco-cities; environmental protection plans such as prevention of pollution, preservation of plant and animal diversity, waste decomposition, reduction of nonrenewable energy consumption, and encouragement of public transportation are being conducted. Moreover; in social issues such as health, security, and ease of access to resources, it is ensured that the living standards of the people are improved, the culture of the people is preserved, and transferred over generations. In economic terms, employment is provided to the public, and local output production and consumption are encouraged to contribute to income growth (De Jong, Wang, & Yu, 2013). A number of smart technological systems are used to save time in taking sustainability measures in eco-cities. These systems are utilized not only for the comfort of the residents of the city but also for the comfort and safety of the tourists visiting the city and for the improvement of urban tourism. Systems described as smart city tourism are the systems that can make various travel decisions such as transportation, accommodation, and activities in a desired tourism destination along with smart systems such as the Internet of Things (IoT), big data, smartphones, and mobile applications (Lee, Hunter, & Chung, 2020). Through these systems, tourists are offered services such as easy access to information, virtual tours, and contactless and online transactions.

The damage to cities and urban tourism due to population growth, unplanned urbanization, and unconscious use of resources cause the depletion of resources. For this reason, it is important to use smart systems in cities. Smart eco-city tourism aims to reduce the negative environmental effects of tourism thanks to the technological systems used. In this way, it is planned to contribute to the sustainable development of urban tourism, to provide quality service to tourists and to increase the quality of life in cities. In this study, the concept of eco-city tourism, which is a sustainable city model, would be perceived as a whole with its environmental, sociocultural, and economic dimensions and would be explained along with smart city tourism.

2. THE CONCEPT OF SUSTAINABILITY AND SUSTAINABLE TOURISM

Issues such as urbanization, population growth, climate change, and environmental pollution, which have been going on for years, lead to excessive and unconscious consumption of resources in the world. The depletion of resources necessary for human life tends to endanger life and the ecosystem. The idea of minimization or, if possible, prevention of such hazards has brought forth the concept of sustainability, which expresses continuity, was first coined at the United Nations Environment Conference held in Stockholm in 1972 and was described in the Brundtland (*Our Common Future*) Report in 1987 as "fulfilling the needs of the present without compromising the future generation's ability to meet their needs" (WCED, 1987, p. 15). Sustainability, in which it is aimed for individuals to live under equal conditions in each generation, expresses the establishment of natural, sociocultural, and economic equilibria (Middleton & Hawkins, 1998, p. 247). The concept of sustainability has principles such as preventing future anxiety, ensuring equality across generations, encouraging participation in decisions, ensuring balance in the ecosystem, and integrating economic and environmental factors (Tuna, 2007, pp. 27–28). The concept aims to establish a balance among environmental, sociocultural, and economic variables (Andersen, Blichfeldt, & Liburd, 2018). The tourism sector would become effective in ensuring sustainability and making the majority of contributions (Budeanu, Miller, Moscardo, & Ooi, 2016). The tourism sector is supported by the natural, environmental, and sociocultural components that are the source of the activities. Therefore, once the concept of sustainability is viewed from the perspective of the tourism sector, the concept of sustainable tourism (ST) emerges. According to UNWTO (2022), ST is described as "a type of tourism that aims to protect and develop future opportunities; fulfills the needs of today's tourists and host regions; preserves cultural integrity, obligatory ecological processes, and biological diversity; and meets economic, social, and aesthetic needs." According to the description, ST aims to ensure that the individuals participating in tourism benefit from tourism resources under equal conditions by providing the preservation-usage balance. Because as the number of individuals participating in tourism activities increases, the level of resource usage also increases, and as a result of excessive consumption, tourism carrying capacity is exceeded and the resources deteriorate. In order to minimize such deterioration, it is necessary to preserve the resources and ensure their conscious utilization. ST, although preserving environmental resources, also prevents cultural degeneration and maintains economic development. According to the World Tourism Organization (WTO) and the United Nations Environment Program (UNEP), the objectives of ST are listed as follows (UNEP, 2009):

- Economic Growth,
- Local Welfare,
- Employment Quality,
- Social Equality,
- Visitor Satisfaction,
- Local Control,
- Social Peace,
- Cultural Diversity,
- Physical Integrity,
- Biodiversity,
- Environmental Cleaning,
- Resource Efficiency.

According to these objectives, the welfare level of both businesses and local life would increase in economic terms by ensuring continuity in tourism centers. Also, employment in the tourism sector would be provided to the local residents. One of the aims of ST is to mitigate poverty and ensure that people benefit from tourism services under equal conditions. By doing this, ST, which promises a safe and peaceful experience for tourists, also preserves the quality of life in society and prevents social deterioration. ST, although preserving the physical destinations of the region, also protects its cultural structure. ST, which aims to minimize the detrimental impacts on the flora and fauna structure, air, water, and soil structure along with the natural attractions that generate resources for the tourism sector, encourages the use of renewable energy systems by reducing the use of nonrenewable resources. In order to put all these into practice, the opinions of both the local community and tourism stakeholders can be sought.

Garrod and Fyall (1998) identified 10 principles related to ST. These principles can be stated as:

(1) Sustainable use of resources,
(2) Reduction of excessive consumption and waste,
(3) Conservation of diversity,

(4) Integration of tourism with planning,
(5) Supporting local economies,
(6) Involvement of local residents,
(7) Consulting the public and relevant groups,
(8) Training of personnel,
(9) Responsible tourism marketing,
(10) Research initiative.

Within the framework of the principles, it can be claimed that ST is an environmental, sociocultural, and economic entity. Besides, the concept prioritizes the region and the local residents and acts jointly with the local government in the decision-making process. To ensure sustainability, the consciousness of the tourists and the local residents as well as the training and knowledge of the personnel are taken into consideration. For the sake of sustainability in tourism, there are a number of applications developed for the conscious usage of resources, the usage of renewable resources, the reduction of excessive consumption, and waste decomposition. These applications are described as *smart* to ensure sustainability in cities, tourism destinations, and tourism enterprises. In the next part of the study, the concepts of smart cities and smart tourism are explained in detail.

3. SMART CITIES AND SMART TOURISM DESTINATIONS

3.1 Smart Cities

Cities have a great deal of social and economic impacts on the environment (Mori & Christodoulou, 2012). Population growth and unplanned construction, pollution, global warming, climate change, resource depletion, and degradation in cities adversely affect the sustainable life of cities. A smart city has been put forward to be used in the solution of these problems of cities. The concept, in general, is associated with technology and defined as city models that provide new services to cities and urban residents by employing smart technologies and real-time data sources, and produce solutions for integrated infrastructure, innovation, digital entrepreneurship, sustainable city development, and renewal opportunities (Caird, Hudson, & Kortuem, 2017). Its main objective is to provide solutions to fulfill the daily needs of smart cities, to ensure environmental protection and public safety, to serve cities, and to engage in commercial activities (Qin, Li, & Zhao, 2010). In order to realize these goals, technology is employed so that the natural, historical, and cultural resources of the cities can be utilized consciously and efficiently. In smart cities, activities are conducted to ensure economic, environmental, and sociocultural sustainability by courtesy of technology (Hollands, 2020, pp. 307–308). Because smart cities increase the quality of life of the city residents and generate a sustainable urban structure (Vasseur & Dunkels, 2010). Dameri (2013) has set four basic goals for smart cities:

(1) *Environmental sustainability*, since it is sustainable;
(2) *Creation of smart intellectual capital* since it is technology-based and digital;
(3) Since it is democratic, *citizen participation*;
(4) *The welfare of the residents*, since it is inclusive.

In order to put the smart city goals into practice, the process needs to be planned and managed accurately. Cohen (2014) stated that smart cities had six important components.

These components are *smart economy, smart people, smart governance, smart transportation/mobility, smart environment,* and *smart life*. With the smart economy, local production can seize opportunities and save money, whereas it is possible to educate people and develop their creativity with smart human resources. Intelligent governance refers to all web-based data and infrastructure systems, whereas intelligent transportation systems aim to mitigate the environmental hazard caused by motor vehicles by using public transportation. Effective use of resources with a smart environment and welfare in terms of safety and health with smart living is considered. In this context, some of the smart cities can be listed as Tokyo, Vienna, Copenhagen, New York, London, Toronto, Berlin, Hong Kong, Paris, and Barcelona (Kaygısız & Aydın, 2017).

Smart cities not only affect the development of cities and the life of the city residents but also affect the tourism of the city and the experiences of the tourists. The fact that the attractiveness of the cities generates a source for tourism and a supply and demand effect for tourists can render cities crucial tourism destinations. Smart tourism applications are conducted in smart cities in order to use these destinations for tourism purposes, to preserve them while using them, and to ensure their sustainability. Smart tourism constitutes the sub-branch of the smart economy, as one of the smart city components (Yalçınkaya, Atay, & Karakaş, 2018).

3.2 Smart Tourism Destinations

The concept of smart tourism was first coined by Gordon Philips in the 2000s and was expressed as a sustainable approach with the objectives of managing and developing tourism enterprises, as well as marketing tourism products (Kıvılcım, 2021). With this approach, upon ensuring continuity in tourism, good managerial planning and development in the sector are aimed. Gretzel, Sigala, Xiang, and Koo (2015) defined smart tourism as "efforts for integrating data obtained from physical infrastructure, social connections, state/organizational resources, and human bodies/minds into business activities, with a clear concentration on efficiency, sustainability and experience enrichment, by making use of advanced technologies." Smart tourism applications, in which tourism services are offered from a technological and innovative perspective, provide services such as information regarding destinations, online reservation and purchase, contactless payment, digital guidance, and 360-degree virtual tours. In destinations where smart tourism services are offered, it is possible to provide savings in terms of cost and time, and sustainability in terms of resources, by courtesy of convenient technology.

Smart tourism destinations are explained in the literature as smart tourism services in smart cities (Boes, Buhalis, & Inversini, 2016). The concept connects multiple stakeholders through a dynamic platform such as information and communications technology (ICT) to bring smartness to tourism destinations, improve the decision-making processes of destinations and support rapid information exchange regarding tourism activities (Buhalis & Amaranggana, 2015). With the development of technology, it is possible to attain information within seconds in tourism destinations. It also requires coordination among all stakeholders in managerial decisions. The aim of smart tourism destinations is to generate a better environment for tourists, local residents, and tourism enterprises in the destination and ecosystem, and to improve tourism systems/processes (Xiang, Stienmetz, & Fesenmaier, 2021, p. 5). To ensure the welfare of the city residents and the tourists visiting the cities, it is aimed to reduce the negativities such as environmental pollution, destruction, and depletion of resources by using technological systems such as transformable energy and smart waste management systems. At the same time, internet services such as the IoT, cloud computing, and big data are also utilized in favor of smart tourism destinations (Zhang, Li, & Liu, 2012). There are a number of features that smart tourism destinations should acquire. These features are presented in Table 8.1.

Smart tourism destinations require cooperation among local residents, tourists, tourism enterprises, local government, and the environment. Smart tourism destinations, which provide a quality service toward tourists by using technology, also save money and provide environmental preservation. Once the right planning is done and good management is

Table 8.1. Smart Tourism Destinations' Characteristics.

No. Stakeholders	Characteristics of Outcome
1. Tourism organizations	• Function as a smart hub that coordinates all relevant information and makes it easily accessible for users to access real-time information • Digitization of core business processes • Optimize their energy use • Engage with local communities, tourists, and government in cocreating tourism experience • Organizational agility, speed decision-making and responsive to customers' needs based on just-in-time insights • Precision targeting and personalized service
2. Governments	• Information governance that supports data openness • Regulate data privacy • Establish public–private partnership
3. Local residents/local communities	• Constantly connected • Sufficiently creative and empowered • Technology savvy • Citizen journalism • Actively involved in developing smart heritage/e-Culture
4. Tourists	• Well-connected and well-informed • Active critics and buzz marketers • Demand highly personalized service • Engaged both socially and technologically • Dynamically discuss through social media • Cocreate experience • Contribute to content • Utilize end-user devices in multiple touch-points
5. Environment	• Interconnected through IoT • Presence of cloud computing services • Innovation ecosystem • Sensor networks throughout the environment • Combine digital information and social contexts which will augment geophysical reality • Interoperable social platforms

Source: Buhalis and Amaranggana (2014).

provided in the destinations, employment and economic development in the tourism sector would be contributed (Cohen, 2014).

4. SMART ECO-CITY TOURISM

4.1 Sustainable City Tourism

In order for the resources, local culture, and economic structure of the cities to persist and develop, it is necessary to ensure sustainability in the cities. This requirement emphasizes the importance of the concept of a sustainable city. Sustainable cities are cities in which socioeconomic interests are combined with environmental and energy concerns to ensure change and development on the condition of persistence (Van Geenhuisan & Nijkamp, 1994). The Ministry of Environment and Urbanization (2016) stated that sustainable cities are the ones that reduce the consumption of natural resources and try to render resources renewable upon realizing socioeconomic development. Sustainable cities improve the quality of life in ecological, economic, and social terms, without burdening future generations (Hall, 2000). Upon evaluating the explanations for the concept, it can be asserted that the purpose of sustainable cities is to protect environmental resources, provide economic development, and prevent degeneration in the sociocultural structure of cities. Examples of sustainable urban practices that aim to improve the quality of life of cities and residents, such as preventing environmental pollution in cities, recycling with smart waste management systems, using renewable energy, ensuring equality in the welfare of generations by reducing poverty, producing and consuming local products, and providing local employment can be exemplified. The city of Gothenburg in Sweden, certain parts of Amsterdam and Vienna, Bristol City, and Canadian cities are examples of sustainable cities (Altuntaş, 2012). One of the factors that influence the sustainability of cities involves the tourism activities of the city. The tourism sector, upon utilizing the tourism resources of the cities, also contributes to economic development. Therefore, it is also crucial to ensure the sustainability of urban tourism.

Sustainable urban tourism can be described as activities that fulfill the needs of present and future generations by providing the balance of preservation and use of the cities' touristic values with the condition of continuity (Kıvılcım, 2021a). In this context, the objective is to spread tourism to four seasons and to achieve environmental, sociocultural, and economic improvement of tourism. Sustainable urban tourism has four major goals:

(1) Economic goals: long-term profit, tourism growth, and funding;
(2) Environmental goals: restoring the old buildings of the city, preserving natural areas, reducing the use of automobiles to prevent air pollution and traffic congestion;
(3) Sociocultural goals: to achieve long-term employment, to give the city residents a voice in tourism, to provide a ST industry;
(4) Experiential goals: to provide tourists with unforgettable experiences and motivate them to revisit the city (İçellioğlu, 2014; Timur & Getz, 2009).

Some urban models have been established to achieve sustainable urban tourism goals. These models are listed as follows (Akgün, 2012; Akkar & ve Belge, 2017; Keskin, 2012; Lıtfın, 2017; Mayer & ve Knox, 2016; Meydan Yıldız, 2016; Tosun, 2013):

- Smart Cities.
- Slow (Calm) Cities.
- Entrepreneurial Cities.
- Eco-Villages.
- Compact Cities.
- Eco-Cities.

4.2 Eco-city Tourism

The term "eco-city" was first coined in 1975 by Paolo Seleri, Mark Roseland, Peter Newman, Jeff Kenworthy, and Isabelle Jennings under the leadership of Richard Register, and defined as cities built on the principles of environmental living (De Jong et al., 2013; Roseland, 1997).

An eco-city is a human settlement modeled on the self-sustaining resilient structure and function of natural ecosystems (Ecocity Builders, 2010). An eco-city:

- Seeks to provide healthy abundance to its inhabitants without consuming more renewable resources than it replaces.
- Seeks to function without producing more waste than it can assimilate or recycle for new uses or that nature can dilute and absorb harmlessly, and without being toxic to itself or neighboring ecosystems.
- Its inhabitants' ecological impacts reflect planetary supportive lifestyles; its social order reflects fundamental principles of fairness, justice, reasonable equity, and consensus at ample levels of happiness (Ecocity standards, 2022).

Eco-city standards are categorized into four basic groups such as urban design, bio-geo-physical conditions, sociocultural features, and ecological imperatives. These categories are:

- Urban Design: Access According to Proximity, Safe and Affordable Housing, Green Building, Eco-Friendly Transportation.
- Bio-Geo-Physical Conditions: Clean Air, Clean and Safe Water, Healthy Soil, Liable Resources/Materials, Clean and Renewable Energy, Healthy and Accessible Food.
- Sociocultural Features: Healthy Culture, Community Capacity/Governance, Healthy, and Fair Economy, Lifelong Education, Quality of Life.
- Ecological Requirements: Healthy Biodiversity, World's Carrying Capacity, Ecological Integrity.

Eco-cities aim to renew their resources without consuming them and offer healthy living conditions in abundance to the city residents. Moreover, it is planned to prevent the hazard to the ecosystem by implementing smart waste management systems and the ability to decompose waste and cares about rights, justice, equality, and order in social life. Eco-cities are characterized by the adoption and implementation of renewable energy systems (Newman, 2010). In order for a city to become an eco-city, it should appeal to all living things, be socially neutral, reduce the use of motor vehicles, encourage pedestrian or bicycle use, contribute to socioeconomic development, protect biological diversity, and reduce energy use (De Jong et al., 2013). Eco-cities in the world include Kazakhstan-Astana Eco-city, Germany-Freiburg, Australia-Moreland Eco-city,

China-Singapore, China-Tianjin Eco-city, Denmark-Kalundborg Industrial Park, Germany-Vauban, India-Karakal Eco-city, Hong Kong-Hung Shui Kiu Eco-city, Cleveland-USA, Curitiba-Brazil, Toronto-Canada, Vaxjo-Sweden, Sydney-Australia, Kampala-Uganda, Puerto Princesa-Philippines, Reykjavik-Iceland, Masdar Eco-city-Abu Dhabi (Kaya & Susan, 2020).

4.3 Smart Eco-city Tourism

The need of utilizing information and communication technologies for the development of cities has led to the emergence of smart eco-cities (Huang, Gao, & Liu, 2019). A smart eco-city involves cities integrated with information and communication technologies (Liu, Wang, Ma, & Yang, 2017). Smart eco-cities, managed by combining digital elements and green planning, concentrate on economic growth and social sustainability (Bibri, 2021). Smart systems are used by utilizing technology to preserve the environmental resources of the cities, save energy, and ensure recycling. These systems consist of big data, the IoT, cloud computing, Wi-Fi, and mobile application. The systems used in smart eco-cities render it easier for city residents to live a comfortable life in areas such as transportation, education, health, and tourism. However, these systems should be used with the right planning through the city administration. It is required to enhance management efficiency and improve the quality of life of residents through information circulation, sustainable development of energy and environment, complete infrastructure, governance, and urban security, rendering it the goal of sustainable urban operation development (Bibri & Krogstie, 2017).

Ahvenniemi, Huovila, Pinto-Seppä, and Airaksinen (2017) emphasized that four basic components of a smart eco-city existed:

- Technology and evidence: includes information and communication technology and data and information, among other technologies.
- Physical environment: includes the building environment and key elements of urban infrastructure, natural environment, and ecological sustainability.
- Society: includes the knowledge economy and pro-business environment, human capital and creativity, governance, cohesion, and cooperation.
- Government: included utilities, urban management, policies, and other institutional regulations.

Smart eco-cities are cities that prevent air, water, and soil pollution; aim to reduce the negative impacts of climate change; ensure waste management; encourage the use of public transportation; and save energy (Liang, Zhang, Chen, & Deng, 2016). These cities, besides ensuring the ecological sustainability of the physical environment, also provide harmony and cooperation between society and the city administration. It is aimed to give society the right to have a say in decisions and to encourage them toward creativity. Renewal of cities, development of tourism in cities, supports the sustainable development of smart eco-cities. In ecologically developed cities, besides the healthy living standards of the city residents, quality service is also provided for the tourists visiting the city. They can seize the opportunity to stay in a clean environment where noise, visual, and air pollution are not present, and to visit socioeconomically developed destinations at which cultural values are preserved. Tourists visiting eco-cities can be described as ecotourists with advanced environmental consciousness. Therefore, tourists try to utilize resources efficiently and save money instead of excessive consumption. Once eco-city tourism becomes associated with ecotourism, it is possible to state that both aim at sustainability and are environmentally-friendly practices, whereas ecotourists are conscious of preserving resources. There are some practices that contribute to the sustainability of tourism in smart eco-cities. These practices involve systems that provide resource-saving, transformation and reuse with smart technological systems in tourism enterprises. According to Ercoşkun and Karaaslan (2009, p. 286), eco-cities should also become smart cities for the sake of their sustainability. In this case, it is thought that evaluating eco-city tourism combined with smart tourism would contribute to the sustainable development of tourism.

5. CONCLUSION

The tourism sector affects the sustainable development of cities in terms of environmental, sociocultural, and economic aspects. The environmental and cultural elements of cities generate a resource for tourism; upon utilizing resources, it also requires preservation to prevent deterioration, destruction, and depletion. Cities host more tourists and operate more tourism facilities. Along with the rapid population growth, the intense tourist demand for the city causes the excessive use of destinations and excess tourism carrying capacity in cities. Furthermore; air, water, soil, noise, and visual pollution occur in cities as a result of excessive usage, the flora and fauna structures deteriorate, and problems such as climate change, greenhouse gas emissions, and global warming occur. The deterioration of the balance of the ecosystem leads to the danger of extinction for all life forms and planets. To prevent such circumstance, certain sustainability practices have been developed for cities. For instance, measures such as saving electricity and water, preventing the spread of smoke from the exhaustion of motor vehicles to the environment by encouraging public transportation, waste decomposition, reducing its usage and ensuring its reuse, and conscious hunting would be mentioned. This situation encourages tourism enterprises to implement smart energy systems, public transportation systems, and waste management systems. Besides taking measures such as conservation of national parks, preventing food waste in hotel enterprises, smart energy systems at rooms and public transportation in travel enterprises, digital systems such as smart technological applications, namely, Wi-Fi, mobile applications, big data, IoT, and cloud computing are also used. These systems save enterprises both time and money. The use of smart technologies to ensure the sustainability of urban tourism has revealed the concept of a smart eco-city. Eco-city tourism, which is an environmental, sociocultural, and economic whole, is associated with smart tourism applications and expressed as a smart eco-city. Smart eco-city tourism, besides maintaining a high level of city residents' welfare, ensuring the healthy and safe living of the people, and providing employment to the city residents, also ensures that the tourists visiting the city spend their holidays in comfortable, healthy, and safe conditions. It aims to transfer the environmental, social, and cultural values, which date back thousands of years ago, to the next generations unharmed. Smart systems used for solving problems such as infrastructure, superstructure, and transportation required for the tourism sector also contribute to the economic improvement of smart cities.

REFERENCES

Ahvenniemi, H., Huovila, A., Pinto-Seppä, I., & Airaksinen, M. (2017). What are the differences between sustainable and smart cities? *Cities*, *60*, 234–245.

Akgün, A. A. (2012). Akıllı kent. In M. Ersoy (Ed.), *Kentsel planlama-ansiklopedik sözlük içinde* (ss. 7). Ninova Yayınları.

Akkar, E. M., & ve Belge, Z. S. (2017). Daha yaşanabilir kentler için mikro ölçek bir yürünebilirlik modeli. *Metu Journal of the Faculty of Architecture*, *34*(1), 231–265. doi:10.4305/metu.jfa.2016.2.11

Altuntaş, A. (2012). Sürdürülebilir Toplumlar ve Metropollerin Baskılarından Kurtulmak İçin Alternatif Bir Yol: Sürdürülebilir Kentler/Sustainable communities and an alternative way to escape the pressures of metropolis: Sustainable cities. *Mustafa Kemal Üniversitesi Sosyal Bilimler Enstitüsü Dergisi*, *9*(17), 135–148.

Andersen, I. M. V., Blichfeldt, B. S., & Liburd, J. J. (2018). Sustainability in coastal tourism development: An example from Denmark. *Current Issues in Tourism*, *21*(12), 1329–1336.

Bibri, S. E. (2021). Data-driven smart eco-cities and sustainable integrated districts: A best-evidence synthesis approach to an extensive literature review. *European Journal of Forest Research*, *9*(1), 1–43.

Bibri, S. E., & Krogstie, J. (2017). Smart sustainable cities of the future: An extensive interdisciplinary literature review. *Sustainable Cities and Society*, *31*, 183–212.

Boes, K., Buhalis, D., & Inversini, A. (2016). Smart tourism destinations: Ecosystems for tourism destination competitiveness. *International Journal of Tourism Cities*, *2*(2), 108–124.

Budeanu, A., Miller, G., Moscardo, G., & Ooi, C. S. (2016). Sustainable tourism, progress, challenges and opportunities: An introduction. *Journal of Cleaner Production*, *111*, 285–294.

Buhalis, D., & Amaranggana, A. (2014). Smart tourism destinations. *Information and Communication Technologies in Tourism*, *2014*, 553–564. doi:10.1007/978-3-319-03973-2_40

Buhalis, D., & Amaranggana, A. (2015). Smart tourism destinations enhancing tourism experience through personalisation of services. In *Information and communication technologies in tourism 2015* (pp. 377–389). Cham: Springer.

Caird, S., Hudson, L., & Kortuem, G. (2017). Communication on smart city evaluation and reporting in UK cities: Pilots, demos and experiments case. In A. Aalto & L. veMontonen (Eds.), *Smart cities in smart regions 2016: Conference proceedings the publication series of Lahti University of applied sciences, part 27* (pp. 20–28). Lahti University of Applied Sciences.

Çevre ve Şehircilik Bakanlığı. (2016). Türkiye'de Çevre Yönetimi için Kurumsal Kapasitenin Geliştirilmesi Projesi (ÇEKAP). *Sürdürülebilir Şehirler Konferansı*. 16 Mayıs 2016.

Cohen, B. (2014). *The smartest cities in the world 2015: Methodology*. Retrieved from https://opus-hslb.bsz-bw.de/frontdoor/deliver/index/docId/642/file/Anlagen9-21.pdf

Dameri, R. P. (2013). Searching for smart city definition: A comprehensive proposal. *International Journal of computers & technology*, *11*(5), 2544–2551.

De Jong, M., Wang, D., & Yu, C. (2013). Exploring the relevance of the eco-city concept in China: The case of Shenzhen Sino-Dutch low carbon city. *Journal of Urban Technology*, *20*(1), 95–113. doi:10.1080/10630732.2012.756202

Ecocity Builders. (2010, February). *What is an eco-city?* Retrieved from https://ecocitybuilders.org/what-is-an-ecocity/

Ecocity Standards. (2022, November). *Ecocity*. Retrieved from https://ecocitystandards.org/ecocity/

Ercoşkun, Ö. Y., & Karaaslan, Ş. (2009). Geleceğin Ekolojik ve Teknolojik Kentleri. *Megaron*, *3*(3).

Garrod, B., & Fyall, A. (1998). Beyond the rhetoric of sustainable tourism? *Tourism Management*, *19*(3), 199–212.

Gretzel, U., Sigala, M., Xiang, Z., & Koo, C. (2015). Smart tourism: Foundations and developments. *Electronic Markets*, *25*(3), 179–188.

Hall, P. (2000). Urban: Global conference in Berlin. *Deutschland*, *4*, 23–30.

Hollands, R. G. (2020). Will the real smart city please stand up?: Intelligent, progressive or entrepreneurial? In *The Routledge companion to smart cities* (pp. 179–199). Routledge.

Huang, Z., Gao, J., & Liu, R. (2019). Discussion on the key success factors of developing smart eco-city based on analytic hierarchy process. *Ekoloji*, *28*(107), 1651–1657.

İçellioğlu, C. Ş. (2014). Kent turizmi ve marka kentler: turizm potansiyeli açısından İstanbul'un swot analizi. *Sosyal Bilimler Dergisi*, *1*, 37–55.

Kaya, H. E., & ve Susan, A. T. (2020). Sürdürülebilir bir kentleşme yaklaşımı olarak, ekolojik planlama ve eko-kentler. *İDEALKENT*, *11*(30), 909–937. doi:10.31198/idealkent.533730

Kaygısız, Ü., & Aydın, S. Z. (2017). Yönetişimde yeni bir ufuk olarak akıllı kentler-smart cities as a new dimension in governance. *Mehmet Akif Ersoy Üniversitesi Sosyal Bilimler Enstitüsü Dergisi*, *9*(18), 56–81.

Keskin, E. B. (2012). Sürdürülebilir kent kavramına farklı bir bakış: yavaş şehirler (Cittaslow). *Paradoks Ekonomi Sosyoloji ve Politika Dergisi*, *8*(1), 81–99.

Kıvılcım, B. (2021). Akıllı otel uygulamalarının sürdürülebilirliğe etkileri: kış turizmi otellerine yönelik nitel bir araştırma. *Balıkesir Üniversitesi Sosyal Bilimler Enstitüsü Dergisi*, *24*(46–1), 1401–1413.

Kıvılcım, B. (2021a). Destinasyon yönetiminde sürdürülebilir kent turizmi. In S. Karasakal & O. Doğan (Eds.), *Destinasyon yönetiminde yeni yaklaşımlar, içinde* (pp. 265–292). Nobel Yayınevi.

Lee, P., Hunter, W. C., & Chung, N. (2020). Smart tourism city: Developments and transformations. *Sustainability*, *12*(10), 3958.

Liang, X., Zhang, W., Chen, L., & Deng, F. (2016). Sustainable urban development capacity measure—A case study in Jiangsu Province, China. *Sustainability*, *8*(3), 270.

Lıtfın, T. K. (2017). *Eko Köyler*. Alfa Yayıncılık.

Liu, J., Wang, F., Ma, X., & Yang, Z. (2017). Recent advances in wireless communication protocols for Internet of Things. *Wireless Communications and Mobile Computing*.

Mayer, H., & ve Knox, P. L. (2016). Slow cities: Sustainable places in a fast world. *Journal of Urban Affairs*, *28*(4), 321–334. doi:10.1111/J.1467-9906.2006.00298.X

Meydan Yıldız, S. G. (2016). *Çevre bilinci ve eko-kent planlaması: Gölbaşı Özel Çevre Koruma Bölgesi örneği* (Yayımlanmamış Doktora Tezi), Ankara Üniversitesi.

Middleton, V. T., & Hawkins, R. (1998). *Sustainable tourism: A marketing perspective*. Routledge.

Mori, K., & Christodoulou, A. (2012). Review of sustainability indices and indicators: Towards a new city sustainability index (CSI). *Environmental Impact Assessment Review*, *32*(1), 94–106.

Newman, P. (2010). Sustainable cities of the future: The behavior change driver. *Sustainable Development Law & Policy*, *11*, 7.

Qin, H., Li, H., & Zhao, X. (2010). Development status of domestic and foreign smart city. *Global Presence*, *9*, 50–52.

Roseland, M. (1997). Dimensions of the eco-city. *Cities*, *14*(4), 197–202. doi:10.1016/S0264-2751(97)00003-6

Timur, S., & ve Getz, D. (2009). Sustainable tourism development: How do destination stakeholders perceive sustainable urban tourism? *Sustainable Development*, *17*(4), 220–232. doi:10.1002/sd.384

Tosun, K. E. (2013). Sürdürülebilir kentsel gelişim sürecinde kompakt kent modelinin analizi. *Dokuz Eylül Üniversitesi Sosyal Bilimler Enstitüsü Dergisi*, *15*(1), 103–120.

Tuna, M. (2007). *Turizm, Çevre ve Toplum* (pp. 27–28s). Ankara: Detay Yayıncılık.

UNEP. (2009). *Sustainable coastal tourism: An integrated planning and management approach*. Environment Programme United Nations.

UNWTO. (2022, November). *'Overtourism'? Understanding and managing urban tourism growth beyond perceptions*. Retrieved from https://www.e-unwto.org/doi/pdf/10.18111/9789284420070

Van Geenhuisan, M., & Nijkamp, P. (1994). Sürdürülebilir kenti nasıl planlamalı. *Toplum ve Bilim Dergisi*, *64*(65), 129–140.

Van der Borg, J. (1992). Tourism and the city: Some guidelines for a sustainable tourism development strategy. In H. Briassoulis ve & J. Van Der Straaten (Eds.), *Tourism and the Environment* (pp. 121–131). Springer.

Vasseur, J. P., & Dunkels, A. (2010). *Interconnecting smart objects with ip: The next internet*. Morgan Kaufmann.

WCED. (1987). *Our common future, Brundtland report*. Vol. 17 – doc. 149, 1–91.

Xiang, Z., Stienmetz, J., & Fesenmaier, D. R. (2021). Smart tourism design: Launching the annals of tourism research curated collection on designing tourism places. *Annals of Tourism Research*, *86*, 103154.

Yalçınkaya, P., Atay, L., & Karakaş, E. (2018). Akıllı turizm uygulamaları. *Gastroia: Journal of Gastronomy and Travel Research*, *2*(2), 85–103.

Yin, S. (2016). *Sustainable city tourism in developing countries: Malaysia experience. Malaysia sustainable cities program*. Working Paper Series.

Zhang, L. Y., Li, N., & Liu, M. (2012). On the basic concept of smart tourism and its theoretical system. *Tourism Tribune*, *27*(5), 66–73.

SMART TOURISM DESTINATIONS AND DIGITALIZATION OF CULTURAL HERITAGE FOR SUSTAINABILITY

Nil Sonuç and Seda Süer

ABSTRACT

The conceptualization of smart emerged by technological advancements penetrated the tourism industry with the pace of globalization transformed the destinations providing digitalized products. Even though smart tourism destinations are initiated by advanced technologies, the notion evolved in embracing Sustainable Development Goals (SDGs) project economic, social, and environmental proliferation. A tourism destination is counted as "smart" which provides advanced technologies to improve the quality of tourists' experiences and enhance the residents' quality of life. The structure of a smart tourism destination is established on four basic pillars that cover technology, innovation, accessibility, and sustainability issues. Designating the notion of smart into tourism destinations is extremely vital since the shifting paradigm of tourists' demands concern sustainability. The development of smart in a tourism destination is a crucial concern for destination management organizations (DMOs) integrating key destination components with the dimensions of a smart city to enhance the tourists' satisfaction and competitiveness of the destination. Therefore, one aim of this chapter is to elucidate the association between key destination components with the dimensions of a smart city to reveal the related smart tourism destination applications. Moreover, cultural heritage in smart tourism destinations forms an important part of tourism both with its tangible and intangible resources which have been involved in the emergent era of digitalization inevitably with all parties and processes. Hence, another aim of this chapter is to examine the dimensional shift in cultural heritage tourism within the framework of digitalization. Sharing cases of digitization of cultural heritage from different parts of the world, this chapter also reflects that it is inevitable to benefit from digitization and ICTs in order to reach the SDGs on the scale of smart tourism destinations. Analysis of academic publications and the national and international reports of the related authorities sums the methodology used to conclude the study with suggestions of future research paths to develop the field.

Smart Cities for Sustainability
Advanced Series in Management, Volume 32, 151–168

Published under exclusive licence by Emerald Publishing Limited
ISSN: 1877-6361/doi:10.1108/S1877-636120230000033010

Keywords: Smart tourism destinations; digitalization; cultural heritage; SDGs; sustainable tourism

1. INTRODUCTION

Rapidly developing advanced technologies and digitalization has become a part of daily life and even a necessity for societies. The cities that applied advanced technologies and digitalization are associated with the notion of "smart" comprising information and communication technologies (ICTs). ICTs infrastructure enacts smart technologies such as "Internet of Things" (IoT) (Buhalis & Amaranggana, 2014), "Cloud Computing" (Boes, Buhalis, & Inversini, 2016), "Big Data and Data Mining" (Thomaz, Biz, Bettoni, Mendes-Filho, & Buhalis, 2017), "virtual reality (VR), augmented reality (AR), artificial intelligence (AI)" (Femenia-Serra & Neuhofer, 2018), "near field communication (NFC)", "radio-frequency identification (RFID)", and "mobile applications (apps)" (Femenia-Serra, Neuhofer, & Ivars-Baidal, 2019) in improving the quality of life within the societies. The ICTs synergize with the social components of societies in delivering assorted services to the residents by reducing energy usage and increasing efficiency. Considering the intensive usage of information in tourism and its dependency on ICTs (Benckendorff, Sheldon, & Fesenmaier, 2014; Koo, Gretzel, Hunter, & Chung, 2015), it is inevitable for destinations to benefit from digitalization and advanced technologies. In essence, the notion of "smart" for destinations is derived from the integration of ICTs into touristic products to improve the tourist experience and satisfaction. Moreover, a smart destination enhances the travel experience and satisfaction of tourists by effectively integrating the physical areas of destinations with virtualization through the ICTs infrastructure in the travel region (Lee, Hwang, & Jang, 2018; Neuhofer, Buhalis, & Ladkin, 2012; Zatori, Smith, & Puczko, 2018). Thus, the ultimate goals of creating a smart tourism destination are to gain an advantage in destination competitiveness by ensuring the efficiency of resource management and sustainability while maximizing tourist satisfaction through a favorable tourism experience (Buhalis & Amaranggana, 2014).

2. SMART CITY AND SMART TOURISM CONCEPTS

The fact that technology reserves a substantial place in daily life caused the demands of tourists for benefiting from many technological opportunities regardless of time and place. The concepts of "smart city" and "smart tourism", which can be described as a response to this demand, are considered interrelated and complementary concepts. With the industrial revolution, technology affects the development of cities and the quality of city life in the community (Akbaş, 2018). In essence, smart cities are expressed as modern cities that meet individuals' needs in the community instantly by utilizing technology effectively. A smart city refers to sustainable and livable cities where smart solutions are produced by using information technologies in daily city life (Jasrotia & Gangotia, 2018). Harrison et al. (2010) comprehensively defined the concept of smart cities as long-term applications where the physical and social infrastructure of cities is combined with ICTs, thereby increasing the operational effectiveness and quality of city life.

Essentially, the concept of a "smart city" is explained as a business model that expresses how innovative systems and processes are used by cities on behalf of stakeholders in order to fulfill certain goals. The basis of this approach depends on managing the transition with innovation and technological solutions for societies. The transition toward smart applications in cities is managed by using technological systems in an innovative way to achieve the determined goals for cities (Khan, Woo, Nam, & Chathoth, 2017). Based on the view that the concept of a smart city also creates the infrastructures of smart concepts for the tourism industry (Yalçınkaya, Atay, & Karakaş, 2018), it is acknowledged that smart city-related structures provide the fundamentals of smart tourism infrastructure. Thus, developing smart tourism infrastructure would be easier and more cost-efficient by enriching the contents and smart city applications that are available.

The tourism industry which has a vital role in the development of countries is experiencing significant transitions with advanced technologies. One of the significant transitions related to tourism is the concept of "smart tourism" implied by technological innovations and developments. Unlike the smart city, smart tourism targets not only the residents' quality of life but also the tourists' experiences, as well as the visit quality, mobility, accessibility, and sustainability (Gretzel, Sigala, Xiang, & Koo, 2015). Expressing that the concept of "smart tourism" is a broader concept than providing Free Wireless Internet (Wi-Fi) service or Mobile Applications, Gretzel, Sigala et al. (2015) comprehensively defined the concept of "smart tourism" as the acquisition of tourism-related data by combining the infrastructure, social relations, resources, and tourist attitudes/behaviors in destinations with advanced technologies, targeting efficiency, sustainability, and enhancement of tourists' experience by utilizing these data in tourism to create value.

Nowadays, smart tourism technologies are given priority as a strategic tool in the tourism management organizations of destinations; thus, projects for smart air service, smart accommodation, and smart transportation structures are on the agenda and are being implemented (Khan et al., 2017). Yavuz (2019) states that "smart tourism" refers to a tourism operation that utilizes the internet and smart technologies, and smart tourism destinations as places where smart technologies are experienced and closely related to "smart tourism" concepts. Therefore, the concept of a "smart destination" includes regions with touristic attractions where smart tourism technologies are used effectively. It is acknowledged that the "smart tourism destination" is considered as a concept that emerged together with the concepts of "smart city" and "smart tourism" in the related literature (Civelek, 2018, pp. 323–332; Jasrotia & Gangotia, 2018; Khan et al., 2017).

3. SMART TOURISM DESTINATIONS

It is admitted that the concept of "smart tourism" developed as a new approach toward sustainability, quality, and the environment as a consequence of the widespread usage of technology in the field of tourism. In the relevant literature, the visiting regions that utilize smart technologies with the purpose of tourism are mentioned with notions as a "smart destination" (Yalçınkaya et al., 2018), "smart tourism city" (Guo, Liu, & Chai, 2014), or "smart tourism destination" (Khan et al., 2017). Da Costa Liberato, Alen-Gonzalez, and De Azevedo Liberato (2018) state that ICTs are acceleratingly embedded into the tourism industry, consequently cities become more attractive and accessible for both residents and tourists, referring to the notion of a "smart tourism destination". Although a smart tourism destination utilizes the combination of smart city and smart tourism applications, it is also the integration of ICTs with the physical infrastructure of cities. In "smart tourism destinations", smart city applications are not only benefited by residents but also tourists, as a factor that supports mobility, accessibility, and sustainability increasing the quality of residents' life, tourists' visits, and tourists' experiences (Gretzel, Sigala et al., 2015).

The notion of "smart tourism destination" expresses a narrower context than the notion of "smart tourism". A "smart tourism destination" implies a tourism destination where new technologies are heavily deployed in order to enrich tourists' experiences, increase the competitiveness of the destination, and promote tourism development plans (Buonincontri & Micera, 2016). Smart tourism destinations generate high level of data and utilize integrated smart systems through the digitalization of services providing an interface between tourists and the destination with the involvement of residents (Khan et al., 2017). Essentially, a tourism destination is configured on the supply and demand forces, and a favorable destination is based on key destination components "attractions, accessibility, amenities, available packages, activities, and ancillary services" acknowledged as the six A's. These six key destination components are developed to add value to the tourists' experience meanwhile raising the profits and the benefits for the destination (Buhalis, 2000).

According to Buhalis (2000), attractions indicate the features such as natural like mountains, artificial like amusement parks, or cultural like music festivals. Accessibility encompasses the entire transportation structure at the destination, including accessible routes, existing terminals, and adequate public transportation. Amenities embrace all services that facilitate favorable hospitality, such as lodging, catering, and leisure activities. Available packages imply the convenience of service packages provided by agencies to attract tourist attention and promote the unique characteristics of the relevant destination. Activities embody all convenient activities that motivate tourists to travel to the destination. Ultimately, ancillary services are day-to-day services such as banking, postal services, and patient services that are not primarily intended for tourists (Buhalis & Amaranggana, 2014).

Integrating smart technologies into tourism destinations is a process that requires the active participation of stakeholders in a technological platform that allows instant information sharing between individuals involved in tourism activities. This platform consists of technological tools and applications that have multiple connection points, support and enable the creation of instant tourist experiences, and increase the effectiveness of tourism resources at the micro and macro level. The main motivation for the development of smart tourism destinations is to benefit from the advanced and smart technologies to enrich tourist experiences and resource management efficiency to maximize consumer satisfaction and destination competitiveness along with sustainability (Buhalis & Amaranggana, 2014).

Table 9.1 represents the combination of the key destination components defined as six A's "attractions, accessibility, conveniences, available packages, activities, and ancillary services" (Buhalis & Amaranggana, 2014) and dimensions of a smart city "smart, economy, smart people, smart governance, smart mobility, smart environment, and smart living" (Giffinger et al., 2007) with the related smart tourism destination applications. While mobile applications used via smartphones are included in the table as the commonly used type of smart technology in destinations, technologies such as NFC, QR codes, Global Positioning System (GPS), sensor, wireless internet networks (Wi-Fi) are other frequently used technologies in smart tourism destinations. Moreover, the table also indicates that these smart technologies are used in public transportation, information acquisition and communication, instant visitor data collection, museums, historical monuments, and similar place visits in destinations.

Table 9.1. Combination of Tourism Applications in Smart Tourism Destinations.

Destination Components (Buhalis, 2000)		Dimensions of a Smart City (Giffinger et al., 2007)		Smart Tourism Destination Applications
Available packages	"Packages prearranged by agents and administrators"	Smart Economy (Competitiveness)	By using information and communication technologies, touristic services such as low-cost e-guidings are provided.	"Wi-Fi, Mobile Applications, NFC, QR Codes, E-Commerce System (Examples: E-Tour Guide, Online Translation, Mobile Phone Recharge Points, Mobile Payment)" (Buhalis & Amaranggana, 2014; Da Costa Liberato et al., 2018; Ghaderi, Hatamifar, & Henderson, 2018; Gretzel, Reino, Kopera, & Koo, 2015; Wang, Li, Zhen, & Zhang, 2016)"
Amenities	"Accommodation and catering services, retail, and other tourism services"	Smart People (Social and Human Capital)	This dimension aims at the contribution of tourists to invention and innovation by using ICTs.	"Mobile Applications, Recommendation Systems (Examples: Online Room Booking, Online Flight Booking, Online Car Booking, Online Rental Taxi, Tourists Forum Sites) (Benckendorff, Xiang, & Sheldon, 2019)"
Ancillary services	"Services benefited by tourists such as banks, communications, post, newsagents, hospitals, and other services"	Smart Governance (Participation)	Effective and efficient communication with tourists is ensured by utilizing the ICTs of the public administration.	"Internet of Things (IoT), RFID, NFC, Mobile Applications, Beacon Technology, Sensors (Examples: E-Maps, Obtaining Information and Communication, Shopping Centers Information, Crowd Management, Tourist Locating, and Tracking) (Divya, 2016; Gretzel, Sigala et al., 2015)"
Accessibility	"Roads, terminals, and vehicles covering the entire transport system"	Smart Mobility (Transport and ICTs)	This dimension includes information and communication-supported transportation systems.	"Wi-Fi, Sensors, Mobile Apps, Smart Cards, NFC, QR Codes (Examples: Instant Public Transport Information, Tourist Bus Services, Bus Access Points and Departure Time Information, City Traffic Information, E-City Guide, E-Taxi Services,

Table 9.1. *(Continued)*

Destination Components (Buhalis, 2000)		Dimensions of a Smart City (Giffinger et al., 2007)		Smart Tourism Destination Applications
				Parking Area Information, Public Transportation, Traffic Control and Management, Bicycle and Vehicle Sharing Systems) (Civelek, 2018, pp. 323–332; Divya, 2016; Gretzel, Sigala et al., 2015)"
Attractions	"Natural, unnatural, artificial, purpose-built, heritage, and specific events"	Smart Environment (Natural Resources)	The green tourism needs of tourists are met with renewable energy, air pollution monitoring technologies, environmentally friendly buildings, energy-efficient lighting systems, and smart water management.	"Mobile Applications, NFC, SMS, Wearable Technologies, VR, AR, Global Positioning System (GPS), Recommendation Systems (Examples: Traffic Control and Management, Bicycle and Vehicle Sharing Systems, Electric Vehicle Stations Smart Buildings, Museum, Historical Monument Visits and Location Information) (Civelek, 2018, pp. 323–332; Da Costa Liberato et al., 2018; Wang et al., 2016)"
Activities	"All available activities at the destination and what customers will demand during their visit"	Smart Living (Quality of Life)	Visits of tourists are facilitated by ICTs.	"Wi-Fi, Mobile Apps (Event Information) (Divya, 2016)"

Source: Adopted by the authors.

4. THE SIGNIFICANCE OF SUSTAINABILITY IN SMART TOURISM DESTINATIONS

Within the acceleratingly competitive environment that prevails in the tourism industry, destinations with explicit competitive and sustainability strategies are more attractive. There beside, advancements in technology have occupied expeditious developments in the tourism industry as well. Correspondingly to these developments, the notion of "smart tourism destination" has originated with crucial aspects. Since smart tourism accelerates the competitive issues of destinations, it also places emphasis on sustainability with technological advancements and practices to maintain resources efficiently. Thus, the tourism industry, which is based on natural and cultural resources, ironically causes damage to these resources, and thus the concept of sustainability for the tourism industry becomes even more crucial. The sustainable tourism approach, which aims to minimize the unfavorable impacts of tourism while trying to maximize its favorable impacts, depends on the premise of caring for the environment, society, and economy.

With the overcrowding in destinations, advancements in ICTs related to the demand and competence of tourists to use developing smart technologies have urged the concept of sustainability a crucial issue in smart tourism destinations (Silik & Dündar, 2020). Today, destinations that adopt Sustainable Development Goals (SDGs) are more successful by gaining a competitive advantage. As the significance of sustainability increases in the development of competitive destinations, technology-oriented smart tourism destinations come to the fore with their sustainability goals and strategies. Successful smart tourism destinations focus heavily on sustainability. Precisely, it is not enough for a smart tourism destination to just have sustainable practices. Therefore, sustainable practices should be supported by technologies that enable the realization of these practices by smart applications. Smart tourism destinations apply smart technologies to improve sustainability practices such as reducing their carbon footprints, adopting eco-friendly approaches, and considering the local residents and their needs.

Considering smart tourism with sustainability emplaces destinations in the scope of social relations harmonized by the cohesion of technology and tourism experience (Hunter, Chung, Gretzel, & Koo, 2015). Therefore, the needs and expectations of the new generation tourists, who are prone to smart technologies, sophisticated and demanding tourists would be satisfied. The smart tourists are more informed about a destination as more attached to social media in sharing information, making recommendations, and benefiting user-generated contents (Ivars, Solsona, & Giner, 2016). Smart tourists mostly depend on information technologies, booking tools, and value easier accessibility to technology, time, money, diversity, flexibility, personalization, and security (Buhalis & ve Amaranggana, 2014).

Perles and ve Ivars (2018) aimed to develop a conceptual model by connecting sustainability and smartness. The model specifically formulated six key functions for sustainable tourism policy. These functions are: "planning and scenario creation in a long-term period, more efficiently usage of resources, monitoring and real-time administration, public-private participation and innovation, transparency and customization of tourist services" (Perles & ve Ivars, 2018). The smart tourism paradigm is based on the cross-use of sustainability because a destination cannot be considered smart if it is not also sustainable. Therefore, smart measures implemented in a tourism destination should target the fundamental issues of local sustainability aimed at improving the quality of life of all its residents while assuring tourism competitiveness (Gonzalez-Reverte, 2019). According to Gonzalez-Reverte (2019), for a smart tourism destination to be considered sustainable, the production and consumption of tourism require three basic dimensions collaborating with technology. The first of these three basic dimensions is social and economic equality, efficiency and reduction in energy consumption, the second is governance and information management, and finally, the satisfaction of the tourists experience.

5. CULTURAL HERITAGE TOURISM AND DIGITALIZATION – ADVANTAGES AND CHALLENGES

A new era has emerged in tourism where the historical and cultural heritage sites of a destination attract tourists offering new and enhanced experiences, through the combined perceived benefits of digitalization and accessibility together with the competitive advantages (Katsoni & Segarra-Oña, 2019). Thus, nowadays cultural heritage tourism is evolving with the growing impetus of digitalization and ICTs as a smart tourism destination appealing to the attraction of tourists. Moreover, the developments in cultural heritage tourism as a smart tourism destination promote SDGs contributing to the preservation of the natural environment and resources (European Commission, 2019). Therefore, cultural heritage tourism in smart tourism destinations needs the support of ICTs for the purposes of promotion, preservation, and support for innovative solutions during the visits of tourists to tangible or intangible heritages (Guttentag, 2010; Pardo Abad, 2019). As exemplified by Pardo Abad (2019) in cases of Industrial Heritage Zones in Spain, all the human resources on the supply and demand sides should be interconnected in order for creating a sustainable structure to enable the consumption of these places for cultural heritage tourism.

As in the example of Ceuti in Spain, the trial of new user-friendly technologies attracts attention which links the visitor and host sides in terms of cocreation and copreservation, and revaluation of the intangible cultural heritage creating a sustainable ecosystem respecting all three core dimensions of sustainability, social, environmental, and economic (Gomez-Oliva, Alvarado-Uribe, Parra-Meroño, & Jara, 2019). On the other hand, even if digitalization seems like a useful option for the preservation of heritage in destinations, the visitors may not be satisfied with its threatening effect on their lived experiences as shown by the results of a study in Cyprus (Yeniasır & Gökbulut, 2022). Visitors are not the only resisting factor; the local authorities may also share the same role as in the Malaysian case while Chinese investors tried to mitigate the perceived risk of smart destinations with social responsibility tools (He & Tritto, 2022). The focal point is that destinations should self-analyze well before they invest in smartness and try to find out if they contribute to the goals of sustainable development as the technology investment is costly and if it is not human-centered, the decisions of investment would result in a loss, as stimulated by a study carried out in Ljubliana (Johnson, 2022).

6. DIGITALIZATION OF CULTURAL HERITAGE FOR SUSTAINABILITY OF TOURISM

The underlying impact of digitalization on cultural heritage enables its protection and creates the basis for its sustainability as its carried through the next generations.

Cultural heritage tourism with the framework of digitalization for its sustainability constitutes a multidimensional theoretical and practical area of research and projects collaborating the archeology, architecture, art, artisanry, and many other fields including the stakeholder analysis above all. When taking into consideration the use of digitalization in tourism for the sustainability of tangible or intangible cultural heritage, it is essential to ask and reply to the questions of "what, where, in what way(how), why, by whom".

6.1 Archeology and Digitalization of Cultural Heritage Tourism

Research by Fang (2022) on digitalization for cultural heritage showed that digital documentation technologies proved to be more successful than traditional techniques of documentation in terms of educational research, tourism, and conservancy of ancient deteriorations. As a contribution to archeology as well as the protection of the findings, passive geophysical tools (e.g., magnetic and EM methods) are suggested to explore archeological remains without destroying or digging in the area (Abdallatif, Odah, El Emam, & Mohsen, 2019). Digital methods in archeology such as 3D recording are emphasized to ease the protection of the findings as well as the process of excavation without destroying the area which evolves into a process of digitalization (Roosevelt, Cobb, Moss, Olson, & Ünlüsoy, 2015). 3D digital and virtual artifacts in archeology (Neamţu, Comes, & Popescu, 2016) created by multiple virtual and augmented reality options also highlight the past of heritage and give an idea of how the artifacts would appear in the past helping additionally restoration projects.

The implementation of digital techniques such as digital archiving, 3D line drawing, virtual restoration, and virtual display technology facilitates archeology and enhances the preservation of cultural heritage as demonstrated in projects of digital heritage such as Nanyuewang museum, Maijishan, Xiangtangshan and Longmen (Li, Luo, & Zha, 2010). The revolutionary effect of various digitization methods on archeology is mentioned in many respects in another study, especially for deeper analysis of stones dating back even to the Paleolithic periods (Shott, 2014).

Digital close-range photogrammetry is researched in a study in Konya, Turkey, for the protection and sustainability of cultural heritage. Close-range photogrammetry for documentation is mentioned as a key term letting the transmission of cultural heritage to future generations, reviving the heritage building in a contemporary manner, collecting data for future plans, learning the history of the building, determining the matters in an ancient construction in terms of architecture, urban planning, and restoration. The case was including the building's processes before and after the fire (Yilmaz, Yakar, Gulec, & Dulgerler, 2007). AR was used to preserve the architectural heritage of an aqueduct in Kayseri, Turkey (Oke, Kozlu, & Çoruh, 2021).

Similarly, a more recent study described the historical building information model (HBIM) and building information model (BIM) as digitized techniques for the conservation and reconstruction of the cultural heritage buildings in the Transilvania Region as a reply to the present prospective degradation of climate change, natural disasters, or simply indifference of younger generations (Chitonu et al., 2020).

6.2 Acknowledgment of Digitalization for Sustainable Cultural Heritage Tourism Management

The museums started to present themselves on the internet in the 1990s (Navarrete, 2019) which represents the beginning of digitalization for museums. European Commission supported research on theoretical implications till the 1990s; however, it has recently turned into a more practical funded tendency in the 2000s. The trial projects created new opinions and tools to depict digital heritage-related products. Systems of museums and scientific heritage such as "CHOSA" to improve access to St Albans historical region and "E-Islam" in Greece to create a virtual environment for Islamic museum dissemination are an example. "eStage" developed for information management, exploitation, and interfaces of European puppetry and "Virtual" including virtual archive for cathedrals in England. "CTIC" project allows access to costume and textile museum collections. Digitalization has provided reachability to cultural heritage in such alternative ways that today more museums are easily accessible in the virtual world. Moreover, digitalization lets access to sites that are difficult to reach by real means of transportation or because of the risk of damage in case of visiting (Tariffi, Morganti, & Segbert, 2004).

To record the intangible cultural heritage with the help of digitization creates an advantage for sustainability and preservation in the case of ancient manuscripts (handwritten copies). Joanna Lumley's India visit reflects a good example. On her trip to Kashmir, her ancestors' birthplace, Joanna Lumley was surprised when she saw the letters that corresponded between her great-grandfather and her grandfather. Two people from the royal family, Pema and Dechem, were working on this archiving of official letters and documents dated for a period comprising of a hundred years beginning from 1875 (Asianexpress, 2017). Similarly, the digitalization of a medieval manuscript reflects the importance of recognition of the past within the present decoding techniques exhilarate the meaning-making among cultures of the past and present (Mak, 2014). Both the protection of and reachability to ancient valuable documents are possible with the methods of digitalization. Latin for five books, "Quinque Libri" was an example of the digitalization of Sardinian records of the local people on their private lives as a part of a digital library which also formed a digital philological resource about the society for the educational field (Tariffi et al., 2004). A lost experience might be modeled in a 3D environment with the help of digitized techniques as shown by Ahmed, Carter, and Ferris (2014) in the case of virtual recreation of an indigenous site.

Experience creation and even cocreation, feeling and senses while traveling on-site, having more quality exhibition opportunities with the improved quality of the visual technology during online visits, improving access to cultural heritage, augmentation of past evidence through digital reconstructions, creating an emotional connection with digitally created gifts or souvenirs are acknowledged as the outstanding benefits obtained through the digitalization of cultural heritage for tourism (Arnold & Kaminski, 2013).

AR and VR are the two prominent examples of digitalization to improve the preservation of heritage as well as the management of heritage tourism. Along with the enrichment of visitor experience, their supportive impact of historical engagement with the places, artifacts, and activities are undeniable. In an offered model, with the help of AR and VR technology, constructs of historical facts are integrated with the contested heritage to reach unique digital tourism experiences with products for heritage preservation (Bec et al., 2019). The beginning of the guided virtual visits of the virtual twin of the Lascaux Cave at Chapelle Sixtine in France is a good example of sustainable cultural heritage tourism (Sonuç & İşçen, 2022).

On the demand side, a digital tourist is described as a person that uses digital products through the museum website, social media profiles, and collaborated portals such as Europena, Google Art Project, or Wikipedia. The management of online or on-site museums with an extraordinarily unique experience to the real or virtual visitor differs in terms of dissemination in many different ways that are out of the reach of the management of the museums themselves. Many different dimensions of online channels exist, and it is difficult to manage the rights to use especially the visuals of the artifacts. On the other hand, sustainable and reliable provision of global access to heritage is a need (Navarrete, 2019).

Another important issue is the involvement of all the related stakeholders in the creation of digital products for the sustainability of cultural heritage tourism. Community participation in managing, planning, and preserving the heritage is emphasized and cocreation, player-centered design, scenario-based design, iterative development, interdisciplinary teamwork, and living labs are some offered stakeholder participation methods (Calvi, Weber-Sabil, Asmar, & Socías Perez, 2022).

The protection and preservation of cultural and natural heritage is a hard objective challenged by the harsh socioeconomic as well as the environmental changes in the world such as climate change as mentioned by many international authorities such as UNESCO, IUCN, WWF, ICOMOS essential for sustainable development (Rodgers & van Oers, 2011). The SDGs promise an alternative achievement if strictly applied by the smart destination authorities which naturally brings sustainability to tourism. Intangible heritage in terms of local beliefs, traditions, dances, and festivities follows in significance the implications for the sustainability of cultural heritage tourism (Kim, Whitford, & Arcodia, 2019).

Overall, digitalization of cultural heritage is applied in many parts of the world for certain purposes such as research and development of products, conservation, reconstruction, preservation, education, experience cocreation, and marketing. Some examples of digitalization technologies applied in various parts of the world with the aims and benefits of usage are given in Table 9.2.

Besides many opportunities and advantages of digitalization in the heritage, drawbacks exist such as a lack of know-how about preference, determination, and usage of digital appliances. Attempts to benefit from incorrect technology-product matches may result in damaging heritage or its perception accidently or even intentionally. Car, as stated by Gervasi, Perri, Simonetti, and Tasso (2022), the perception of a virtual world may vary from a fake world to an alternative reality. Thus, measures should be taken in terms of the security and reliability of digitized creations of heritage. On the other hand, digital technology is ever-developing and the cost of budget for its adoption may be excessive. To conclude, the digitalization of cultural heritage for the sustainability of tourism is a field that requires intricate research and multidimensional development. Certain research attempts to prove exists such as a model creation effort of Gatelier, Ross, Phillips, and Suquet (2022) for a business model implementing digital heritage experience interpretations.

Table 9.2. Digital Technologies Applied for Sustainability of Cultural Heritage Tourism.

Digitalization Technologies	Purpose of Usage	Cultural Heritage (Tangible of Intangible) Area or Category
"Oblique aerial photography, 3D laser scanning technology, 360 degrees panorama technology" (Cai, Fang, Zhang, & Chen, 2021)	Virtual experience, Tourism marketing	Mount Lushan Heritage Area
"Photogrammetry, aerial 3D mapping, polygon modeling, 3D sculpting, 3D painting, UV Mapping, etc. (The major software/tools used: Agisoft Photoscan, Autodesk Remake, Pixologic ZBrush, xNormal, Autodesk 3ds Max, Unity, SteamVR, HTC Vive" (Esmaeili, Thwaites, & Woods, 2017)	Creation of virtual cultural heritage environments	Virtual Museum Heritage and Tourism (VMHT)
"Augmented Reality Tourism System (ARTS): real-time environment viewing module, a space-switching module, and an Augmented Reality (AR) guide graphic module" (Shih, Diao, & Chen, 2019)	Mobile app, 3D scans and AR scenes used for pedagogical aid for students and tourists	Shih Yih Hall, Lukang
"3D environment produced by the digitization of the archaeological site, the excavation, and archaeological finds" (Vlizos, Sharamyeva, & Kotsopoulos, 2021)	Experiential learning for visitors and students, involvement of communities	Sanctuary of Amyklaion, Sparta, Greece
"Augmented reality, virtual reality and 3D printing" (Jung & tom Dieck, 2017)	Value cocreation, development of products	The United Kingdom
"3D Sound Narratives from Social Platforms, HCI – Human Computer Interaction, PLUGGY (Pluggable Social Platform for Heritage Awareness and Participation), 3D Tune-in Toolkit" (Lim, Khan, & Picinali, 2021)	Inclusivity, safeguarding heritage, meaningful experiences	User-Generated Content (UGC)
"3D printing technology" (Du, 2020)	Product development	Intangible Cultural Heritage

"Laser scanning and 3D visualizations" (Tait, Laing, Grinnall, Burnett, & Isaacs, 2016)	Cultural resilience, community engagement resulting in coproducing, gaining a new perspective to the heritage	The United Kingdom
"3D immersive cultural heritage visualization" (Tufail et al., 2022)	Increasing value perception, contributing to sustainability	Seokguram Grotto, South Korea
"Crowdsourcing; web-mapping; 3D modeling" (Dhonju, Xiao, Mills, & Sarhosis, 2018)	Large-scale heritage documentation and sharing, enabling management, exploitation, and education of cultural heritage resources	online geo-crowdsourcing system, termed Share Our Cultural Heritage (SOCH),
"Crowdsourcing 3D cultural heritage, mass photogrammetry" (Ch'ng, Cai, Zhang, & Leow, 2019)	Preservation of heritage, global sharing for experience and educational purposes, and avoiding theft	East Asia Museums
"Mobile 3D Laser Mapping System (Zebedee)" (Zlot et al., 2014)	3D model construction of complex heritage sites for documentation, conservation, continuity of research, and education for tourism	Queensland
"VR (Virtual Reality) based digitization" (Hajirasouli, Banihashemi, Kumarasuriyar, Talebi, & Tabadkani, 2021)	For the survival of endangered heritage sites	Kandan İran- exhibition in Australia
"VR, AR, and Mixed Reality (MR)" (Jin & Liu, 2022)	Image viewing, production, and dissemination of heritage	VR Restored Image of the Old Summer Palace
"AI (Artificial Intelligence), 3D reconstructions, film footage, laser scans and visualizations, archaeological and architectural data, historical photographs, and user-generated content" (Terras et al., 2021)	Revaluation, reimagery, cocreation, innovative investment strategy	The digitalization of Galleries, Libraries, Archives, and Museums (GLAM)

Source: Adopted by the authors.

7. CONCLUSION

This study clarifies the process from smart cities to smart tourism, with its impact on daily life as a result of the development of advanced and smart technologies. Smart tourism destinations, which emerged as a result of the concepts of smart city and smart tourism, are equipped with various smart technologies that can meet the demands of today's tourists, increase the tourist experience and the quality of life of the residents while providing tourist satisfaction, and contribute to sustainability. In this study, tourism applications providing benefits for smart tourism destinations are presented by matching key destination components with dimensions of a smart city to express the fields of usage. Especially in recent years, the importance of cultural heritage tourism and the sustainability of cultural heritage areas come to the fore within the scope of smart tourism destinations.

Smart tourism not only develops advanced technological tools for destinations but also enables sustainable development. For this reason, the concept of sustainability comes to the fore as a crucial issue in smart tourism destinations. Destinations need to adapt their sustainability goals to their strategies and goals of being smart; therefore, there is a need to embrace the concepts of sustainability and smartness together. In addition, destinations with smartness in tourism should adopt best practices in the sustainability of their destinations and develop sustainable practices specific to their destinations with diverse smart tourism applications. The recommendations for tourism managers and destination management stakeholders regarding smart tourism destination development can be listed as follows:

- Establishment of smart destination management organizations that deal with the smart destination development process.
- Determining the practices that are suitable for the structure of the destination by considering the previous examples in the world where smart destinations are developed.
- Identifying and applying smart technologies, which are pioneers in smart tourism destination development, in a way that can be integrated into the destination.
- Determining a suitable road map by revealing the resources, advantages, and disadvantages of the destination where a smart tourism destination could be developed.
- When the smart tourism destination development process is completed, it should be ensured to follow the constantly developing smart technologies.

Tourism and tourist experiences evolve with the emergence of digitalization. The production and consumption as well as the marketing and distribution channels of tourism also evolve. Virtual and augmented reality, AI, and metaverse begin to play a more prominent role in the dissemination of tourism products. Digitalization enriches the cultural heritage tourists' experience of the destination without giving harm to cultural heritage sites, thus it is advantageous for sustainability. The differentiation of tourist products also increases even more with the digitization of heritage and experiences. Disadvantages might be the evolving nature of the traveling experience. When comparing the on-site experience versus the virtual experience, which is more costly, or more enjoyable, time will show which of the two will be higher in popularity. Or one will feed the other and increase the popularity of the other reciprocatively.

As a result of the sustainability and digitalization efforts, the concept of tourism as well as the tourist will need to be redefined. Touristic products will not be perishable anymore; their experience will stay more memorable. Touristic products are not necessarily owned at the destination point. Some touristic products are already available at the departure point in the virtual environment. In fact, this does not create a disadvantage but rather, enhances an enrichment of the touristic product. On the other side, the producers of tourist products should adapt to this new world of sustainable, inclusive, and cocreative tourism and should invest in the know-how of digitization.

Does it bring innovation? In fact, it is an effort to be indelible that exists since the first footprint of humans on the planet. Drawing on the walls of the caves, writing on clay tablets, inventing of papyrus, parchment, paper, pen, pencil, printing, recording on floppy disks, and in the end, on the internet creating a digital world. What comes next? Even though humans evolved in terms of creating languages and means to communicate, does it still look for ways to reinvention the sixth, seventh, or eighth senses or more? Without using any intermediary tools, return to telepathic communication as their ancestors did in caves? Will science be able to keep the heritage recording digitally forever overcoming climate change and other possible disasters? Will virtual visits to tourism destinations and museums augment, thanks to their budgetary advantage? Or will the tourists still require kinesthetic visits for the best experiences? Time and science will unveil the answers to these questions.

REFERENCES

Abdallatif, T., Odah, H. H., El Emam, A. E., & Mohsen, A. (2019). Geomagnetism exploration of the Egyptian archaeology: Thirty-years of success and challenges. In G. El-Qady & M. Metwaly (Eds.), *Archaeogeophysics. Natural science in archaeology*. Cham: Springer. doi:10.1007/978-3-319-78861-6_7

Ahmed, N., Carter, M., & Ferris, N. (2014). Sustainable archaeology through progressive assembly 3D digitization. *World Archaeology*, *46*(1), 137–154.

Akbaş, İ. (2018). Akıllı kentler: AB ve Türkiye analizi. *Sosyal Bilimler Dergisi*, *5*(26), 139–163.

Arnold, D., & Kaminski, J. (2013). Cultural heritage tourism and the digital future. In J. Kaminski, A. M. Benson, & D. Arnold (Eds.), *Contemporary issues in cultural heritage tourism* (pp. 289–310). Routledge.

Asianexpress. (2017, July 7). *Darling… I'm home: Joanna Lumley takes a trip back to Kashmir,- the country of her birth*. Retrieved from https://www.asianexpress.co.uk/2017/07/darling-im-home-joanna-lumley-takes-a-trip-back-to-kashmir-the-country-of-her-birth/

Bec, A., Moyle, B., Timms, K., Schaffer, V., Skavronskaya, L., & Little, C. (2019). Management of immersive heritage tourism experiences: A conceptual model. *Tourism Management*, *72*, 117–120.

Benckendorff, P., Sheldon, P., & Fesenmaier, D. R. (2014). *Tourism information technology*. Oxford: CABI.

Benckendorff, P., Xiang, Z., & Sheldon, P. (2019). *Tourism information technology* (3rd ed.). Boston, MA: CABI.

Boes, K., Buhalis, D., & Inversini, A. (2016). Smart tourism destinations: Ecosystems for tourism destination competitiveness. *International Journal of Tourism Cities*, *2*(2), 108–124. doi:10.1108/IJTC-12-2015-0032

Buhalis, D. (2000). Marketing the competitive destination of the future. *Tourism Management*, *21*(1), 97–116.

Buhalis, D., & Amaranggana, A. (2014). Smart tourism destinations. In Z. Xiang & I. Tussyadiah (Eds.), *Information and communication technologies in tourism* (pp. 553–564). Cham: Springer.

Buonincontri, P., & Micera, R. (2016). The experience co-creation in smart tourism destinations: A multiple case analysis of European destinations. *Information Technology and Tourism*, *16*(3), 285–315.

Cai, Z., Fang, C., Zhang, Q., & Chen, F. (2021). Joint development of cultural heritage protection and tourism: The case of Mount Lushan cultural landscape heritage site. *Heritage Science*, *9*(1), 1–16.

Calvi, L., Weber-Sabil, J., Asmar, D., & Socías Perez, X. (2022). A framework for stakeholders' involvement in digital productions for cultural heritage tourism. *Multimodal Technologies and Interaction*, *6*(52), 1–20.

Chiţonu, G. C., Rolfsen, C. N., & Deaconu, O. (2020, March). Accessibility to Transylvania's cultural heritage through BIM-heritage and community involvement. *IOP Conference Series: Materials Science and Engineering*, *789*(1), 1–8.

Ch'ng, E., Cai, S., Zhang, T. E., & Leow, F.-T. (2019). Crowdsourcing 3D cultural heritage: Best practice for mass photogrammetry. *Journal of Cultural Heritage Management and Sustainable Development*, *9*(1), 24–42.

Civelek, A. (2018). Smart cities and smart tourism: Smart city projects and applications in Turkey. In *Multidisciplinary academic conference on management, marketing, and economics, 7–9 December Prague*.

Da Costa Liberato, P. M., Alen-Gonzalez, E., & De Azevedo Liberato, D. F. V. (2018). Digital technology in a smart tourist destination: The case of Porto. *Journal of Urban Technology*, *25*(1), 75–97. doi:10.1080/10630732.2017.1413228

Dhonju, H., Xiao, W., Mills, J., & Sarhosis, V. (2018). Share our cultural heritage (SOCH): Worldwide 3D heritage reconstruction and visualization via web and mobile GIS. *ISPRS International Journal of Geo-Information*, *7*(9), 360. doi:10.3390/ijgi7090360

Divya, K. (2016). Study and reviews of smart city based tourism mobile app. *International Journal of Computer Trends and Technology*, *33*(5), 226–230.

Du, Y. (2020, March). The promotion of intangible cultural heritage tourism creative products' development through 3D printing technology. In *4th international conference on culture, education and economic development of modern society (ICCESE 2020)* (pp. 354–357). Atlantis Press.

Esmaeili, H., Thwaites, H., & Woods, P. C. (2017). Workflows and challenges involved in creation of realistic immersive virtual museum, heritage, and tourism experiences: A comprehensive reference for 3D asset capturing. In *13th international conference on signal-image technology and internet-based systems (SITIS), Jaipur, India* (pp. 465–472). doi:10.1109/SITIS.2017.82.

European Commission. European Capital of Smart Tourism. (2019). Retrieved from https://smarttourismcapital.eu. Accessed on October 1, 2019.

Fang, X. (2022). Research on the development path of cultural heritage information visualization from the perspective of digital humanities. *Mobile Information Systems*. Article ID 2652920, 9 pages. doi:10.1155/2022/2652920

Femenia-Serra, F., & Neuhofer, B. (2018). Smart tourism experiences: Conceptualisation, key dimensions and research agenda. *Journal of Regional Research*, *42*, 129–150.

Femenia-Serra, F., Neuhofer, B., & Ivars-Baidal, J. (2019). Towards a conceptualisation of smart tourists and their role within the smart destination scenario. *Service Industries Journal*, *39*(2), 109–133. doi:10.1080/02642069.2018.1508458

Gatelier, E., Ross, D., Phillips, L., & Suquet, J. B. (2022). A business model innovation methodology for implementing digital interpretation experiences in European cultural heritage attractions. *Journal of Heritage Tourism*, *17*(4), 391–408.

Gervasi, O., Perri, D., Simonetti, M., & Tasso, S. (2022). Strategies for the digitalization of cultural heritage. In *Computational science and its applications–ICCSA 2022 workshops: Malaga, Spain, July 4–7, 2022, proceedings, part VI* (pp. 486–502). Cham: Springer International Publishing.

Ghaderi, Z., Hatamifar, P., & Henderson, J. C. (2018). Destination selection by smart tourists: The case of Isfahan, Iran. *Asia Pacific Journal of Tourism Research*, *23*(4), 385–394. doi:10.1080/10941665.2018.1444650

Giffinger, R., Fertner, C., Kramar, H., Kalasek, R., Pichler-Milanović, N., & Meijers, E. (2007). *Smart cities: Ranking of European medium-sized cities, centre of regional science (SRF)*. Vienna: University of Technology.

Gomez-Oliva, A., Alvarado-Uribe, J., Parra-Meroño, M. C., & Jara, A. J. (2019). Transforming communication channels to the co-creation and diffusion of intangible heritage in smart tourism destination: Creation and testing in Ceutí (Spain). *Sustainability*, *11*(14), 3848.

González-Reverté, F. (2019). Building sustainable smart destinations: An approach based on the development of Spanish smart tourism plans. *Sustainability*, *11*(23), 6874.

Gretzel, U., Reino, S., Kopera, S., & Koo, C. (2015). Smart tourism challenges. *Journal of Tourism*, *16*(1), 41–47.

Gretzel, U., Sigala, M., Xiang, Z., & Koo, C. (2015). Smart tourism: Foundations and developments. *Electronic Markets*, *25*, 179–188.

Guo, Y., Liu, H., & Chai, Y. (2014). The embedding convergence of smart cities and tourism Internet of Thıngs in Chına: An advance perspective. *Advances in Hospitality and Tourism Research*, *2*(1), 54–69.

Guttentag, D. A. (2010). Virtual reality: Applications and implications for tourism. *Tourism Management*, *31*(5), 637–651.

Hajirasouli, A., Banihashemi, S., Kumarasuriyar, A., Talebi, S., & Tabadkani, A. (2021). Virtual reality-based digitisation for endangered heritage sites: Theoretical framework and application. *Journal of Cultural Heritage*, *49*, 140–151.

Harrison, C., Eckman, B., Hamilton, R., Hartswick, P., Kalagnanam, J., Paraszczak, J., & Williams, P. (2010). Foundations for smarter cities. *IBM Journal of Research and Development*, *54*(4), 1–16.

He, Y., & Tritto, A. (2022). Urban utopia or pipe dream? Examining Chinese-invested smart city development in Southeast Asia. *Third World Quarterly*, *43*(9), 2244–2268.

Hunter, W. C., Chung, N., Gretzel, U., & Koo, C. (2015). Constructivist research in smart tourism. *Asia Pacific Journal of Information Systems*, *25*(1), 105–120.

Ivars, J., Solsona, F. J., & Giner, D. (2016). Gestión turística y tecnología de la ınformación y la comunicación (TIC): El nuevo enfoque de los destinos ınteligentes. Tourism management and ınformation and communication technologies (ICTs): The new smart destinations approach. *Documents d'Analisi Geografica*, *6*, 327–346.

Jasrotia, A., & Gangotia, A. (2018). Smart cities to smart tourism destinations: A review paper. *Journal of Tourism Intelligence and Smartness*, *1*(1), 47–56.

Jin, P., & Liu, Y. (2022). Fluid space: Digitisation of cultural heritage and its media dissemination. *Telematics and Informatics Reports*, *8*, 100022.

Johnson, A.-G. (2022). Why are smart destinations not all technology-oriented? Examining the development of smart tourism initiatives based on path dependence. *Current Issues in Tourism*. doi:10.1080/13683500.2022.2053071

Jung, T. H., & tom Dieck, M. C. (2017). Augmented reality, virtual reality and 3D printing for the co-creation of value for the visitor experience at cultural heritage places. *Journal of Place Management and Development*, *10*(2), 140–151.

Katsoni, V., & Segarra-Oña, M. (2019). *Smart tourism as a driver for culture and sustainability*. Springer International Publishing.

Khan, M. S., Woo, M., Nam, K., & Chathoth, P. K. (2017). Smart city and smart tourism: A case of Dubai. *Sustainability*, *9*, 1–24.

Kim, S., Whitford, M., & Arcodia, C. (2019). Development of intangible cultural heritage as a sustainable tourism resource: The intangible cultural heritage practitioners' perspectives. *Journal of Heritage Tourism*, *14*(5–6), 422–435.

Koo, C., Gretzel, U., Hunter, W. C., & Chung, N. (2015). The role of IT in tourism. *Asia Pacific Journal of Information Systems*, *25*(1), 99–104.

Lee, C. S., Hwang, Y. K., & Jang, H. Y. (2018). Moderating effect of growth mindset on the relationship between attitude toward tourism and meaning in life. *International Journal of Pure and Applied Mathematics*, *120*(6), 5523–5540.

Li, R., Luo, T., & Zha, H. (2010). 3D digitization and its applications in cultural heritage. In M. Ioannides, D. Fellner, A. Georgopoulos, & D. G. Hadjimitsis (Eds.), *Digital heritage. EuroMed 2010. Lecture notes in computer science, 6436*. Berlin, Heidelberg: Springer. doi:10.1007/978-3-642-16873-4_29

Lim, V., Khan, S., & Picinali, L. (2021). Towards a more accessible cultural heritage: Challenges and opportunities in contextualisation using 3D sound narratives. *Applied Sciences*, *11*(8), 3336. doi:10.3390/app11083336

Mak, B. (2014). Archaeology of a digitization. *Journal of the Association for Information Science and Technology*, *65*(8), 1515–1526.

Navarrete, T. (2019). Digital heritage tourism: Innovations in museums. *World Leisure Journal*, *61*(3), 200–214.

Neamţu, C., Comes, R., & Popescu, D. (2016). Methodology to create digital and virtual 3D artefacts in archaeology. *Journal of Ancient History and Archaeology*, *3*(4), 65–74.

Neuhofer, B., Buhalis, D., & Ladkin, A. (2012). Conceptualising technology enhanced destination experiences. *Journal of Destination Marketing and Management*, *1*(1–2), 36–46.

Oke, A., Kozlu, H., & Çoruh, L. (2021). Use of augmented reality in the preservation of architectural heritage: Case of the aqueduct Kuru Kopru (Kayseri, Turkey). *International Journal of Architecture and Planning*, *9*(1), 110–130.

Pardo Abad, C. J. (2019). Application of digital techniques in ındustrial heritage areas and building efficient management models: Some case studies in Spain. *Applied Sciences*, *9*(20), 4420.

Perles, F., & ve Ivars, J. (2018). Smart sustainability: A new perspective in the sustainable tourism debate. *Journal of Regional Research*, *42*, 151–170.

Rodgers, A. P., & van Oers, R. (2011). Editorial: Bridging cultural heritage and sustainable development. *Journal of Cultural Heritage Management and Sustainable Development*, *1*(1), 5–14.

Roosevelt, C. H., Cobb, P., Moss, E., Olson, B. R., & Ünlüsoy, S. (2015). Excavation is destruction digitization: Advances in archaeological practice. *Journal of Field Archaeology*, *40*(3), 325–346.

Shih, N.-J., Diao, P.-H., & Chen, Y. (2019). ARTS, an AR tourism system, for the integration of 3D scanning and smartphone AR in cultural heritage tourism and pedagogy. *Sensors*, *19*(17), 3725. doi:10.3390/s19173725

Shott, M. (2014). Digitizing archaeology: A subtle revolution in analysis. *World Archaeology*, *46*(1), 1–9.

Silik, C. E., & Dündar, Y. (2020). Sürdürülebilirlik kapsamında akıllı turizm destinasyonları. In B. Ilgaz (Ed.), *Akıllı kent akıllı turizm* (pp. 117–137). Ankara: Gazi Kitapevi.

Sonuç, N., & İşçen, M. (2022). Digitalisation in the tourism and hospitality industry: Perspectives of the supply and demand sides. In *Optimizing digital solutions for hyper-personalization in tourism and hospitality* (pp. 40–63). Hershey, PA: IGI Global.

Tait, E., Laing, R., Grinnall, A., Burnett, S., & Isaacs, J. (2016). (Re)presenting heritage: Laser scanning and 3D visualisations for cultural resilience and community engagement. *Journal of Information Science*, *42*(3), 420–433. doi:10.1177/0165551516636306

Tariffi, F., Morganti, B., & Segbert, M. (2004). Digital cultural heritage projects in Europe: An overview of TRIS and the take-up trial projects. *Program*, *38*(1), 15–28.

Terras, M., Coleman, S., Drost, S., Elsden, C., Helgason, I., Lechelt, S., ... Speed, C. (2021). The value of mass-digitised cultural heritage content in creative contexts. *Big Data & Society*, *8*(1). 20539517211006165.

Thomaz, G. M., Biz, A. A., Bettoni, E. M., Mendes-Filho, L., & Buhalis, D. (2017). Content mining framework in social media: A FIFA world cup 2014 case analysis. *Information and Management*, *54*(6), 786–801. doi: 10.1016/j.im.2016.11.005

Tufail, M., Park, J., Kim, H., Kim, S., Kim, K. M., & Lee, J.-H. (2022). How do visitors perceive the significance of tangible cultural heritage through a 3D reconstructed immersive visual experience at the Seokguram Grotto, South Korea? *Journal of Heritage Tourism*, *17*(4), 409–430.

Vlizos, S., Sharamyeva, J. A., & Kotsopoulos, K. (2021). Interdisciplinary design of an educational applications development platform in a 3D environment focused on cultural heritage tourism. In M. Shehade & T. Stylianou-Lambert (Eds.), *Emerging technologies and the digital transformation of museums and heritage sites. RISE IMET 2021. Communications in computer and information science* (Vol. 1432). Cham: Springer. doi:10.1007/978-3-030-83647-4_6

Wang, X., Li, X. R., Zhen, F., & Zhang, J. (2016). How smart is your tourist attraction? Measuring tourist preferences of smart tourism attractions via A FCEM-AHP and IPA approach. *Tourism Management*, *54*, 309–320. doi:10.1016/j. tourman.2015.12.003

Yalçınkaya, P., Atay, L., & Karakaş, E. (2018). Akıllı turizm uygulamaları. *Gastroia: Journal of Gastronomy and Travel Research*, *2*(2), 34–52.

Yavuz, M. C. (2019). Akıllı destinasyon: Turizm, inovasyon, girişimcilik ve çözüm bekleyen konular. *Journal of Tourism Theory and Research*, *5*(2), 203–211.

Yeniasır, M., & Gökbulut, B. (2022). Effectiveness of usage of digital heritage in the sustainability of cultural tourism on islands: The case of northern Cyprus. *Sustainability*, *14*(6), 3621.

Yilmaz, H. M., Yakar, M., Gulec, S. A., & Dulgerler, O. N. (2007). Importance of digital close-range photogrammetry in documentation of cultural heritage. *Journal of Cultural Heritage*, *8*, 428–433.

Zatori, A., Smith, M. K., & Puczko, L. (2018). Experience-involvement, memorability and authenticity: The service provider's effect on tourist experience. *Tourism Management*, *67*, 111–126.

Zlot, R., Bosse, M., Greenop, K., Jarzab, Z., Juckes, E., & Roberts, J. (2014). Efficiently capturing large, complex cultural heritage sites with a handheld mobile 3D laser mapping system. *Journal of Cultural Heritage*, *15*(6), 670–678.

RECENT DEVELOPMENTS IN THE EVALUATION OF RENEWABLE ENERGY RESOURCES IN TOURISM BUSINESSES: A BIBLIOMETRIC ANALYSIS

Metin Sürme and Dilara Bahtiyar Sari

ABSTRACT

Energy use occupies an important place among the service activities offered to tourists that guide the tourism industry. The realization of basic needs such as heating, cooling, ventilation, lighting and decontamination in these enterprises are among the important factors that directly affect energy use. In order to obtain the energy needed for the sustainability of services at a more affordable cost, renewable energy sources should be put into operation. In this direction, it makes it more advantageous for businesses in the tourism sector to invest by turning to renewable energy sources in order to maintain their activities more economically. In this context, the main purpose of the study carried out in this part of the book is to reveal the latest developments in the field of evaluation of renewable energy sources in tourism enterprises. Bibliometric analysis was carried out by using the Web of Science (WoS) database in the research and with the findings obtained, it was concluded that the field is new and up-to-date and needs to be studied more. When looking at the WoS categories of studies titled renewable energy in tourism enterprises; it was concluded that more energy fuel, green sustainable science technology, science themes were given weight. According to the network analysis, the most cited authors and countries' densities were determined and the intensive expressions in the network keywords in their studies titled renewable energy in tourism enterprises are renewable energy, renewable energy sources, sustainable tourism, sustainability, carbon dioxide (CO_2) emissions, green marketing, blue economy, and energy efficiency.

Smart Cities for Sustainability
Advanced Series in Management, Volume 32, 169–178

ISSN: 1877-6361/doi:10.1108/S1877-636120230000033011

Keywords: Renewable energy sources in tourism enterprises; recent developments; bibliometric analysis; content analysis; tourism; renewable energy sources; sustainability

1. INTRODUCTION

The origin of the universe is energy. Light, mass, motion, heat, sound, chemical, nuclear, etc., are different forms of energy. Human beings have absolutely needed the consumption of energy-giving foods other than vitamins and protein in order to survive since the day they existed on the planet we live in. Man, who met his first energy needs thanks to his metabolism, started to search for the energy needed outside of his own power after a while as a result of social and technological development, and found the forces around him and benefited from the power of fire, wind, water, and domesticated animals. The differentiation of energy needs showed itself with the Industrial Revolution in Europe in the 18th century. Steam in the seventeenth century, coal in the eighteenth century, oil and electric power in the nineteenth century were used. In the first half of the twentieth century, these energy sources began to be used in all areas of life. Energy sources developed with the addition of nuclear energy to these resources have enabled the advancement of technology (Inan, Akbulut, & Aslan, 2018). Energy resources are generally classified according to their use and recyclability. According to the classification made for their use, energy sources are renewable and nonrenewable; they are classified as primary and secondary energy sources according to their recyclability (Koç & Kaya, 2015).

In today's world, energy is accepted as one of the most significant elements. In the labor-intensive tourism sector, energy expenses have a ratio of approximately 8%–12% in total expenses (Öztürk, Öztürk, & Dombaycı, 2018). The relationship between tourism and energy consumption takes place through several channels. Firstly, the tourism industry is a high energy demanding industry. Energy use in the tourism sector is demanded for hotels and transportation in the host country or for transportation between countries. The factor that increases the carbon dioxide (CO_2) emission, the most among tourism activities, is transportation due to the intense use of fossil fuels. In particular, air transport is thought to have the greatest impact on energy consumption. In addition, energy is demanded for food, heating, cooling, sea, and other travel transportation and other needs within the country. In summary, tourism activities can increase CO_2 emissions directly by using fossil fuels or indirectly by the consumption of electrical energy mostly produced from the use of coal, natural gas, and oil. For this reason, the increase in the number of tourists coming to the country may increase the energy demand and lead to an increase in pollution and greenhouse gas emissions (Arı, 2021). Considering this aspect, the importance of taking measures against environmental pollution, which is one of the biggest problems of the world, can be better understood. It is estimated that around 7 million deaths occurred in the world due to air pollution in 2012 alone. It is estimated that the increase in greenhouse gas emissions will be 52% by 2050 without effective measures (Khan, Yaseen, & Ali, 2019). In the light of this information, the importance of the use of renewable energy sources in tourism enterprises is understood more.

2. CONCEPTUAL FRAMEWORK

Sustainability of energy is supposed to be ensured in terms of sustainable development, natural environment, and continuity of tourism. Sustainability of energy, on the other hand, will be achieved by utilizing renewable energy sources that are unlimited in nature instead of scarce fossil fuels. In this context, the tourism sector, which hosts an international movement, and the accommodation enterprises, which are the most important organizations of this sector, will attract more attention than other sectors on a global scale by offering people the opportunity to experience these new energy applications (Sevinc, 2019). In addition to this, as an energy-intensive sector, there is a strong relationship between the use of renewable energy in terms of achieving higher tourism revenues and increasing competitiveness, which encourages renewable energy investments (Ben Jebli, Ben Youssef, & Apergis, 2019; Sofi, Bhat, Ahmad, Aara, & Aswani, 2022). When the relevant literature is examined, it shows that the use of renewable energy sources should be encouraged in order to ensure sustainability in tourism. Ben Jebli et al. (2019) conducted a research to determine the relationship between CO_2 emissions, economic growth, renewable energy consumption, foreign direct investment, and tourist arrivals. As a result of the research, there is evidence of bidirectional causality between renewable energy, tourism, foreign direct investment, and CO_2 emissions. Similar results were seen in the research results of Isik, Kasımati, and Ongan (2017). Katircioglu, Feridun, and Kilinc (2014) examined the relationship between international tourism, energy use, and environmental pollution. As a result of the research, it has been seen that tourism has effects on both energy use and carbon dioxide pollution. In a similar study, Solarin (2014) investigated macroeconomic variables and tourist arrivals as determinants of Malaysia's CO_2 emissions and found a long-term relationship between tourist arrivals and CO_2 emissions. These results reveal a strong relationship between the tourism sector and CO_2 emissions. However, Sevinç (2019) stated that any zero-energy hotel project similar to the project of the European Union for the application of zero-energy buildings in the hotel sector has not yet been realized in our country, and with the realization and proliferation of zero-energy hotels, technologies for the use of renewable energy sources in other sectors working in the accommodation sector. They stated that in this way, high-tech companies will emerge and a new employment area will be created, and environmental pollution can be reduced by reducing the use of fossil fuels at the same time.

Tourism businesses around the world are increasingly competing to become more effective and sustainable. However, it can be seen from the relevant literature that specific steps have not been taken regarding renewable energy management planning in many countries. For example: Calderón-Vargas, Asmat-Campos, and Chávez-Arroyo (2021), in their study, found that Peru has potential in relation to its renewable resources, but the legislation is not specific, less relevant to the tourism sector, and the legislation on this subject is very general. They have stated.

One of the biggest expenses of accommodation establishments is energy expenditures. When the relevant literature is examined, there are a limited number of studies examining the use of renewable energy in accommodation establishments (Dalton, Lockington, & Baldock, 2008; Dhirasasna & Sahin, 2021; Karagiorgas et al., 2006; Meschede, Dunkelberg, Stöhr, Peesel, & Hesselbach, 2017; Sardianou & Kostakis, 2020; Yenişehirlioğlu & Türkay, 2021). It is seen that green marketing practices are mostly the subject of studies within the scope of sustainable tourism (Gökdeniz, 2017; Hong, Kim, & Kim, 2003; Ibnou-Laaroussi, Rjoub, & Wong, 2020; Lütfi & Dilek, 2013; Meler & Ham, 2012; Seyhan & Yılmaz, 2010). On the other hand, it is seen that there has been an increase in the number of research articles on the general development of interest in renewable energy sources in recent years. As well as this, only 7.6% of these articles were about environmental impacts, whether positive or negative (Zhan et al., 2021). In the light of all this information, it is valuable for the future and sustainability of tourism to determine the status of the existing literature on the use of renewable energy resources in tourism enterprises and to see which areas are strong and/or weak. Because the results obtained from this study are in line with the existing literature on the use of renewable energy in tourism enterprises. It's like being photographed. It is thought that this information will lead to future studies.

3. METHOD

3.1 Research Model

A bibliography consists of publications and works related to a particular subject, place, and period. The purpose of the bibliography is to create documents and bibliography by making research and analysis based on choosing the best among these works. Bibliometrics is the numerical analysis of bibliographic features on a topic, of this context, the concept of bibliometry was first used in 1969 to numerically analyze the bibliographic information of works published as bibliographies the purpose of bibliometric analysis is to numerically evaluate the relationships between publications on a science or subject. Therefore, the bibliometric analysis method includes the analyzes made in order to determine the nature and development process of a discipline as a result of combining, interpreting, and analyzing all the resources (articles, books, and all related documents) and statistics available in the literature. Especially in recent years, bibliometric analysis has become an important part of research evaluation method and is widely used in scientific and applied fields. Researchers use this method to investigate the impact of their field in various fields, to determine the level of influence of a number of researchers, to identify the impact of a particular article or to reveal studies that are particularly effective in a certain period of time (Vishwakarma & Mukherjee, 2019).

The study was carried out according to the case study method, one of the qualitative research methods. A case study is a research method that provides an in-depth examination of a phenomenon or event. In this study, this method was preferred because it was desired to examine the articles published on desert tourism in terms of bibliometric parameters and to reveal the current situation. The research process was developed as follows (Fig. 10.1). The methodological part of the research was adapted from the bibliometric analysis on geoparks by Herrera-Franco, Montalván-Burbano, Carrión-Mero, Jaya-Montalvo, and Gurumendi-Noriega (2021).

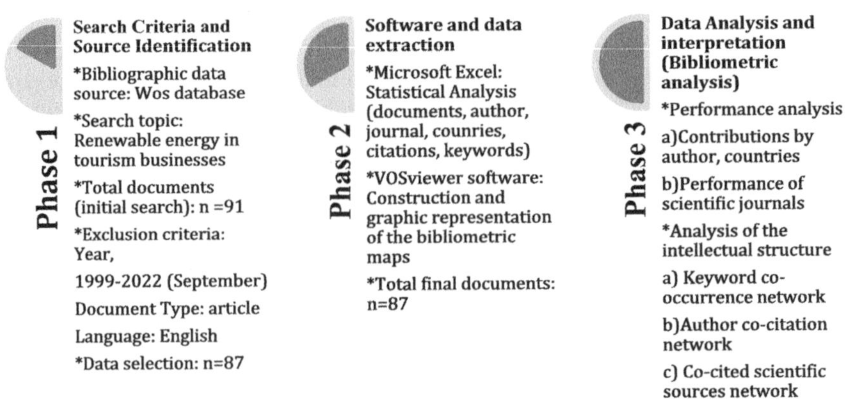

Fig. 10.1. Methodological Process of the Bibliometric Analysis.

3.2 Data Collection Process

The bibliometric data in this study was taken from the WoS database produced by Clarivate Analytics. The WoS database is one of the world's most crucial scientific citation search and analytical information platforms. This database offers comprehensive data content from different disciplines (Li, Rollins, & Yan, 2018). For this reason, the specified database was preferred for the study. The range of reviews of the literature covers the period from 1999 to September 9, 2022. The bibliometric analysis technique was used in the study. VOSviewer (Version 1.6.16) package program was used to create and display bibliometric maps for network analysis of journals. The program used is VOSviewer (Van Eck & Waltman, 2009) an open-source tool developed by Nees Jan van Eck and Ludo Waltman at Leiden University Science and Technology Research Center (CWTS) in the Netherlands (Akkaya & Ertekin, 2021).

3.3 Findings of the Research

In the light of the findings obtained, it was understood that the majority of academic studies published as document types were articles with a rate of approximately 63%. It has been determined that the number of studies titled "renewable energy in tourism enterprises" has increased according to the years. As a general evaluation by years, in the last five of the year when viewed; There were 9 studies in 2022, 14 studies in 2021, 6 studies in 2020, 6 studies in 2019 and 8 studies in 2018. Fig. 10.2 shows the number of studies by years. Due to the new and up-to-date nature of the subject area, it has been revealed that the studies have increased over the years. It is foreseen that the topics related to this field will increase in the future.

According to another finding obtained from the research, it was determined that the dominant language was English (97%). The top 10 countries contributing the most to the

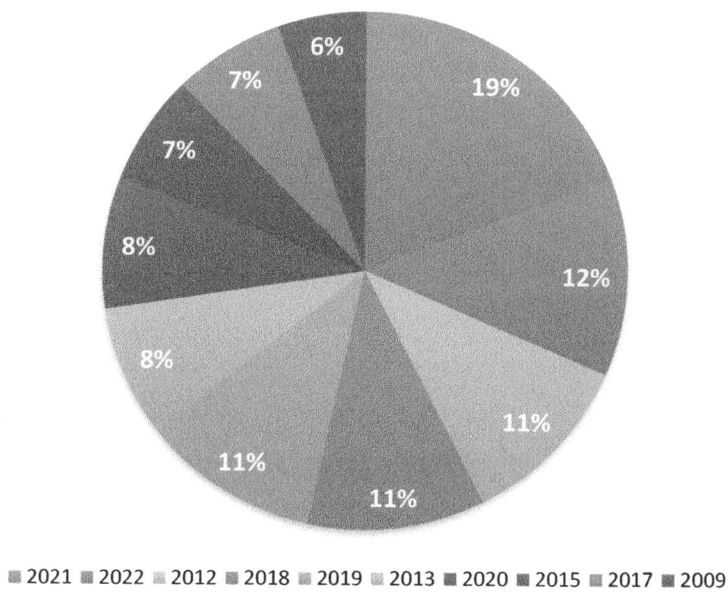

Fig. 10.2. Distribution of Studies on Renewable Energy in Tourism Businesses by Years.

field emerged as Peoples R China, Croatia, Italy, Germany, England, the United States, Greece, Spain, Australia, and Austria, respectively (Fig. 10.3).

Similarly, the top 10 researchers, editors, funding organizations, and universities contributing to the field were obtained.

In Table 10.1, in the distribution of studies on desert tourism by universities, the first three universities with the highest study density are University of Juraj Dobrila Pula, University of Rijeka, Lappeenranta University of Technology.

In Table 10.2, the first ten countries that contribute to the field of renewable energy in tourism enterprises are included. The top ten countries contributing the most to the field emerged as Italy, Germany, England, USA, Greece, Spain, Australia and Austria respectively.

As seen in Fig. 10.4, the intense expressions in the keyword network in the studies titled renewable energy in tourism enterprises; renewable energy, renewable energy sources, sustainable tourism, sustainability, CO_2 emissions, green marketing, blue economy, energy efficiency.

When looking at the WoS categories of the studies titled renewable energy in tourism enterprises; According to the network analysis, the densities of the most cited countries are determined and presented below.

In Fig. 10.5, the networks of countries working on renewable energy sources in tourism enterprises are included. As seen in the above network, the concentrated countries are Peoples R China, Germany, Croatia, Italy, and the United States.

4. CONCLUSION

At the beginning of the inevitable principles of the concept of sustainability, there are processes such as increasing the share of renewable energy resources, conducting and managing energy efficiency and saving studies. Katırcıoğlu (2014), in many researches in the field of energy economy, tried to determine the connection between energy,

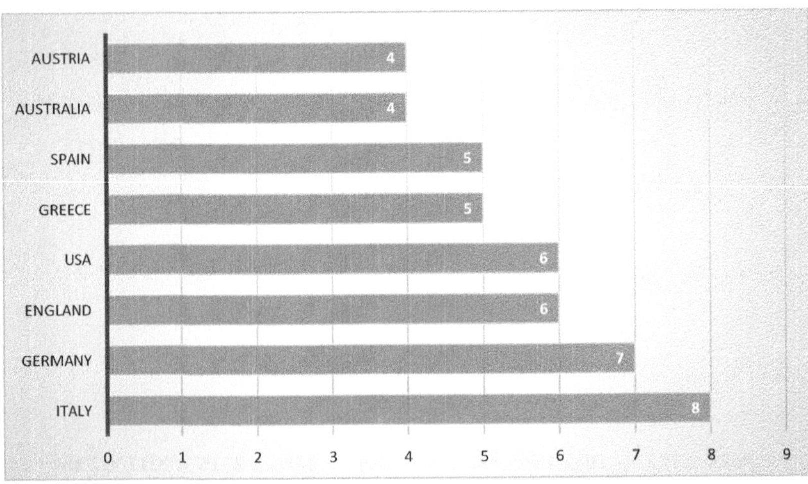

Fig. 10.3. Top 10 Countries Contributing to the Field of Renewable Energy in Tourism Businesses.

Table 10.1. Top 10 Institutions That Made the Most Scientific Contribution to the Field of Renewable Energy in Tourism Businesses.

Affiliations	Record Count	% of 87
University of Juraj Dobrila Pula	3	3.448
University of Rijeka	3	3.448
Lappeenranta University of Technology	2	2.299
National Chung Hsing University	2	2.299
Universidad Privada Del Norte	2	2.299
University College London	2	2.299
University of London	2	2.299
University of Oradea	2	2.299
University of Southern Queensland	2	2.299
2C Ecosolutions	1	1.149

environmental pollution, and economic growth, and according to the literature formed as a result of these researches; that climate changes are among the important factors in the pollution and deterioration of the environment; the main factor causing climate change is CO_2 emissions; that certain segments of the economy and some sectors need to be more interested in the relationship between energy and climate change, and that one of these

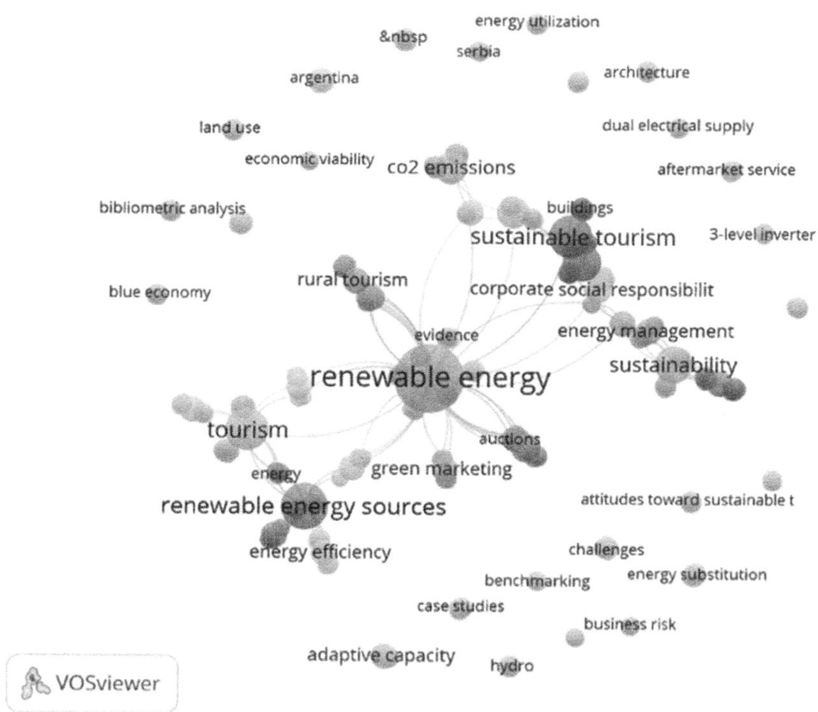

Fig. 10.4. Keyword Network on Renewable Energy in Tourism Businesses.

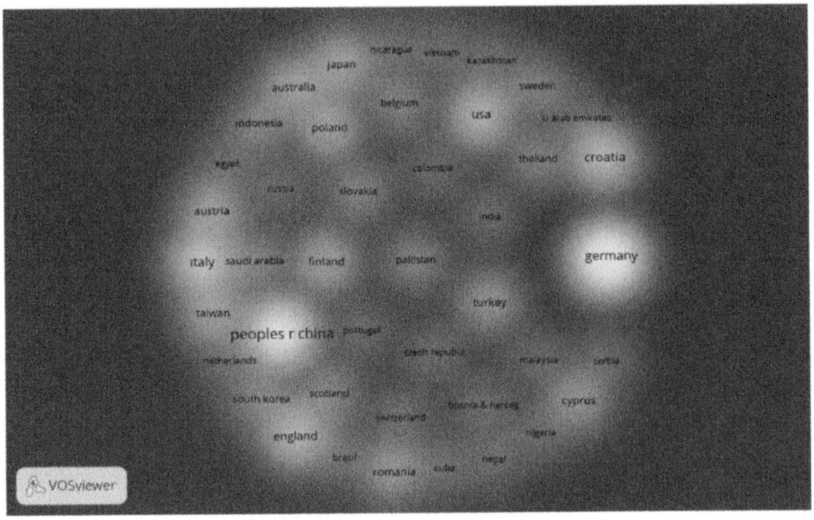

Fig. 10.5. Country Level Network for the Citation of the Studies on Renewable Energy in Tourism Businesses.

sectors is the tourism sector; draws attention to the fact that although international tourism movements and the increase in the number of international tourists contribute to the economic development of countries, they cause an increase in energy consumption.

Becken (2013) states that the current climate changes cause concern in the tourism sector and that the stakeholders in the sector set targets to reduce greenhouse gas emissions, and in this direction, the World Travel and Tourism Council's greenhouse gas emissions will be 25%–30% by 2020; It is working to reduce it by 50% until 2035; Although energy expenses seem to be insignificant with a rate of 3%–6% among the operating costs of the accommodation sector, they are still the second most important cost factor after labor costs; the energy used in the accommodation sector is mostly used in hot water, cooling, and lighting, and the energy used in these is electricity, gas, and diesel, which are secondary energy sources that are shown as the main source of CO_2 emissions; states that these energy sources cause climate changes.

According to research on energy; currently, most of the energy consumed in the world is met from fossil fuel resources and these resources are decreasing day by day. Published reports show that billions of tons of carbon dioxide and millions of tons of sulfur compounds, lead and toxic chemicals are released into the atmosphere as a result of industrial activities. The use of fossil fuels as an energy source causes countries that do not have these resources to become foreign-dependent on fuel and the economic values they produce are exported due to energy imports. This situation causes problems in terms of economic development of countries that are poor in terms of energy resources. When we evaluate tourism activities in terms of the use of energy resources, although the tourism activities experienced today make great contributions to the economies of the countries, it is understood that various functions (transportation, accommodation, catering, etc.) used in the fulfillment of tourism activities cause an increase in energy consumption. As a result of this research, it has been determined that the number of studies on renewable energy in tourism enterprises has increased over the years. Most of the academic studies are in the form of articles (63%), the dominant language is English (97%), and the top 10 countries contributing the most to the field are Peoples R China, Croatia, Italy, Germany, England, USA, Greece, Spain respectively Australia, Austria originated. Similarly, the top 10 researchers, editors, funding organizations, and universities contributing to the field were obtained. When looking at the WoS categories of the studies titled renewable energy in tourism enterprises; more energy fuels, green sustainable science technology. It can be said that he concentrates on sciences themes. According to the network analysis, by determining the densities of the most cited authors and countries, the intense expressions in the keyword network in their studies titled renewable energy in tourism enterprises; renewable energy, renewable energy sources, sustainable tourism, sustainability, CO_2 emissions, green marketing, blue economy, energy efficiency.

As a limitation of the study, the findings obtained could not be compared or discussed with the findings of other studies due to the insufficient and few published studies in the field of study.

The actual study is a bibliometric analysis study. For future studies, the subject of the study can be searched with different methods, or a similar study can be done in other databases with the same method, except the WoS database. The obtained results can be compared with the findings of this study.

REFERENCES

Akkaya, G., & Ertekin, P. (2021). İki Kere Farklı Bireylere Yönelik Literatürün Görsel Olarak İncelenmesi: Bibliyometrik Bir Calısma. *Pamukkale Üniversitesi Eğitim Fakültesi Dergisi, 52*, 492–517. doi:10.9779/pauefd.706012

Arı, A. (2021). Yenilenebilir enerji, turizm, CO_2 ve GSYH ilişkisinin Türkiye için analizi. *Akademik Yaklaşımlar Dergisi, 12*(2), 192–205.

Becken, S. (2013). Operators' perceptions of energy use and actual saving opportunities for tourism accommodation. *Asia Pacific Journal of Tourism Research, 18*(1–2), 72–91.

Ben Jebli, M., Ben Youssef, S., & Apergis, N. (2019). The dynamic linkage between renewable energy, tourism, CO_2 emissions, economic growth, foreign direct investment, and trade. *Latin American Economic Review, 28*(1), 1–19.

Calderón-Vargas, F., Asmat-Campos, D., & Chávez-Arroyo, P. (2021). Sustainable tourism policies in Peru and their link with renewable energy: Analysis in the main museums of the Moche route. *Heliyon, 7*(10), e08188.

Dalton, G. J., Lockington, D. A., & Baldock, T. E. (2008). Feasibility analysis of stand-alone renewable energy supply options for a large hotel. *Renewable Energy, 33*(7), 1475–1490.

Dhirasasna, N., & Sahin, O. (2021). A system dynamics model for renewable energy technology adoption of the hotel sector. *Renewable Energy, 163*, 1994–2007.

Gökdeniz, A. (2017). Konaklama sektöründe yeşil yönetim kavramı, eko etiket ve yeşil yönetim sertifikaları ve otellerde yeşil yönetim uygulama örnekleri. *International Journal of Social and Economic Sciences, 7*(2), 70–77.

Herrera-Franco, G., Montalván-Burbano, N., Carrión-Mero, P., Jaya-Montalvo, M., & Gurumendi-Noriega, M. (2021). Worldwide research on geoparks through bibliometric analysis. *Sustainability, 13*, 1175, 1–32.

Hong, S. K., Kim, S. I., & Kim, J. H. (2003). Implications of potential green tourism development. *Annals of Tourism Research, 30*(2), 323–341.

Ibnou-Laaroussi, S., Rjoub, H., & Wong, W. K. (2020). Sustainability of green tourism among international tourists and its influence on the achievement of green environment: Evidence from North Cyprus. *Sustainability, 12*(14), 5698.

İnan, İ, Akbulut, İ., & Aslan, E. (2018). Enerji sorununun çözümünde yenilenemez ve yenilenebilir enerji kaynaklarının yeri ve önemi. *Türk Dünyası Araştırmaları, 120*(237), 11–40.

Işik, C., Kasımatı, E., & Ongan, S. (2017). Analyzing the causalities between economic growth, financial development, international trade, tourism expenditure and/on the CO_2 emissions in Greece. *Energy Sources, Part B: Economics, Planning and Policy, 12*(7), 665–673.

Karagiorgas, M., Tsoutsos, T., Drosou, V., Pouffary, S., Pagano, T., Lara, G. L., & Mendes, J. M. M. (2006). HOTRES: Renewable energies in the hotels. An extensive technical tool for the hotel industry. *Renewable and Sustainable Energy Reviews*, *10*(3), 198–224.

Katırcıoglu, S. T. (2014). International tourism, energy consumption, and environmental pollution: The case of Turkey. *Renewable and Sustainable Energy Reviews*, *36*(2014), 180–187.

Katircioglu, S. T., Feridun, M., & Kilinc, C. (2014). Estimating tourism-induced energy consumption and CO_2 emissions: The case of Cyprus. *Renewable and Sustainable Energy Reviews*, *29*, 634–640.

Khan, M. T. I., Yaseen, M. R., & Ali, Q. (2019). Nexus between financial development, tourism, renewable energy, and greenhouse gas emission in high-income countries: A continent-wise analysis. *Energy Economics*, *83*, 293–310.

Koç, E., & Kaya, K. (2015). Enerji kaynakları–yenilenebilir enerji durumu. *Mühendis ve Makine Dergisi*, *56*(668), 36–47.

Li, K., Rollinsve, J., & Yan, E. (2018). Web of science use in published research and review papers 1997–2017: A selective, dynamic, cross-domain, content-based analysis. *Scientometrics*, *115*, 1–20.

Lütfi, A., & Dilek, S. E. (2013). Konaklama işletmelerinde yeşil pazarlama uygulamalari: Ibis otel örneği. *Süleyman Demirel Üniversitesi İktisadi ve İdari Bilimler Fakültesi Dergisi*, *18*(1), 203–219.

Meler, M., & Ham, M. (2012). Green marketing for green tourism. In *Faculty of tourism and hospitality management in Opatija. Biennial international congress tourism & hospitality industry* (p. 130). University of Rijeka, Faculty of Tourism & Hospitality Management.

Meschede, H., Dunkelberg, H., Stöhr, F., Peesel, R. H., & Hesselbach, J. (2017). Assessment of probabilistic distributed factors influencing renewable energy supply for hotels using Monte-Carlo methods. *Energy*, *128*, 86–100.

Öztürk, H. K., Öztürk, H. M., & Dombaycı, Ö. A. (2018). Turizm sektöründe enerji tüketimi ve enerji tasarruf olanakları. *Güncel Turizm Araştırmaları Dergisi*, *2*(1), 17–28.

Sardianou, E., & Kostakis, I. E. (2020). Perceived barriers to invest in renewable energy sources in the Cretan hotel industry. *International Journal of Sustainable Energy*, *39*(3), 240–249.

Sevinç, Y. (2019). *Turizm mühendisliğinde enerji yönetimi: bir saha çalışması*. Sakarya Üniversitesi Fen Bilimleri Enstitüsü. Yüksek Lisans Tezi.

Seyhan, G., & Yilmaz, B. S. (2010). Sürdürülebilir turizm kapsamında konaklama işletmelerinde yeşil pazarlama: Calısta luxury resort hotel. *Dokuz Eylül Üniversitesi İşletme Fakültesi Dergisi*, *11*(1), 51–74.

Sofi, A. A., Bhat, M. Y., Ahmad, L., Aara, R. R., & Aswani, R. S. (2022). Renewable energy and transitioning towards sustainable tourism: Inferences from kernel density and nonparametric approach. *Renewable Energy*, *193*, 963–975.

Solarin, S. A. (2014). Tourist arrivals and macroeconomic determinants of CO_2 emissions in Malaysia. *Anatolia*, *25*(2), 228–241.

Van Eck, N. J., & Waltman, L. (2009). How to normalize cooccurrence data? An analysis of some well-known similarity measures. *Journal of the American Society for Information Science and Technology*, *60*(8), 1635–1651.

Vishwakarma, P., & Mukherjee, S. (2019). Forty-three years journey of tourism recreation research: A bibliometric analysis. *Tourism Recreation Research*, *44*(4), 1–16. doi:10.1080/02508281.2019.1608066

Yenişehirlioğlu, E., & Türkay, O. (2021). Yenilenebilir Enerji Teknolojilerinin Konaklama İşletmelerinde Kullanımı ve Turizm Arzındaki Paradigma Kırılması: Avusturya Stadhale Hotel İncelemesi. *Türk Turizm Araştırmaları Dergisi*, *5*(1), 623–641.

Zhan, Z., Ali, L., Sarwat, S., Godil, D. I., Dinca, G., & Anser, M. K. (2021). A step towards environmental mitigation: Do tourism, renewable energy and institutions really matter? A QARDL approach. *Science of the Total Environment*, *778*, 146209.

URBAN 6.0 AND UTOPIA CONCEPTS VIA SUSTAINABLE, CLEAN, INCLUSIVE, INNOVATIVE, AND U-MOBILITY THROUGH THE THEORY OF COMPREHENSIVE EVERYTHING

Hamid Doost Mohammadian

ABSTRACT

Nowadays, sustainable, clean, inclusive, innovative, and smart mobility in addition to urban transformation is required to achieve sustainable development as a path to preserve the world for future concerns and improve quality of life at the present, even to be kept up with growing citizens' needs. Mobility as an infrastructure component plays fundamental roles in urban transformation, and economic development. In this chapter, and based on the 5th wave theory, related theories, models, and concepts, modern, clean, and inclusive mobility founded on high future of 4th technologies (which is called 5th technologies), digitalization, smartness, sustainability, and CSR 2.0 strategies is declared as proper clean mobility technologies to create sustainable and smart cities. Such smart cities are able to deal with challenges made by rapid, unplanned urbanization and globalization to achieve sustainable development. In this research, roles of inclusive and smart mobility systems as path to create modern and sustainable urban areas to make the world more sustainable and livable for living are declared. Literature reviews, case studies, interviews, and questionaries are applied as main methods to recognize inclusive and modern mobility and its roles in urban transformation to achieve sustainable development. This chapter is based on know-how and do-how of the author Prof. Hamid Doost on sustainability such as cooperating with Danish Sustainable Platforms Company, working with Erasmus Plus as an academic leader in Germany since 2017, cooperating with Copenhagen's former mayor and researching on sustainability. In this chapter, impact of sustainable mobility, sustainable buildings, and smart cities on CSR 2.0 and social responsibility, how these parameters improve sustainable development and sustainability in social responsibility, corporate social responsibility (CSR), and how social responsibility could influence humanities are explored.

Smart Cities for Sustainability
Advanced Series in Management, Volume 32, 179–209

ISSN: 1877-6361/doi:10.1108/S1877-636120230000033012

Keywords: Mobility; i-sustainability plus theory; sustainability and sustainable development; modern cities; high technologies; ubiquitous (U); smartness; the 5th wave/tomorrow age theory; Urban 6.0; sustainable; clean; inclusive; innovative and smart mobility

1. INTRODUCTION

Today, sustainable development of cities is the most important issue in urban planning. Information and communication technologies (ICTs) have brought tremendous changes and developments by entering different urban areas. In other words, the movement of the world's cities toward high and clean technologies has many advantages; one of the most important of these advantages is sustainable smart mobility, because it can make a major contribution to the sustainability of cities. For example, the intelligent transportation system tries to reduce the volume of traffic and the duration of the trip by using and maximizing the productivity of the growth and development of technology, and as a result, reducing fuel consumption and air pollution, which ultimately moves toward sustainable development in the long run. One of the most basic infrastructures required for the development and increase of the level of social and environmental welfare of any country is the existence of sustainable smart mobility in that country. Cities inherently face complex and extensive challenges that can only be solved through a systematic approach. As a result, urban planners around the world are trying to develop models based on the fifth wave theory for the development of twenty-first century cities in order to respond to the new demands and expectations of today's world, with an integrated view of all dimensions of urbanization, so that urban communities are ready to face the challenges, in order to reduce the limitations and effects of global shocks, to survive for future generations, and to improve livability and quality of life. Therefore, one of the new concepts to deal with the current challenges of cities is the development of a smart city so that, along with a sustainable smart mobility system, it can have the most efficiency and serve the citizens, which is also proposed as the main axis of development. It can be said that smart mobility is one of the areas that can be affected by information technology and urban smarting processes, and its results can be predicted, forecasted, and then planned in the path of urban sustainability (Report of the World Commission on Environment and Development: Our Common Future, 1987; Valcárcel-Aguiar, Murias, & Rodríguez-González, 2019).

2. LITERATURE REVIEW/BACKGROUND

2.1 Mobility

Smart mobility, as one of the dimensions of the smart city, provides benefits for the environment, and it focuses on new projects and initiatives and sustainable transportation systems, to support urban traffic and city mobility, using an integrated ICT infrastructure. It can be said that sustainable mobility is an efficient form of transportation that greatly reduces polluting energy. And the cities that have sustainable mobility are classified as sustainable smart cities. Some examples of smart mobility services include providing real-time public and multipurpose transportation information, optimizing traffic lights to attend to traffic demand in real time (Ardnt et al., 2014; Benevolo, Dameri, & D'Auria, 2016; Finnveden & Gunnarsson-Ostling, 2016; Gillis, Semanjski, & Lauwers, 2015; Jeekel, 2017; Kayal, Singh, & Kumar, 2014; Narang, 2007).

2.2 Sustainability and Sustainable Development

Sustainability in its broadest sense refers to the ability of society, ecosystem, or any current system to continue functioning in the unlimited future, without being forced to weaken as a result of the analysis of the resources that the system depends on or due to excessive load on them. On the other hand, the ability of the system for endurance and reliability is inevitably dependent on the success that the system gains in establishing a connection with the external environment. In other words, the stability of the system is completely dependent on the ability of the system to adapt, change, and respond to the environment, and since the environment is always changing, this system adaptation process must be a dynamic and sensitive process. In fact, an example that can be talked about as one of the dimensions of sustainability is the environment. This issue is in complete contradiction with the views of the past two centuries of man. In the past two centuries, the environment has been considered as an external and separate issue to humans. In this view, the relationship between humans and the environment can be seen in the belief that humans must live by completely conquering nature. It has been believed that man should overcome all the environmental obstacles with his thinking and knowledge and rule the nature for himself. Also, this view continued with influences on the development of capitalism and the industrial revolution of the fourth generation. Also, the concept of sustainable development reflects the undeniable fact that ecological considerations can and should be used in economic activities. These considerations include the ideas of creating a logical environment in which the claim of development in order to promote the quality of all aspects of life is challenged (Barbier & Burgess, 2017; Basiago, 1998; Our Common Future Brundtland Report, 1987).

To help and accelerate the creation of sustainable smart cities, strengthening peace and sustainable development is a priority. Therefore, based on the seven-pillar sustainability model (7Ps) of Professor Dr. Dost Mohammadian, innovative sustainable strategies by moving toward cultural, educational, technological, political sustainability in addition to environmental, social, economic dimensions, along with strengthening peace and love are the way forward for this key idea (Feil & Schreiber, 2017; Kuhlman & Farrington, 2010; Klarin, 2018) (Fig. 11.1).

2.3 Livability and Quality of Life

Livability and quality of life are the main indicators concerned on satisfaction of humans about their lives.

Quality of life is not a new concept. The history of this concept should be seen in Greek philosophy. Aristotle has mentioned it in the discussion of happiness, and philosophers such as Jean-Paul Sartre and others have paid attention to it in the new era. As Javanahaj (1994), one of the theoreticians of this field, with emphasis on the character and individual and social identity of human beings and individual and social values, philosophically considers the quality of life to be the result of the combination of utilitarianism and existentialism. Quality of life has been conceptualized in different ways. Some researchers believe that this concept is a set of psychological, social, and service needs of life. Some consider it as a feeling that a person has toward his well-being. Therefore, it can be said that a set of social, psychological and physical, cultural, environmental, and economic aspects are in a subset of quality of life.

Livability is a general concept that is related to a number of other concepts and terms, such as sustainability, quality of life and quality of place, and healthy communities. Although the definition of livability is different from one community to another, social

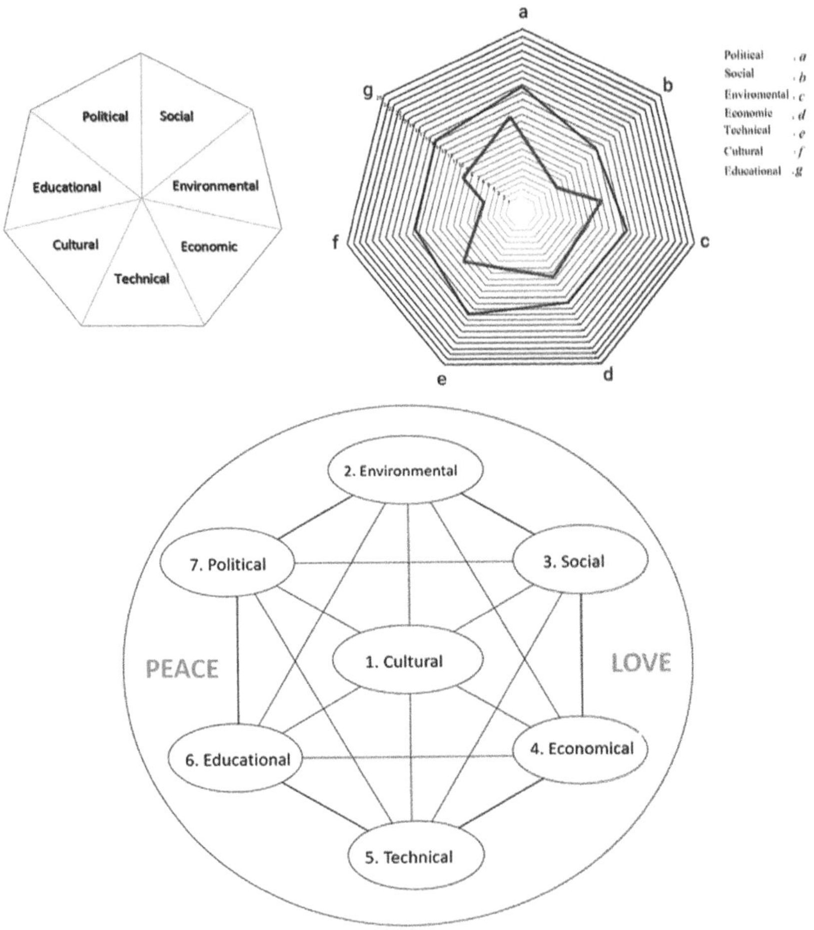

Fig. 11.1. 7Ps Model With the Pillars' Priority, Connections, and PEACE/LOVE [Doost. H, 2017] (Author).

planning goals can be used to create indigenous criteria of livability and are often used to define different dimensions of the community and the shared experiences that shape it. It can be said that livability focuses on the human experience of the place and considers these experiences in a specific time and place. Livability as a concept can be very broad or limited according to the context in which it is defined. Nevertheless, the quality of life in any place is in the center of attention of this concept and includes many different measurable indicators, which usually include density, transportation, security, and sustainability (Berger, Feindt, Holden, & Rubic, 2014; Bullock, Brereton, O'Neill, Clinch, & Russell, 2008; Giap, Thye, & Grace, 2014; Ruen, 2011; Zavrl & Zeren, 2010).

2.4 Smart and Sustainable Infrastructure

Infrastructure of each city plays significant roles in creating sustainable and smart cities that are required to supply growing and modern citizens' needs and sustainable development (Ogie, Perez, & Dignum, 2017; What is Sustainable Infrastructure, 2018). In particular, different services such as transportation, supply of drinking water and electricity, the disposal and treatment of waste water, the mobility of people and goods, the provision of ICTs, industries, businesses, and citizens play roles in developing proper infrastructure (Smart Cities and Infrastructure, 2016). Sustainable and smart technical infrastructure such energy, transportation, water supply, information networks, and sewage are needed to make modern cities that are able to make high quality of livability and life for their citizens (Ogie et al., 2017).

Sustainable infrastructure is a kind of infrastructure that is planned, designed, constructed, operated, and decommissioned in a path to align with economic and financial, social, environmental, climate change, and generally sustainability (Smart Cities and Infrastructure, 2016). In other words, sustainable infrastructure is a tool to optimize the environmental, social, and economic impact.

These days, it is needed to create smart infrastructure to supply citizens' need, develop sustainability, and create sustainable urban areas with high quality of life (Al-Hader & Rodzi, 2009). Smart infrastructure has multiple actors such as smart citizen, smart economy, smart building and construction, smart mobility, smart governance, smart life, and smart environment. The key idea of these components is the optimal use of energy and resources and ultimately progress toward sustainability and performance effectiveness (Aralu, 2015; Idowu & Esere, 2013).

Generally, sustainable and smart infrastructure could create modern urban areas based on sustainability, quality of life, and livability toward sustainable development.

2.5 Technologies

As it was mentioned, high technologies, digitalization, smartness, and ubiquitous are required to achieve smart services in cities toward sustainability. The main technologies that could be applied to make smart infrastructure are mentioned in this sector.

2.5.1 Technologies Based on Internet and Information

One of modern technologies are internet and information-based technologies such as ICT, Internet of Things (IoT), and Internet of Energy (IoE).

Information Communication Technology (ICT)

In the last 50 years, the occurrence of extensive developments in the field of computers and communications has led to major changes in different areas of human life. Humans have always used technology, and the history of human life is full of ICTs that are known as new or excellent technologies and have had the greatest impact on human life. The world of communication and information production is rapidly changing, and today, we have witnessed their convergence with each other more than in the past, in such a way that data and information are transferred to all parts of the world quickly and in an unimaginable time and are available to users. But in the definition of ICT, it can be said, technology is the collection, organization, storage, and dissemination of information, including sound, image, text, or number, which is done using computer and telecommunication tools. Regardless of the various definitions and the wide range of application of ICT in different parts of human life, quick access to information and doing things regardless of geographical distances and regardless of time limits is the most central achievement of this technology. Reliable and accessible communication can be used efficiently as a part of the tool for raising global problems. ICTs alone may not feed the hungry, eradicate hunger, or reduce child mortality, but they are increasingly important factors that drive economic growth and social equality. It enables efficient production, diversification, and distribution of agricultural products and makes it possible to provide basic health services to people in need in areas deprived of health facilities. It also allows educators to bring their knowledge to the most remote parts of the planet. Access to information technology can foster and assist the creation of small enterprises and clusters in the world's poorest and most remote regions, linking the bulk of domestic and global markets. Information technology has made it possible for poor infrastructures to jump so that distance from markets is no longer a cause of loss, and inefficient distribution channels will disappear and become a thing of the past. It should be noted that ICT is the beating heart of progress and development, and we should make it the center of all development strategies in the improvement of all aspects of society, including economic, social, and cultural (Hussein, 2019; Information and Communication Technology Development Indices, 2003; Nureni; Patel, Patel, Scholar, & Salazar, 2016; Ratheeswar, 2018).

Internet of Things (IoT)

Today's world consists of a set of physical objects, including machines, infrastructures, and devices that are equipped with a network of sensors and intelligent agents. This enables them to perform activities such as monitoring the environment, reporting the situation, receiving instructions, and even showing feedback based on the collected information. This is a concept that is referred to as the IoT today.

The IoE is a new generation of energy with more power, greater compatibility with the environment, and also has the most features of renewable energy, which plays an essential role in social, economic development, and other things. Connectivity anywhere, anytime, on any network with anything is the main goal of the IoT, which can be applied to various businesses and processes to improve their services and performance (Sarhan, 2018).

Applying IoT in different businesses and services has many benefits such as: using the IoT in various businesses and services has many advantages, including saving energy, saving costs, reducing time, improving smart services, and all the benefits. There are challenges such as data management, high cost for specific components, interoperability, security challenges, data management, technical challenges such as appropriate technology infrastructure, sensors, etc., and educational barriers to train knowledgeable, expert, and intellectual capital, which make it difficult to use the IoT. Therefore, it is necessary to deal with these challenges in order to use the IoT and benefit from its benefits (Kafle, Mahmud, Morsalin, & Town, 2012; Kranenburg & Bassi, 2019; Mahmoud, Yousuf, Aloul, & Zualkernan, 2015).

In the era of advanced technologies such as the IoT, the possibilities to improve the ways of life and human interaction in cities are endless. Traditional urban areas are transformed into modern areas through smart jobs, smart infrastructure and mobility, and overall smart living (Clemente, Ottorino, & Dario, 2018; Sarhan, 2018).

Internet of Energy (IoE)

IoE as a modern kind of ICT is able to achieve changes in sources, loads, developing clean energy, renewable energy, plugging in electric vehicles, etc., toward improving processes, performances, and services (High Technology Indicators, 2016). The IoE is a concept that combines the IoT and energy management with the aim of distributing and sharing energy. All types of energy including renewable and nonrenewable energy and all energy consumers will interact through IOT. Based on this approach, energy providers and consumers will have a two-way interaction in real time so that production adjustments, consumption savings, and optimal distribution are possible. The purpose of the IoE is to improve the use of energy by emphasizing the use of renewable energies and the correct distribution of energy to meet all the needs of energy sources (Ahmed, 2018; Artificial Intelligence and Big Data: Innovation Landscape Brief, 2019; Artificial Intelligence and Machine Learning 101, 2019; Kersting, 2018; Kuhl, Goutier, Hirt, & Satzger, 2019).

2.5.2 High Technologies

Technology is a tool needed to produce new products and processes. High technology is a type of technology founded on rapid renovation of knowledge, very superior to other technologies and due to its degree of complexity that requires continued effort in research and a solid technological base (Machine Learning, 2017). In order to compete and continue to survive in complex global conditions, cities are forced to become intelligent and provide infrastructure to use high technologies. One of the most effective ways to achieve technologies is advanced knowledge and research and development, which provides the path for sustainability (Perez, Deligianni, Ravi, & Yung; Craig, 2005; Mosavi & Varkonyi, 2017). Machine learning (ML) and big data are main types of high technologies.

Machine Learning (ML)

ML is an important research field in artificial intelligence (AI), which enables computers to model based on experiences and accurately predict future events. Also, as pillars of information technology, it is another type of advanced technology that has an important role and fundamental importance in human life. The main approaches to ML can be classified into two main categories: supervised learning and unsupervised learning. A typical problem in supervised learning is classification, while unsupervised learning is relatively common in clustering problems. Common classification methods include neural networks, support vector machines, and decision trees, and the most widely used clustering method is K-means. ML methods have been widely and successfully used in many different fields such as health, education, wireless sensor networks, finance and international relations, manufacturing, etc. Recently, semi-supervised learning is another method of ML approach that learns knowledge through labeled and unlabeled data (Alexandru, Alexandru, Coardos, & Tudora, 2016; Etzion & Aragon-Correa, 2016; Gorostidi, 2014; Mukred & Jianguo, 2017; Thabet & Soomro, 2015; Victor & Fekete Farkas, 2018).

Big Data

Big data is introduced in 2011 by International Data Corporation as "a new generation of technologies and architectures, designed to economically extract value from very large volumes of a wide variety of data, by enabling high-velocity capture, discovery, and/or analysis" (Alter, 2019). Big data and its application could solve challenges and problems of organization in different fields (Digital Sustainability; Tregua, D'Auria, & Bifulco, 2014). Big data is about collection of data sources, technologies, and methodologies applied to develop data creation as a path to deal with organizational challenges and problems (Digital Sustainability).

The main characters of big data are variety, velocity, volume, veracity, variability, visualization, and value (Alter, 2019; Sustainable and Digital Sustainable Development as the Framework for Digital Transformation, 2018).

Big data is a new era confront with challenges and barriers to be applied. Therefore, it needs to be discovered and explored more in order to apply in efficient and productive path. Data integration complexity, architecting big data system, lack of qualified and knowledgeable capitals, lack of business sponsorship, data security, privacy and governance, data ownership and other political issues, real time data, poor data quality, immaturity of new data types and sources, inadequate infrastructure, high cost, lack of metadata and scheme for big data, managerial challenges, and cultural issues are main challenges related to apply big data (Seele & Lock, 2017; Sustainable and Digital Sustainable Development as the Framework for Digital Transformation, 2018). Big data is too big, too fast, and modern; therefore, it needs special sophisticated systems to be able to deal with its challenges.

Big data is a digital revolution that could not only influence on businesses performances but also on sustainable development.

Modern technologies such as AI, ML, robotic, ICT, information technology (IT), IoT, big data, and so on could be utilized as techniques to develop digitalization, smartness, and ubiquitous as techniques to make sustainable and smart urban areas. In particular, digitalization is a concept applied in western countries and in eastern ones ubiquitous are popular as new technology applied to make services aligned with rapid urbanization and modern demands.

2.6 Smartness, Digitalization, and Ubiquitous (U)

Smartness is linked to competitiveness and sustainability. The smart city is the basis for the formation and expansion of the future city. Also, smart cities are of fundamental use in order to facilitate development through infrastructures and new ICTs, to improve the quality of development based on the preservation and promotion of human and social capital, and, subsequently, to solve problems and challenges facing development. They provide a more peaceful environment by providing new and innovative solutions. Smart growth as a priority and smart city as a destination or expected model compared to the current state of the city, along with two other main priorities, i.e., inclusive growth and sustainable growth, are considered. Therefore, the intelligent vision of urban development and subsequently smart cities, as a response to the pervasive challenges in the field of urbanization, is the destination (Ho Lee, Hoon Han, Taik Leem, & Yigitcanlar, 2008; Mühlhäuse & Gurevych, 2008; Quality of city life multiple criteria analysis - Scientific Figure on ResearchGate; Sassi, 2016; Sen, 2012).

Digitalization is an idea in eastern world and ubiquitous applied in western countries. Both of them are about applying modern technologies to demand citizen needs and deal with urbanization challenges.

ICT has been proposed as the main axis of transformation and development of the third millennium. In this area, the city is also changing due to the evolution of the concept of urban space and the emergence of new ICT infrastructure. In other words, today, the role of information technology and communication infrastructures are extremely important; the reason for this is the emergence of environmental, social, and economic opportunities and the disappearance of time and place limitations.

Ubiquitous is derived from the noun ubiquity. Its meaning is to be present everywhere at the same time (Ilhan, Möhlmann, & Stock, 2015). Ubiquitous creates condition to overstate for things and people that seem to turn up everywhere, has been more popular than ubiquity (Ilhan et al., 2015; Lichá, 2015; Yeon Mee, Hyun Soo, Soo Young, & So-yeon, 2009).

Ubiquitous = everything everywhere at the same time

Ubiquitous = Reality + virtual reality + future of high technologies (5th technologies)

2.7 The 5th Wave/Tomorrow Age Theory or Theory of Comprehensive Everything

The fifth wave theory was invented and improved by the author between 2010 and 2017. This theory considers the threshold of tomorrow from 2020 to 2030 and states that it is necessary for cities to prepare themselves for tomorrow's challenges. Indeed, cities must become smart and sustainable places that anticipate the needs of their citizens, and move toward high livability. In other words, cities should become a peaceful environment that provides effective services to its citizens. In recent decades, clean technologies have become an important way to fight social, economic, environmental, educational, cultural, and political risks and crises. Recently, the challenge of the Covid-19 pandemic has created profound changes in all the mentioned dimensions in societies at a very high speed, which the author, in his theory, mentions as a great accelerator. It can be said that the cities that benefit from the power of clean and smart technologies are flexible and continue to exist. Figs. 11.2 and 11.3 show the theory of wave 5/the era of tomorrow, the eras, and technologies that we have passed and the point where we are:

In addition to the concept of aliveness, livability is related to quality of life, sustainability, and healthy communities, so as another aspect of the fifth wave theory, livability is strongly emphasized. In other words, since clean technologies play a significant role in decision-making and policymaking strategies of smart and sustainable cities, it can be accepted that smartness and clean technologies are a way to improve livability and sustainability (Fig. 11.4).

2.8 i-Sustainability Plus Theory for Technological Society and innovative Ecosystem

The constant changes and developments that are taking place with great speed are one of the most prominent features of the present age that can be seen in many aspects of life. These constant changes and developments along with the development of high technologies and smartness have made the achievement of organizations their goals in a complex and difficult path. Today, organizations need to be able to quickly adapt to developments in various fields in order to remain competitive and meet the needs of society in order to be sustainable and to survive. In this regard, creativity and innovation as a determining factor in the performance of the organization, its success, and long-term survival are considered. i-Sustainability Plus is a theory of high-stability work, innovation, and digitalization. In other words, the theory is made up of a trinity that, with today's rapid technological

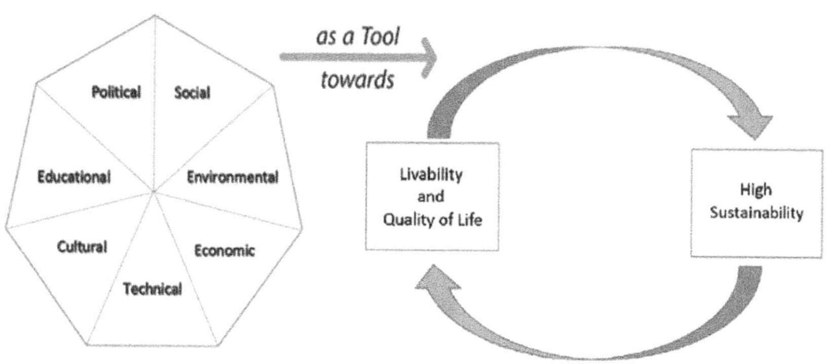

Fig. 11.2. Relation Among Quality of Liveability and Life and 7 Pillars of Sustainability (Author).

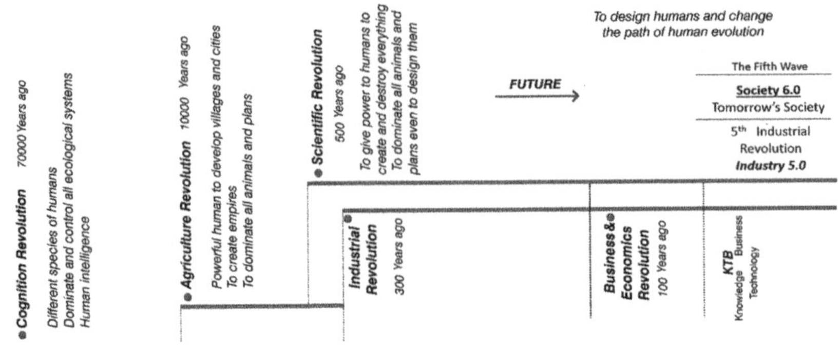

Fig. 11.3. Revolutions in the 5th Wave/Tomorrow Age Theory (Author).

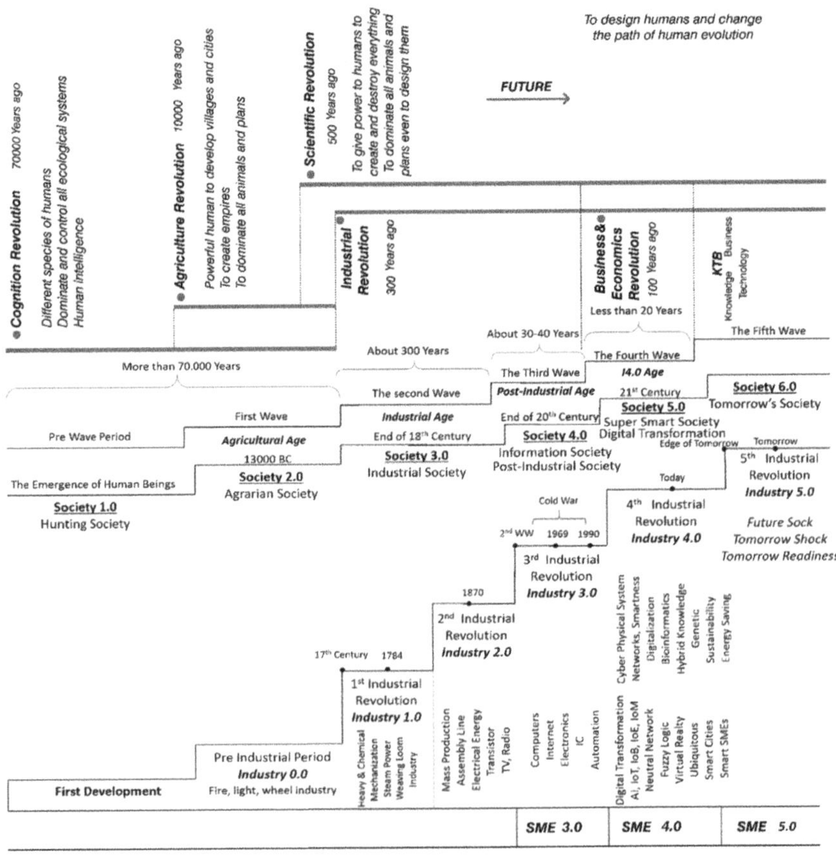

Fig. 11.4. The 5th Wave Theory (Author).

advances, makes businesses more innovative for greater durability and stability in the face of the future of the fourth industry, and places great emphasis on the importance of digital learning. In this theory, innovative management is the innovation in an organization's business, marketing, and technology systems to achieve success and best results.

Fig. 11.5 presents the i-Sustainability Plus theory, which is made of the trinity open innovation, seven pillars of sustainability (7Ps model), and 4.0 smart high technologies such as IoT-education:

2.9 Urban 6.0 and Utopia

Utopia is a symbol of an ideal and flawless reality. In recent times, with the advancement of technologies and the phenomenon of globalization, it is a question of a smart and sustainable society and urban space that meets the needs of its citizens in the best possible way without harming the needs of future generations. In general, these cities provide high livability for their citizens in line with high quality of life.

A utopia is an imagined community, society, or urban space that possesses highly desirable or nearly perfect qualities for its citizens.

The 5th wave theory forecasted a crisis at the first edge of tomorrow (2020–2030), which has led to the creation of communities, societies, cities, and businesses founded on high technologies, IoT, and on appropriate business strategies concerned with sustainability which can create new concepts and new businesses, societies, and cities which are capable of tackling future concerns. These are commonly known as "Tomorrow's Society and Business" and "Super-Intelligent Society", and they function in a super-intelligent business environment.

Another theoretical basis is the "i-Sustainability Plus theory", which is made up of the trinity open innovation, sustainability, and 4.0 smart high technologies, e.g., digitization and ubiquitous and super-intelligent society (Society 6.0).

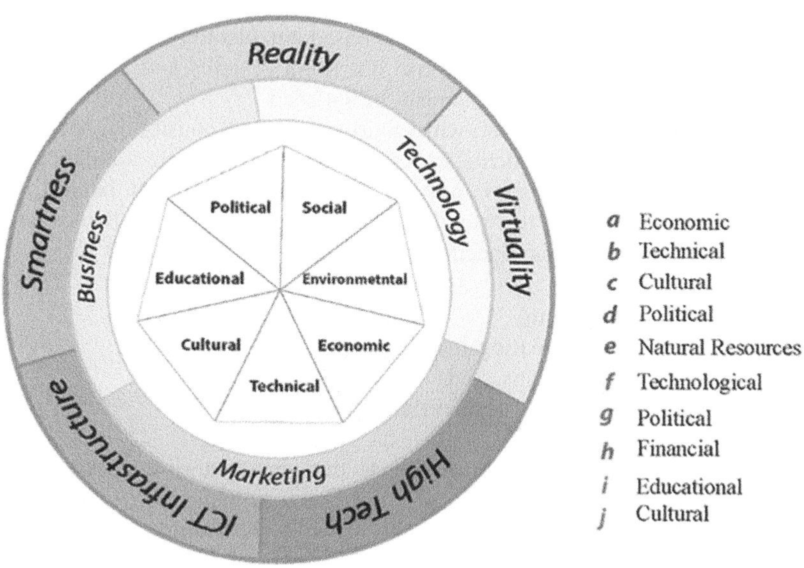

Fig. 11.5. i-Sustainability Plus Theory (Author).

This construct, which includes the idea of a sustainable ubiquitous and smart society and city, is evaluated as a new idea of urban living in tomorrow's business, and societies entitled Urban 6.0 (Prof. Dr. Hamid Doost Mohammadian, 2017).

It is critical to focus not only on profit optimizing but also on sustainability and corporate social responsibility (CSR) for the environment and saving of resources.

The 5th wave theory concentrates on the readiness to change into the new age that we are just entering. Within this project, the applicant aims to implement readiness strategies and design a future scenario to apply these theories and evaluate the results to design empirical investigation from my theoretical approaches (cf. on 5th wave theory). The 5th wave theory is designed to:

- forecast,
- prevent, and
- face the postsustainability impact.

of today's social challenges and tomorrow's urbans shocks using

- open innovation,
- implementation,
- development, and
- application.

of these technologies to provide a blue–green sustainability and digital readiness and recovery strategies with the CSR approach (e.g., for the Urban 6.0, SME 5.0 in Industry 5.0 (future of Industry 4.0 as symbol of west culture) and the super-intelligent society entitled Society 6.0 (future of Society 5.0 as symbol of nonwest culture)).

This urban development aims to make the world a better place to live in with a high blue–green sustainability impact for business and life toward a utopia with a focus on equality in such categories as economics, government, and justice, with the method and structure of proposed implementation varying based on ideology. A utopia's nature is inherently contradictory because societies are not homogeneous and have desires which conflict and, therefore, cannot simultaneously be satisfied.

In getting the world back on its feet, with environmental responsibility, social cohesion, and economic efficiency for investments, cohesion is crucial to ensure a balanced recovery from the coronavirus pandemic and make sure that no one is left behind. With the REACT-EU initiative, we want to extend the crisis response and repair measures to the most affected sectors and regions. Specifically, this initiative will help us to:

- reinforce the health care system,
- support affected citizens by creating jobs for them,
- enhance basic services for citizens, and
- make the world economies more resilient and sustainable.

In fact, smart and successful cities are those that create a sustainable mobility and transportation paradigm for themselves; and by monitoring the environment and being aware of their weaknesses and strengths, they monitor the changes, opportunities, and risks of the external environment in step with the growing needs of their citizens. These cities manage things with the aim of creating a balance between technology and business with social cohesion, technology, and business with environmental responsibility, including focusing on green–blue sustainability and technology and business with long-term economic efficiency (Fig. 11.6).

The scope of discussion of stability in smart moving surfaces is wide. Today, the paradigm of sustainability has largely replaced success in the literature of rail, land, air, and water transportation management, and almost most governments define the achievement of sustainable mobility in their urban value system. In the corona epidemic, as the most dangerous challenge of the century and its adverse effects on communities and cities, the importance of sustainability and sustainable development in general became more of a priority for governments, and cities were forced to move toward greater intelligence and a paradigm of sustainability. In this way, a new concept of smart cities was created with three priorities: (1) environmental responsibility, (2) social cohesion, and (3) economic efficiency (Fig. 11.7)

3. METHODOLOGY

It is possible to discover the scientific truth by thinking and trying at the same time as preparing a plan to anticipate uncertainties and challenges through research. Therefore,

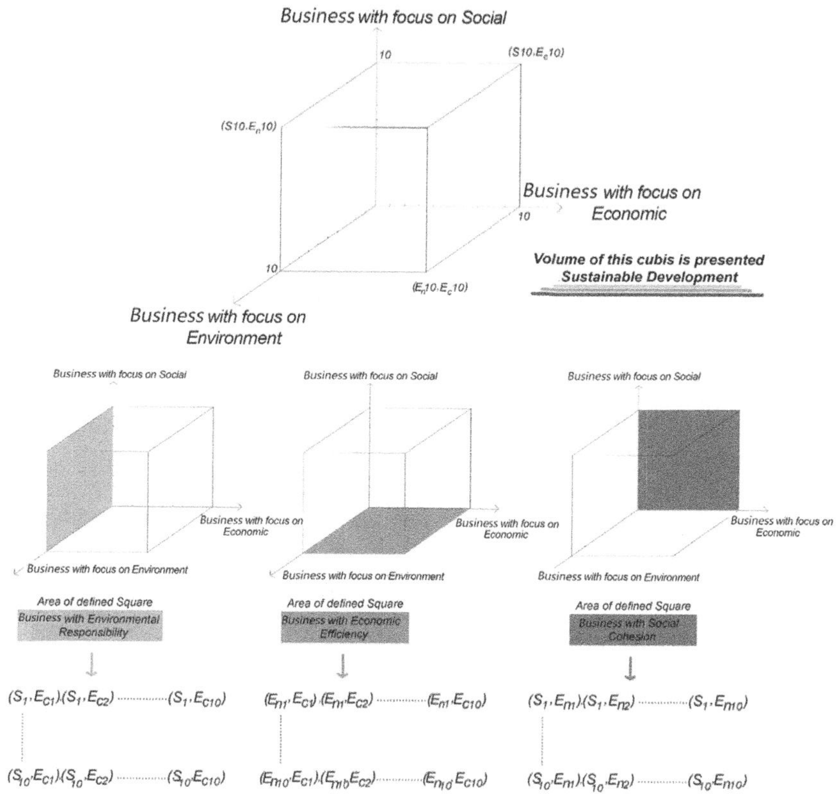

Fig. 11.6. Business With Focus on 3D SocioEcoEnvironment (Author).

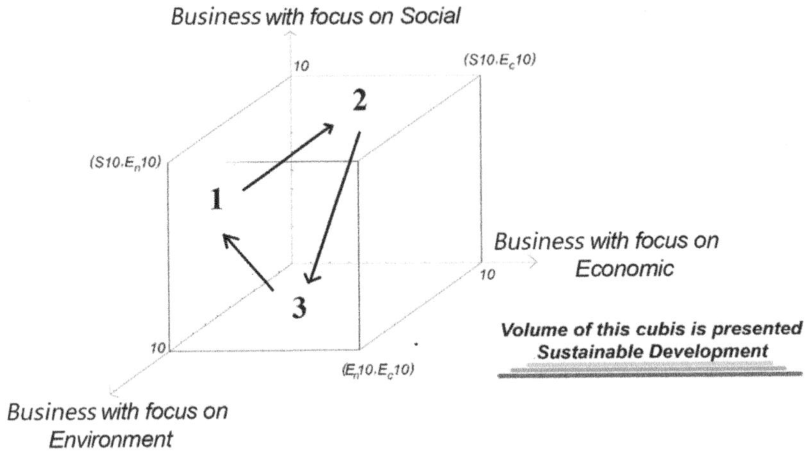

Fig. 11.7. Sustainable Technology and Business in SocioEdu Environment (Doost. H, 2019) (Author).

mapping is necessary to face the future path. The main goal of all research processes is to increase efficiency and effectiveness (productivity of knowledge) through the production and development of science and knowledge. Therefore, in this chapter, we are looking for sustainable, clean, inclusive, innovative, and smart mobility technologies to achieve City 6.0 and sustainability through the fifth wave theory, theories, models, and related concepts. In this regard, in order to achieve the goal of the research, using the doost research method (DRM) and its levels, it is necessary to answer the main question raised below. Also, in this research, an explanatory mixed methods approach is applied that has several steps: literature review, case studies, survey, questionnaire, and analyzing.

- How would sustainable and smart mobility play roles in achieving sustainable development?

In order to find out logical answer to main question and aims of the study, sub questions are proposed:

- How would sustainable and smart mobility be developed? And what is the role of modern technologies such as LCT, IoT, digitalization, and so on on creating sustainable and smart mobility?
- Which dimensions are required to achieve proper mobility?
- Why is it required to design modern urban areas based on sustainability and high quality of life?
- What is the relation among quality of life, livability, and sustainable development?
- How would sustainable and smart mobility influence on livability of urban areas and achieving sustainable development?

In this research, an explanatory mixed methods approach is applied that has several steps: literature review, case studies, survey, questionnaire, and analyzing.

Stage 1: The Development of a Research Method

The purpose of the surveys and case studies was to indicate policies, strategies, and solutions applied in mobility field by developed countries that help them to create sustainable and livable urban areas. Additionally, questionaries among experts were distributed in order to find out roles of mobility in urban plan and quality of citizens' life.

In particular, the impact of appropriate transportation as inclusive and sustainable mobility as one of the necessary components for modern and sustainable cities was investigated. Also, it should be noted that the role of technology in this infrastructure is undeniable.

Stage 2: The Development of a Research Tool.

The methodology of the DRM which is the revised Kiwi and Kampenhood besides of BESQol (built environment sustainability and quality of life) were used as tools for analyzing in this research.

DRM methodology founded on Persian Zarathustra (good thoughts, good words, and good deeds) is used in social science, engineering, and management. The main goal of this method is to identify the key dimensions of human well-being. This method has seven questions and three levels (Sung Hwang, 2013) (Fig. 11.8).

BESQol is the other methodology tool applied in this research that is related to the built environment designed to help develop sustainable low carbon developments that provide capabilities for a high quality of life for all members of the community (Ishida, 2000, 2017).

Step 3: Analysis and collection of study results

Based on the data from the research analysis, results were obtained which are discussed by the author.

4. CASE STUDIES

4.1 South Korea/Seoul

Ubiquitous is an idea introduced and applied in Eastern era. South Korea is the pioneer of designing and creating U-city. In particular, this idea and concept help South Korea to confront with its challenges based on rapid urbanization such as environmental challenges, supplying citizens' needs, contamination, etc. Applying ubiquitous services in different cities could struggle with these challenges and make better urban areas. Fundamentally, ubiquitous services and city help South Korea to deal with urbanization challenges and reduce impacts of them on urban planning in order to create more sustainable urban areas

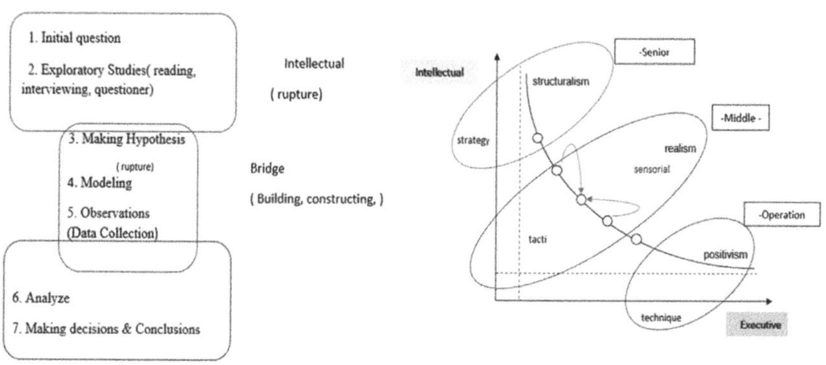

Fig. 11.8. DRM (Doost Research Methodology) Methodology and Its Levels (Sung Hwang, 2013).

where citizens are satisfied about their lives (D'Auria, Tregua, & Bifulco, 2014; Mora & Bolici, 2017).

Fig. 11.9 presents the main features of U-eco city founded on South Korea.

The main goal of this project is sustainable, smart, and future-oriented urban design and planning, in which the combination of high technologies of urban management and ecological system leads to a comfortable urban environment for citizens in line with better livability and higher quality of life. (Noori, Hoppe, & de Jong, 2020).

In many ways, the city of Seoul can be considered both as a map and a guide for how to develop a smart city and a platform for implementing and testing new types of technology. Almost all aspects of life in Seoul are driven by technology, from integrated public transport systems to government emergency alert systems (A Digital City for Everyone, 2019).

Basically, 3 key ideas have been realized to be used in Seoul as a solution:

(1) ICT infrastructure and development, which plays a fundamental and important role in achieving U-city services in line with demand forecasting.
(2) Integrated urban management framework to expand and develop services through standards smart users.
(3) Smart users: In addition to ICT as a tool in U-cities, citizens who are aware of modern technologies and are able to use smart technology as a tool to interact with smart services and achieve U-life.

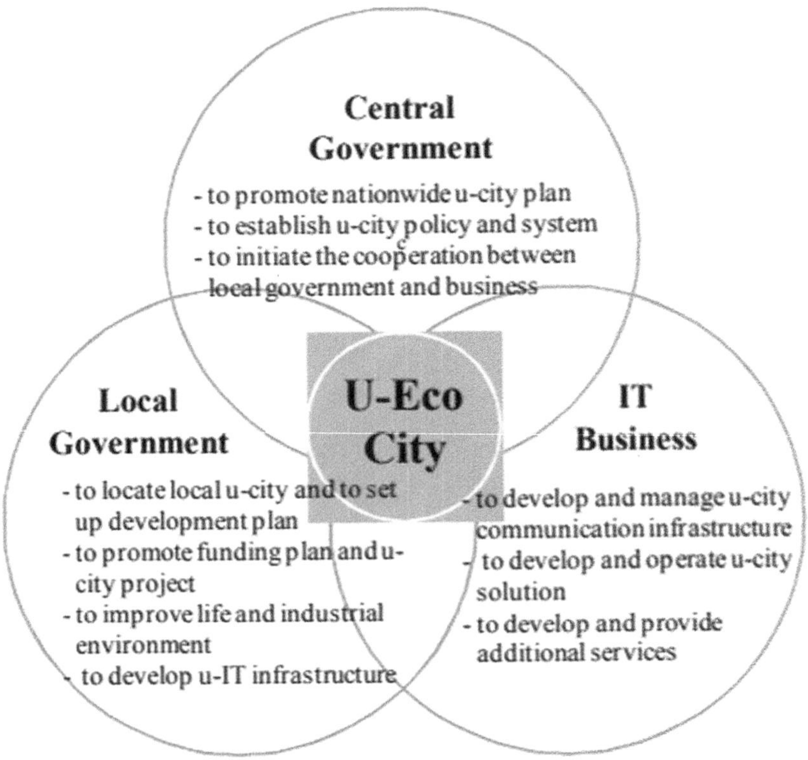

Fig. 11.9. The Main U-City Operation Project in Korea (Mora & Bolici, 2017).

4.2 The Netherlands/Amsterdam

Digital cities based on integrating information (both achievable and real-time), innovation, and technology are developed specifically in western countries to create more sustainable and livable cities (Doost Mohammadian, 2020d). Digital cities have been emerged in the United States for the first. Most of these cities are developed by America Online (AOL). And after that, digital cities have been emerged in Europe from the 1990s to 2000s (Doost Mohammadian, 2020c, 2020d).

Digital city is a kind of city founded on computer network and urban information resources to form a virtual digital space for a city in order to supply citizens' needs and deal with urbanization challenges. Substantively an open, complex, and adaptive system are main features of such city that is able to make an information service marketplace and information resource deployment center. In particular, digital cities could develop sustainability and improve quality of life in urban areas (Doost Mohammadian, 2020d).

Improvement of digital cities creates smart cities. The main differences between digital cities and smart one is related to technology applied by these two kinds of cities. Digital cities are based on web services, and smart cities are founded on sensory services. Additionally, digital cities could be developed to virtual and real cities in recent years. Amsterdam founded on community networks and social interactions, Helsinki based on virtual spaces, and Kyto focused on connecting virtual and real spaces are famous examples of digital cities. In particular, digital cities are based on technology that is needed to apply in different sectors including technology for information integration, public participation, social agents, and information security. Applying technology in different aspects of life could be introduced as main indicators required to create digital city (Doost Mohammadian, 2020a, 2020b). Generally, digital cities have different goals and components (Doost Mohammadian, 2020c). Table 11.1 presents four digital cities, their components, and goals.

Amsterdam is selected as case study in this research to be declared comprehensively.

Amsterdam was introduced as digital city in 1994 in The European Digital Cities Conference. Community networks and social interactions among citizens are main indicators to make this city a digital one (Doost Mohammadian, 2020d).

Table 11.1. Comparison of Digital Cities (Doost Mohammadian, 2020d).

	AOL	Amsterdam	Helsinki	Kyoto
Goal	Vertical Market	Public Communication Specs	Next Generation Metropolitan Network	Social Information Infrastructure for Everyday Life
Architecture	Accumulating Urban Information	Loosely Coupled with the Physical City Platform for Community Network	High-speed Network Tightly Coupled with the Physical City	Tightly Coupled with Physical City Multilayer Architecture: Information Layer, Interface Layer, and Interaction Layer
Technology	WEB Chat	City Metaphor for Public Participation	3D Virtual City Network Technology	3D Virtual City Information Integrated Social Agent
Organization	Profit Organization	Nonprofit Organization	Digital City Consortium Initiated be Helsinki Telephone Company	Digital City Forum (universities, companies, and local governments)

In particular, Amsterdam could also be introduced as smart city besides digital city decade (Doost Mohammadian, 2020c; Skill Development for Industry 4.0, 2016). Strategic urban planning founded on strategic thinking, collaboration, and inclusive criteria plays significant roles to make Amsterdam a smart city. Additionally, ICT infrastructure and digital services are the other indicators playing roles in Amsterdam digitalization and smartness (Skill Development for Industry 4.0, 2016).

Applying technology in different services and all aspects of life, creativity, proper urban plan, and applying smartness in different fields are main indicators made Amsterdam be smart (Doost Mohammadian, 2020c, 2020d). Additionally, smart citizens and who are familiar with technologies and aligned with them impact on achieving smart Amsterdam (Doost Mohammadian, 2020d).

To achieve such city, Amsterdam confronts challenges such as: accessing to internet for everyone, accessibility of public services, education, human resource competencies, security, ICT infrastructure, personal data, cost, increasing digital resilience of Amsterdam, and proper governance. However, the city deals with these challenges and create digital and smart city (Doost Mohammadian, 2020d).

Reducing CO_2 emission by applying smart energy and smart grid is the main goal of smart and digital Amsterdam. Clean environment and sustainable water management are two main results of digitalization and smartness of Amsterdam (Doost Mohammadian, 2020a). Generally, being smart and digital could create sustainable and livable urban area that citizens are satisfied about their lives.

5. INTERVIEWS AND QUESTIONNAIRE

5.1 Questionnaire

In order to measure and evaluate the relationship between sustainable and smart mobility and quality of life and livability, a questionnaire with a 5-point Likert scale and including 10 questions was distributed among 100 experts in the fields of sustainability and mobility, to examine the relationship between universal, smart, and sustainable mobility, smart and sustainable infrastructure, and quality of life and livability (Table 11.2).

Table 11.2. Questionnaire Result.

Quality of Life Indicators	Score
Environment Sustainability	4.5
Improvement of Urban Infrastructure	4.6
Economic Condition	4
Security	3.9
Health and Diseases	4.2
Improving Public Transportation	4.4
Energy Management and Reduction of Energy Consumption	4.5
Water Management	4.1
Reducing Risks and Mobility Challenges	4.3
Increasing Livability of areas	4.6

5.2 Interviews

Interview was conducted with a set of structured questions consisting of seven questions in order to find out how smart, sustainable, and inclusive mobility could be achieved. Additionally, importance of such mobility to design modern urban areas with high quality of life, livability, and sustainability is explored among interviews' questions. These questions were as follows:

- What is sustainable and smart mobility?
- How would sustainable and smart mobility be achieved?
- What are the drivers of sustainable and smart mobility?
- Does risk management play roles in creating smart and sustainable mobility?
- What are the roles of citizens and government in developing smart and sustainable management?
- What is the role of smart and sustainable mobility for environmental sustainability and future of cities?
- Why is it vital to create modern urban areas based on sustainable and smart infrastructure?

6. RESULTS AND DISCUSSION

6.1 Results

Based on survey and analyzing, it is declared that new concepts of urban areas like ubiquitous cities, digital and smart ones, are needed to deal with global challenges and achieve sustainable development. Additionally, these urban areas are able to be kept up with rapid urbanization demands as well as high quality of life and livability.

In particular, there are different paths and tools needed to design modern urban areas based on sustainability and high quality of life. Sustainable infrastructure, sustainable and smart comprehensive strategic urban plan, value engineering, and sustainability are main indicators needed to create modern areas. As we are living in an era, technologies are main accelerators to be applied to find out solutions to design modern urban areas; technologies are introduced as main drivers in designing sustainable and smart cities. Additionally, innovation, management like risk management, project management, citizens, and governments are other drivers to make sustainable and livable urban areas. In particular, there is a relation among sustainable and livable urban areas, sustainable development, and high quality of life.

Mobility as an infrastructure component could influence on creating such cities. Generally, different paths and tools including seven pillars of sustainability, high technologies, risk management, innovation, comprehensive strategic urban plan, and so on are needed to establish proper mobility systems. In this chapter, inclusive and modern mobility is introduced as suitable system to design modern urban areas. In following context, inclusive and modern mobility is comprehensively introduced.

Fundamentally, inclusive and modern mobility could be one of main tools required to create sustainable and smart cities toward sustainable development via its privileges such as environmental sustainability, economic sustainability, social sustainability, smart management, and so on.

6.2 Discussion

It is expected that the world will reach about 10 billion population in 2025. This fact presents that the world confronts challenges based on rapid urbanization and growing population. Environmental challenges, climate change, social instability, insecurity, unemployment, poverty, slum, economic challenges, and health diseases are main challenges derived from rapid urbanization and growing population that threaten the future of the world and humanities. In particular, the environment, nature, and humanities are faced with risks and significant challenges to 2050. Therefore, it is vital to deal with these challenges in order to maintain the world and humanities for future and make the world with less challenges in 2050. Sustainable development is introduced as path to deal with these challenges as well as making more sustainable and livable world for living.

Fundamentally, urban transformation and urban development are required to achieve sustainable development in order to deal with urbanization challenges, improve quality of life at the present, and maintain the world for future. So, new concept of urban areas could be applied as solution for urban transformation and achieving sustainable development. Different players have roles and impacts in urban transformation. Infrastructure, comprehensive strategic urban plan, industries, businesses, education, technical aspect, citizens, and government are main players in urban transformation and city development. Additionally, innovation and high technologies are main accelerators that are applied to improve players' performances in order to achieve urban development. In other words, high technologies and innovation could be utilized to make city indicators aligned with sustainability to create sustainable and smart cities toward sustainable development.

Such modern cities are based on smart services such as smart mobility, smart and sustainable construction, smart businesses, smart industries, smart education, and so on. Additionally, smart citizens who are familiar and aware of such services play significant roles in creating modern urban areas. It means that citizens should be aligned with these services and able to apply them in order to create smart cities based on smart services. In particular, these smart services are able to supply rapid and modern needs of citizens and achieve strategies, services, and performances aligned with sustainability pillars toward sustainable development. Generally, smartness is able to supply modern and rapid demands besides developing sustainability. Therefore, smartness is a path to create modern urban areas that could deal with urbanization challenges, improve quality of life, and achieve sustainable development toward more sustainable and livable world.

Technology mirrors culture, and also technologies play significant roles in creating modern urban areas. In other words, high technologies such as IT, ICT, big data, digitalization, and so on are drivers to design modern urban areas. In Eastern countries ubiquitous services and in Western countries digitalization and smartness are applied as techniques and solutions to deal with urbanization challenges, supply citizens' need, and create livable and sustainable urban areas. In particular, these ideas for designing modern urban areas are utilized to reduce global challenges, maintain the nature for future, improve quality of life at the moment, and generally achieve sustainable development. Fundamentally, designing new concept of cities is needed to develop sustainability in order to make the world a better place for living. Fig. 11.10 presents main services applied by countries to develop modern and smart services:

Also, new concepts for urban areas based on sustainability and smartness are mentioned in this chapter to achieve sustainable development. New concept for urban areas founded on sustainability, smartness, and comprehensive strategic urban plan based on sustainability strategies and smart services is declared to achieve sustainable

Countries	West	Countries	East
Scandinavian Countries	• Infrastructure Technologies • Sustainable Mobility • Waste and Waste Water Management • Renewable Energy Technologies • Bicycle as Main Transport System • Public Transport Technologies • Educational Technologies • Top 10 Innovative Economics in the World	***Japan***	• Society 5.0 • Automotive and less CO2 emissions Technologies • Infrastructure Technologies
Germany	• Industry 4.0 • Automotive Technologies • SME Management • Infrastructure Technology • Digitalization • Vocational Training • Improving Public Transportation like Dutch Bahn as the best train system in Europe • The most Innovative Economics in the World	***South Korea***	• Ubiquitous Technologies • The Second Innovative Economics • Automotive and less CO2 emissions Technologies • Infrastructure Technologies

Developing Smart Services to Supply Citizens' Needs, Create Sustainable and Smart Businesses, Industries and Life

Dealing With Urbanization Challenges and Create Sustainable And Smart Cities such as Digital Cities, Smart Cities and Ubiquitous Ones towards Sustainable Development

Fig. 11.10. Smart Services Applied by West/Non-West Countries Toward Cultural Sustainability (Author, 2017).

development. Such city is needed to develop seven pillars of sustainability including environmental, economic, social, cultural, educational, technical, and political sustainability. Fundamentally, comprehensive strategic urban plan–based sustainability strategies and smart strategies play important roles in creating blue–green cities. Furthermore, value engineering to manage and control cost of projects are needed to develop sustainability and create modern cities. In particular, sustainability strategies, proper comprehensive strategic plan, suitable infrastructure, and value engineering are vital to design sustainable and smart modern urban areas. Additionally, innovation and high technologies are accelerators to find out solutions to develop such areas toward sustainable development. Fig. 11.11 presents how sustainable and smart cities would be achieved toward sustainable development:

In this chapter, roles of mobility in creating sustainable and smart cities are explored and declared.

Mobility as a component of infrastructure could play significant roles on creating sustainable and smart mobilities.

One of main challenges that urban areas and world confront with is environmental challenges. Such challenges are significant threats for nature, environment, and humanities. Many diseases, depletion of resources, extinction of spices, and so on are made by environmental challenges. Therefore, it is vital to deal with these challenges to preserve

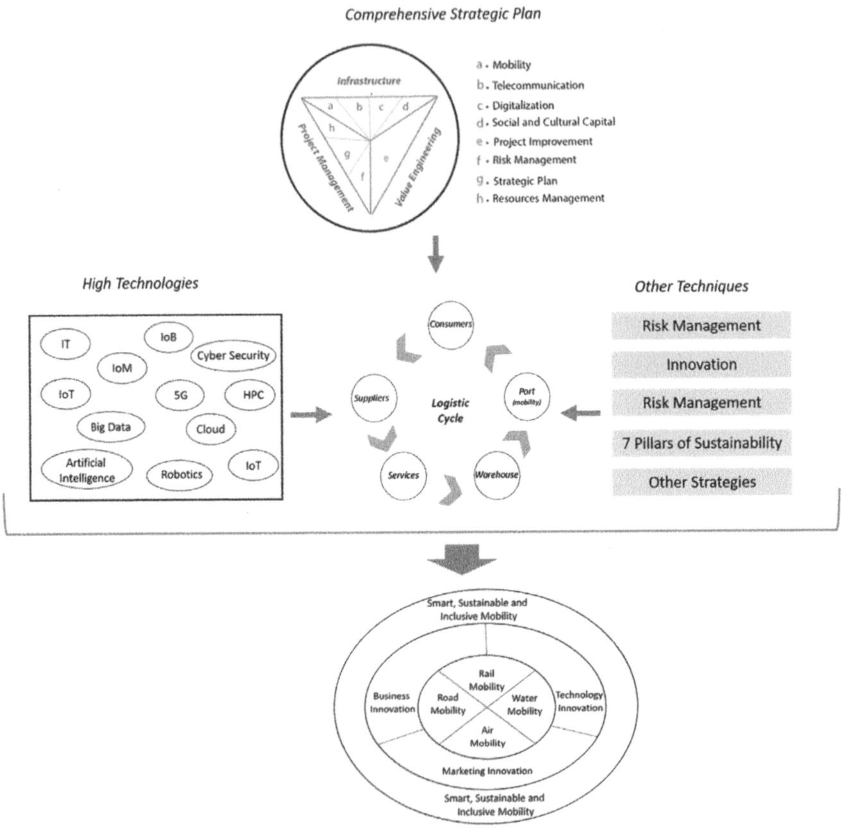

Fig. 11.11. Sustainable, Clean, Inclusive, Innovative, and Smart Mobility Indicators (Author).

nature and humanities. Mobility as an infrastructure indicator could influence on reducing environmental challenges and their impacts on humans' life.

In an urban area, movement as an important dimension and indicator can affect the quality of human life. In other words, economic development and livability and social welfare depend on transportation systems. Transportation has four main categories including water transportation, air transportation, road transportation, and rail transportation. Each of these transportation systems has risks that lead to concern and risks that are considered to be a sustainable threat. Therefore, smart and sustainable mobility is a type of sustainable mobility suitable for green and blue strategies, which is introduced as a way to create sustainable infrastructure. Green strategies to face and deal with environmental challenges are focused on smart policies in line with sustainability. The three traditional pillars of sustainability include social, environmental, and economic sustainability. But at the beginning of the twentieth century and according to the opinion of the authors, sustainability has seven pillars: cultural, social, environmental, technical, political, educational, and, of course, economic. Thus, green strategies act as catalysts to help countries improve these dimensions of high sustainability, which include the seven pillars mentioned above, through smart transportation and mobility systems. For example, strategies compatible with environmental responsibility keep mobility in step with the environment and do not destroy it. Mobility based on clean energy such as hybrid vehicles is a good example of environmentally friendly mobility. Water-based concerns such as water scarcity, water pollution, water purification, etc., are highly effective through mobility related to water management and are in the direction of sustainability.

Additionally, applying risk management to achieve sustainable mobility systems that is aligned with blue–green strategies plays significant roles in developing smart mobility. In other words, risk management to control, manage, and reduce risks related to different transportation systems would increase mobility systems sustainability, even help mobility systems to be aligned with blue and green strategies and has less impacts on environment.

The other factors required for sustainable and smart mobility is smart and sustainable energy management. Sustainable and smart management helps mobility and transportation systems and industries to work in a path with least greenhouse gases emissions, environmental risks besides least costs. In particular, energy management is a technique concerned on utilizing renewable energy management, applying waste as new source of energy, increasing energy efficiency and less environmental impacts, improving productivity, applying clean energies, and so on. Therefore, smart and sustainable energy management is a tool to help mobility and transportation be aligned with sustainability strategies.

Improving public transportation systems could play significant roles in developing sustainable mobility. Smart transportation application and systems could be applied as a technique to encourage citizens to use public transportation. Uber or Moovit are applications help transportation systems to be enhanced and help users to ride easily on public transportation. In particular, proper public transportation systems are one of main indicators required to achieve sustainable mobility systems in countries. In specific, culture plays significant roles in encouraging citizens to use public transportation. Therefore, enhancing cultural norms to ride on public vehicles would help countries to improve public transportation systems.

Fundamentally, clean energy plays significant roles in making mobility sustainable. Transpiration systems based on renewable and clean energies such as hybrid cars, electric vehicles, public transportation working with electric and renewable energies, airplanes with clean energies, and so on are needed to achieve mobility systems aligned with environmental sustainability and sustainable water management systems. In other words, transportation systems based on clean and renewable energies play significant roles in achieving mobility aligned with blue and green strategies.

Creating areas founded on free-vehicles transportation is one of main factors to develop sustainable mobility toward sustainable and smart cities. Designing urban areas in a path have walking areas or bike areas could be applied as techniques to create free-vehicles transportation systems. Such solution could help cities to make sustainable and smart mobility aligned with environment, nature, and water management. In specific, this feature can reduce environmental challenges such as air pollution, water contamination, CO_2 and greenhouse gases emissions, depletion of nonrenewable resources, and so on. Therefore, free-vehicles transportation is one of main features of sustainable and smart mobility.

Besides of all these factors, education and knowledgeable capitals play important roles in gaining sustainable, smart, and inclusive mobility. Proper education to train knowledgeable, expert, and qualified capitals is needed to design comprehensive strategic urban plan concerned on proper mobility. In addition, education toward altering cultural norms could influence on mobility systems. For instance, make children aware of importance of sustainability, using public transportation, etc., through education and making them as culture would improve mobility in blue–green and inclusive path.

Additionally, two accelerators as techniques to develop these indicators play significant roles in developing sustainable, smart, and inclusive mobility: innovation and high technologies.

High technologies such as IT, IoT, and IoE, digitalization, and ubiquitous tools and techniques are required to develop smart services and mobility. For instance, digitalization through improving ubiquitous life and services; smartness via improving E-services; high technologies through hybrid vehicles and electronic vehicles are main solutions to achieve smart and sustainable mobility. In other words, high technologies are tools to make solutions for strategies needed to achieve smart mobility systems. Therefore, improving high technologies play significant roles in developing smart and sustainable mobility and transportation systems. Smart management, vehicles working with clean energies such as hybrid cars, smart transportation, ubiquitous transportation and so on, all are founded on high technologies. Fundamentally, high technologies are vital to create smart and sustainable mobility toward sustainable and liveable urban areas.

The other tool required to develop smart mobility strategies is innovation. In particular, innovation and innovative management could make mobility systems to find out solutions to develop smart and sustainable transportation with the least risks. For instance, ubiquitous idea is founded on innovative approach. Therefore, innovation and creativity are required to develop smart and sustainable mobility systems.

Fundamentally, smart, sustainable, and inclusive mobility based on environmentally friendly strategies and sustainable water management could deal with environmental challenges like air pollution, climate change, water contamination, extinction of spices, disturb environment and nature, depletion of nonrenewable energies such as benzene and gasoline, and so on in order to maintain the world for upcoming generations. Additionally, reducing environmental challenges would make the world and cities more livable and sustainable at present with less challenges, especially health problems for present citizens. Therefore, smart and sustainable mobility systems could reduce impacts of environmental problems on human's life in order to increase quality of life for them.

The other goal of such mobility systems is economic sustainability. Reducing financial risks of mobility systems, making balance among taxes, improving services and price strategies, promoting economic situation of a country by successful and sustainable mobility systems, and so on are main advantages of sustainable and smart mobility that countries could benefit from to develop economic sustainability.

Improving security and safety, reducing health challenges, enhancing social responsibility, etc., are main social aspects of sustainable and smart mobility that would influence on improving social sustainability. So, creating proper context for social sustainability could be named as the other advantage of sustainable and mobility.

Sustainable and smart mobility system needs knowledgeable and expert capitals besides of smart citizens to be popular and accessible in urban areas. Therefore, it is needed to train qualified capitals to develop such mobility. Practical and vocational training play significant roles in achieving sustainable and smart mobility. Additionally, high technologies are fundamental tools to make such mobility systems. In particular, sustainable and updated education systems are needed to achieve smart and sustainable mobility. Therefore, education sustainability is the other privilege of such mobility systems.

Hybrid cars, electronic vehicles, smart services, ubiquitous services, applications like Moovit, and so on are main indicators to develop smart and sustainable mobility. These indicators are based on improving technologies and high technologies. Therefore, technical sustainability is a fundamental tool for developing such mobility systems. It could be concluded that smart and sustainable mobility needs to improve technical sustainability in order to able to find out solutions to achieve its strategies toward sustainability. So, technical sustainability could be derived from sustainable and smart mobility systems.

Smart and sustainable mobility systems would encourage citizens to utilize public transportation as well as riding vehicles based on clean energies in order to develop sustainability. In other words, such mobility would influence on cultural norms and change them. Therefore, cultural sustainability would be achieved by proper mobility systems.

In addition, political sustainability could be gained by improving other pillars of sustainability as well as proper strategies of sustainable and smart mobility such as safety, local–international investment, and policies concerned on global political situation.

Generally, sustainability pillars including environmental, economic, social, cultural, educational, technical, and political aspects could be improved by sustainable and smart mobility systems. Therefore, sustainable development could be gained by smart and inclusive mobility.

Basically, sustainable development can deal with urbanization challenges such as environmental challenges, air pollution, poor drinking water quality, water scarcity, poverty, problems related to waste disposal and high energy consumption, insecurity, economic problems, etc. In particular, sustainable development can provide the needs and indicators required by humans for a high quality of life. For example, we can mention: education, housing conditions, and balance between work and life, etc., in order to increase human satisfaction of about their lives. Therefore, sustainable development is able to improve livability and quality of citizens' life besides struggling with urbanization challenges. Fundamentally, high quality of livability and life besides sustainable development could be achieved by smart and sustainable mobility. Fig. 11.12 presents the impact of inclusive mobility on developing sustainability:

In particular, the main privileges of smart, sustainable, and inclusive mobility are:

- Equality and social cohesion through privacy, improved safety and security, well-being and improved health, etc.
- Clean energy is a type of transportation system to reduce environmental challenges.
- Free-vehicles city to achieve environmental sustainability.
- More efficient and productive traffic management.
- Immediate access to services to meet the needs of citizens in any place and at any time.
- Greener, safer, and cleaner mobility.
- Smart and more sustainable transportation.
- Sustainable and smart energy management to maintain natural resources for future.
- Less disruption in environment and nature to preserve the world for future.

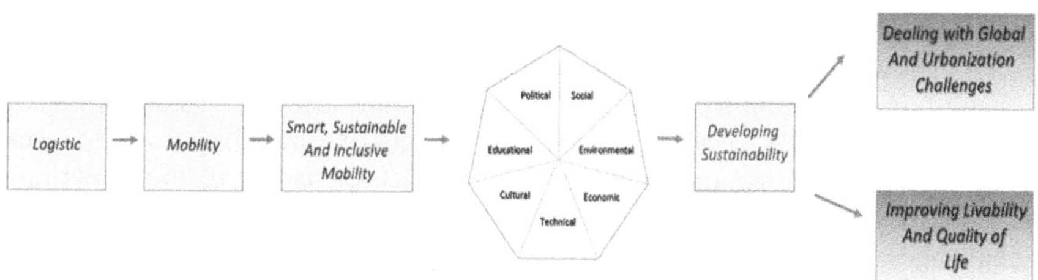

Fig. 11.12. Sustainable, Clean, Inclusive, Innovative and Smart Mobility, and Sustainable Development (Author).

- More smart and clean logistics.
- Achieving sustainable and smart urban planning and design and reducing environmental risks, crises, and challenges through reducing noise pollution, green gas emissions, air pollution, and the overall use of clean energy.
- Improving and developing economic stability through attracting investment and creating employment, improving productivity, improving export transportation, etc.
- In step with the demand of citizens and the growth of their needs.
- Less logistic and mobility risks such as transportation accidents, water contamination, air pollution, and so on toward sustainable development.
- Improving education systems to train capitals aware of familiar with high technologies and smart mobility.
- Improving cultural and social norms such as security, safety, reducing health diseases, and so on.
- Be aligned with global standards.
- Respecting social responsibility in developing sustainability and preserving the world for future.
- A step toward designing a smart, inclusive, and sustainable city by creating inclusive, smart, and sustainable infrastructure.
- Creating more sustainability in urban areas.
- Supplying citizens' needs based on Maslow's hierarchy.
- Making quality of life and livability higher in urban areas through other introduced privileges.
- Creating modern urban areas based on high livability and sustainability that are able to deal with global challenges and achieve sustainable development.
- Making the world more sustainable and livable through sustainable development.

Generally, smart and inclusive mobility could influence on developing sustainable and smart infrastructure that is required for designing modern urban areas.

In particular, smart and sustainable infrastructure not only improves sustainable development but also provides human needs including economic development, housing, security, education, etc. These needs cause livability and quality of life. The term livability refers to the degree of providing the necessities of a society based on the needs and capacities of the people of that society. An unlivable society ignores the needs of the people of that society and does not respect their wishes. Urban livability is one of the fundamentals of sustainable urban development. The livability approach can be seen as the result of an increase in the unsustainable patterns of urban life, which in the long run reduces the power of natural resources. Improving the living conditions in urban environments, in addition to increasing their attractiveness, provides a suitable platform for achieving the goals of sustainable development and a higher quality of life for people.

Fig. 11.13 shows the reality between smart and inclusive mobility and quality of life:

Clean and inclusive mobility based on environmentally friendly strategies, sustainable water management policies, sustainable and smart infrastructure, ubiquitous, low emission greenhouse gases, smartness, and digitalization is indicated as one of the keys that could make sustainable and more livable urban areas. These urban areas are able to reduce global challenges and their impacts on the world as well as sustainable development. Fundamentally, smart and inclusive mobility could make the world a better place for living through its privileges (Fig. 11.14).

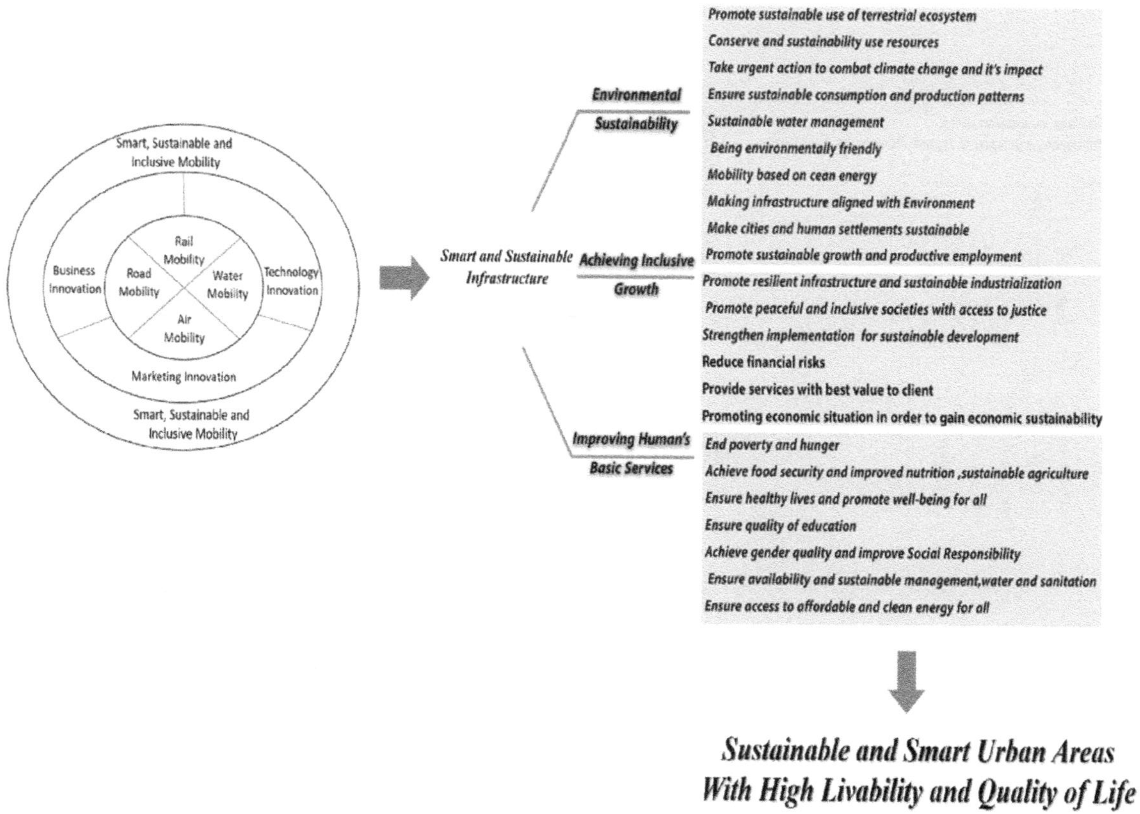

Fig. 11.13. Sustainable, Clean, Inclusive, Innovative and Smart Mobility, and Sustainable Smart Infrastructure (Author).

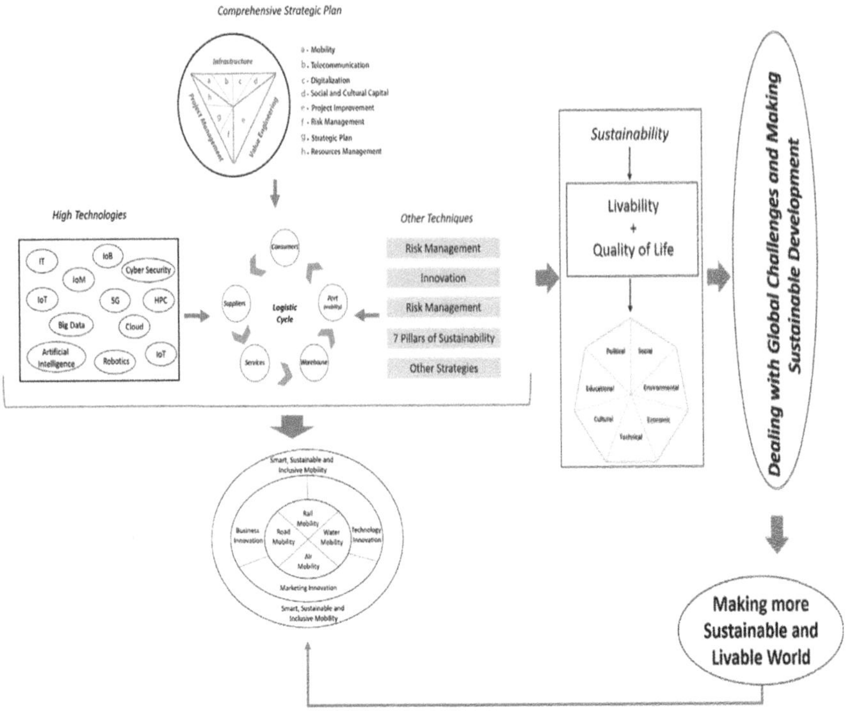

Fig. 11.14. Sustainable, Clean, Inclusive, Innovative and Smart Mobility Make the World as Better Place for Living (Author).

7. CONCLUSION

Globalization has been the dominant concepts of theoretical and practical issues of development in recent decades, especially after the development of digital communications and virtual networks. This phenomenon has a direct relationship with the city, and its main features are manifested and determined on the basis of the city, according to the required infrastructures and requirements and spheres of influence, a way that can simultaneously bring challenges in different dimensions for the city.

It is essential that the planners, policymakers, and finally the government strive to design and create clean and inclusive mobility in order to anticipate, prevent, and face the challenges of urbanization and achieve higher livability and quality of life.

Modern and inclusive mobility based on policies compatible with efficient management of water, environment, innovation, and advanced technologies is the way to modern and inclusive cities.

According to research, the below points are concluded:

- Designing modern sustainable and livable urban settings based on sustainability could be applied a solution to make the world as a better place for living through sustainable development and dealing with global challenges.
- Mobility as a component of urban infrastructure is a driver to design modern, sustainable, and smart urban areas. Mobility could play role in making sustainable and

blue–green infrastructure concerned on environmentally friendly, water management, and smart strategies.

- Clean and inclusive mobility is a mobility system aligned with environmentally friendly, smartness, sustainability, growing demands, and sustainable infrastructure. Indicators such as sustainability, digitalization, smart planning, ubiquitous concept, proper education, knowledgeable, expert and vocational social capitals, risk management, and innovation and innovative procedure are needed to achieve such mobility systems.
- Inclusive and smart mobility could influence on developing other pillars of sustainability, quality of life, and livability toward more sustainable and livable areas for living. Therefore, such mobility would make the world a better place for living.
- Technology is a tool to advance strategies, solutions, and polices needed for inclusive and smart mobility.
- As technology plays fundamental role in creating clean and inclusive mobility, smart citizens are needed to be aligned and familiar with these smart services. Fundamentally, training smart citizens are aware of importance and usage of technology to apply them in their life and benefit from their privileges.
- Besides, of clean and inclusive mobility, the other components of comprehensive sustainable strategic urban plan such as project management, value engineering, and other indicators like smart education, smart businesses, smart citizens, and so on are required to design modern, sustainable, and smart cities to make the world more sustainable and livable.

REFERENCES

A digital city for everyone. 2019. The agenda for the digital city (Vol. 1).

Ahmed, H. (2018). AI advantages and disadvantages. *International Journal of Scientific Engineering and Applied Science (IJSEAS)*, *4*.

Al-Hader, M., & Rodzi, A. (2009). The smart city infrastructure development and monitoring. *Theoretical and Empirical Researches in Urban Management*, *4*.

Alexandru, A., Alexandru, C., Coardos, D., & Tudora, E. (2016). Big data: Concepts, technologies and applications in the public sector. *International Journal of Computer, Electrical, Automation, Control and Information Engineering*, *10*.

Alter, S. (2019). Making sense of smartness in the context of smart devices and smart systems. *Information Systems Frontiers*. doi:10.1007/s10796-019-09919-9

Aralu, U. (2015). Influence of information and communication technology on digital divide. *Global Journal of Computer Science and Technology: H Information & Technology*, *15*.

Ardnt, W., Jain, A., Emberger, G., Lah, O., Sohr, A., & Tomaschek, J. (2014). *Future Megacities2: Mobility and transportation concepts for sustainable transportation*. Berlin: Jovis.

(2019). *Artificial intelligence and big data: Innovation landscape brief*. International Renewable Energy Agency (IRENA).

(2019). *Artificial intelligence and machine learning 101*. Micro Focus.

Barbier, E. B., & Burgess, J. C. (2017). The sustainable development goals and the systems approach to sustainability. *Economics*, *11*. Retrieved from http://www.economics-ejournal.org/

Basiago, A. D. (1998). Economic, social, and environmental sustainability in development theory and urban planning practice. *Environmentalist*.

Benevolo, C., Dameri, R., & D'Auria, B. (2016). *Smart mobility in smart city: Action taxonomy, ICT intensity and public benefits*. Springer International Publishing Switzerland.

Berger, G., Feindt, B., Holden, E., & Rubic, F. (2014). Sustainable mobility-challenges for a complex transition. *Journal of Environmental Policy and Planning*, *16*.

Bullock, C., Brereton, F., O'Neill, J., Clinch, P., & Russell, P. (2008). *Quality of life and the environment*. Environmental Protection Agency.

Clemente, C., Ottorino, V., & Dario, A. (2018). A novel platform for the experimental training on internet of energy. In *Proceeding of IEEE global engineering education conference (EDUCON)*. 17–20 April. Tenerife, Spain.

Craig, J. (2005). *Introduction to robotics: Mechanics and control.* Upper Saddle River, NJ: Pearson Education International.

D'Auria, A., Tregua, M., & Bifulco, F. (2014). Digital city towards smart city: A theoretical overview. In *the proceeding of the 2nd international virtual scientific conference*. October 27–31. London, The United Kingdom.

Digital sustainability: Global sustainability as a driver of innovation and growth. Stockholm: Cyber Group.

Doost Mohammadian, H. (2020a). *Rezaie: Global SMEs; Volume 1 final 201021_sustainable smart innovative global SMEs.* Bielefeld: Fachhochschule des Mittelstands (FHM) – University of Applied Sciences.

Doost Mohammadian, H. (2020b). *Rezaie: Global SMEs; Volume 2 SMEs 4.0 to gain sustainable development globally.* Bielefeld: Fachhochschule des Mittelstands (FHM) – University of Applied Sciences.

Doost Mohammadian, H. (2020c). *SMEs in energy sector Merk- Ubiquitous blue-green energy management; Volume 1 Foundation1.* Bielefeld: Fachhochschule des Mittelstands (FHM) – University of Applied Sciences.

Doost Mohammadian, H. (2020d). Digital transformation in academic society and innovative ecosystems in the World beyond Covid19-pandemic with using 7PS model for IoT. In *The 2020 IEEE conference LWMOOCS VII, learning with MOOCS 2020 in Antigua Guatemala, from 30 September – 02 October 2020.*

Etzion, D., & Aragon-Correa, J. (2016). Big data, management, and sustainability: Strategic opportunities ahead. *Organization and Environment, 29.*

Feil, A. A., & Schreiber, D. (2017). Sustainability and sustainable development: Unraveling overlays and scope of their meanings. *Cadernos Ebape. br, 15*, 667–681.

Finnveden, G., & Gunnarsson-Ostling, U. (2016). Sustainable development goals for cities. In *JPI urban Europe symposium "shaping common ground in urban sustainability" Brussel.*

Giap, T., Thye, W., & Grace, A. (2014). A new approach to measuring the livability of cities: The global livable cities index. *World Review of Science, Technology and Sustainable Development, 11*(2).

Gillis, D., Semanjski, I., & Lauwers, D. (2015). How to monitor sustainable mobility in cities? Literature review in the frame of creating a set of sustainable mobility indicators. *Sustainability Journal, 8.* No one.

Gorostidi, M. (2014). Routes, advantages and disadvantages of robotic para-aortic lymphadenectomy: A review. *OA Robotic Surgery, 10.*

(2016). *High technology indicators.* Instituto Nacional De Estadistica.

Ho Lee, S., Hoon Han, J., Taik Leem, Y., & Yigitcanlar, T. (2008). *Towards ubiquitous city: Concept, planning, and experiences in the Republic of Korea.* Hershey: IGI Global, Information Science Reference.

Hussein, A. (2019). Internet of Things (IoT): Research challenges and future applications. *International Journal of Advanced Computer Science and Applications, 10.*

Idowu, A. I., & Esere, M. (2013). ICT and higher educational system in Nigeria. *Academic Journals, 8.*

Ilhan, A., Möhlmann, R., & Stock, G. (2015). *Citizens' acceptance of U-life services in the ubiquitous city Songdo.* Springer Science+Business Media Singapore.

(2003). *Information and communication technology development indices.* New York and Geneva: United Nation.

Ishida, T. (2000). Understanding digital cities. *Lecture Notes in Computer Science*, 7–17.

Ishida, T. (2017, April). Digital city, smart city and beyond. In *Proceedings of the 26th International Conference on World Wide Web Companion* (pp. 1151–1152).

Jeekel, H. (2017). Social sustainability and smart mobility: Exploring the relationship. In *World conference on transport research – WCTR 2016 Shanghai.*

Kafle, Y. R., Mahmud, K., Morsalin, S., & Town, G. E. (2012). Towards an Internet of Energy. In *Proceeding of IEEE international conference on power system technology (POWERCON).* Oct 30, Nov 2. New Zealand.

Kayal, P., Singh, R., & Kumar, M. (2014). *Defining sustainable urban mobility.* The Energy and Resources Institute. TERI-NFA Working Paper No.11.

Kersting, K. (2018). Machine learning artificial intelligence: Two fellow travelers on the quest for intelligence behavior in machines. *Frontiers in Big Data, 1.*

Klarin, T. (2018). The concept of sustainable development: From it is beginning to the contemporary issues. *Zagreb International Review of Economics and Business, 21*(1).

Kranenburg, R., & Bassi, A. (2019). *IoT Challenges.* Retrieved from http://www.comcjournal.com/content/1/1/9

Kuhl, N., Goutier, M., Hirt, R., & Satzger, G. (2019). Machine learning in artificial intelligence: Towards a common understanding. In *Proceeding of Hawaii international conference on system sciences (HICSS-52).* Maui, Hawaii.

Kuhlman, T., & Farrington, J. (2010). What is sustainability? *Sustainability Journal, 2.*

Lichá, A. (2015). *Songdo and Sejong: Master-planned cities in South Korea.* halshs-01216229f.

(2017). *Machine learning: The power and promise of computers that learn by example.* The Royal Society.

Mahmoud, R., Yousuf, T., Aloul, F., & Zualkernan, I. (2015). Internet of Things (IoT) security: Current status, challenges and prospective measures. In *Proceeding of 10th international conference for internet technology and secured transactions*. 14–16 December, London, United Kingdom.

Mora, L., & Bolici, R. (2017). How to become a smart city: Learning from Amsterdam. In *Book: Smart and sustainable planning for cities and regions: results of SSPCR 2015*. Springer.

Mosavi, A., & Varkonyi, A. (2017). Learning in robotics. *International Journal of Computer Application, 157*.

Mühlhäuse, M., & Gurevych, I. (2008). *Introduction to ubiquitous computing*. Hershey, PA: IGI Global.

Mukred, M., & Jianguo, Z. H. (2017). Use of big data to improve environmental sustainability in developing countries. *International Journal of Business and Management, 12*.

Narang, S. (2007). Inclusive and sustainable urban planning: A guide for municipalities. In *An introduction to urban strategic planning* (Vol. 1). UNON.

Noori, N., Hoppe, T., & de Jong, M. (2020). Classifying pathways for smart city development: Comparing design, governance and implementation in Amsterdam, Barcelona: Dubai, and Abu Dhabi. *Sustainability, 12*. MDPI.

Nureni, Y. Information communication technology (ICT): Concepts and applications.

Ogie, R., Perez, P., & Dignum, V. (2017). Smart infrastructure: An emerging frontier for multidisciplinary research. In *Proceedings of the institution of civil engineers – Smart infrastructure and construction* (pp. 1–9).

(1987). *Our common future, Brundtland report; Chapter 2: Towards sustainable development. United Nation world commission on environment and development*. Oslo: United Nations.

Patel, K., Patel, S., Scholar, P., & Salazar, C. (2016). Internet of Things-IoT: Definition, characteristics, architecture, enabling technologies, application and future challenge. *International Journal of Engineering Science and Computing (IJESC), 6*.

Perez, J., Deligianni, F., Ravi, D., & Yung, J. Artificial intelligence and robotics. UK-RAS Network.

Quality of city life multiple criteria analysis – Scientific figure on ResearchGate. Retrieved from https://www.researchgate.net/figure/Indicators-most-widely-applied-for-analysing-sustainable-city

Ratheeswar, K. (2018). Information communication technology in education. *Journal of Applied Science and Research, 3*.

(1987). *Report of the world commission on environment and development: Our common future*. United Nation.

Ruen, K. (2011). The quality of life and the environment. In *Asia Pacific International Conference on Environment-Behaviour Studies, North Cyprus*.

Sarhan, Q. (2018). Internet of Things: A survey of challenges and issues. *International Journal of Internet of Things and Cyber-Assurance, 1*.

Sassi, P. (2016). Built environment sustainability and quality of life (BESQoL) assessment methodology. In W. Leal Filho & L. Brandli (Eds.), *Engaging stakeholders in education for sustainable development at University level. World sustainability series*. Cham: Springer.

Seele, P., & Lock, I. (2017). The game – Changing potential of digitalization for sustainability: possibilities, perils, and pathways. *Journal of Sustainability Science, 12*.

Sen, J. (2012). Ubiquitous computing: Applications, challenges and future trends. In *Embedded systems and wireless technology: Theory and practical application* (pp. 1–40). Boca Raton, FL: CRC Press, Taylor and Francis Group.

(2016). *Skill development for industry 4.0*. BRICS Skill Development Working Group.

Smart Cities and Infrastructure. (2016). *Commission on science and technology for development nineteenth session*. United Nation.

Sung Hwang, J. (2013). *Smart cities Seoul: Case study*. ITU-T Technology Watch Report.

(2018). *Sustainable and digital sustainable development as the framework for digital transformation*. Berlin: German Council Sustainable Development.

Thabet, N., & Soomro, T. (2015). Big data challenges. *Journal of Computer Engineering & Information Technology, 4*(3).

Tregua, M., D'Auria, A., & Bifulco, F. (2014). Digital city towards smart city: A theoretical overview. In *Proceeding of 2nd international virtual scientific conference. Slovakia. 9–13 June 2014*.

Valcárcel-Aguiar, B., Murias, P., & Rodríguez-González, D. (2019). Sustainable urban livability: A practical proposal based on a composite indicator. *Sustainability, 11*. MDPI.

Victor, V., & Fekete Farkas, M. (2018). *The era of big data and path towards sustainability*. Innovation Management and Education Excellence through Vision 2020.

(2018). *What is sustainable infrastructure? A framework to guide sustainability across the project Cycle*. Inter-American Development Bank.

Yeon Mee, K., Hyun Soo, K., Soo Young, M., & So-yeon, B. (2009). *Ubiquitous eco-city planning in Korea. A project for the realization of ecological city planning and ubiquitous network society*. Real Corp 2009: Cities 3.0 – smart, sustainable, integrative.

Zavrl, M., & Zeren, M. (2010). Sustainability of urban infrastructures. *Sustainability, 2*. MDPI.

MAPPING FUTURE URBAN PLAN – TOWARD BLUE-GREEN SMART CITY AND MOBILITY THROUGH THE 5TH WAVE, I-SUSTAINABILITY PLUS, AND DCT THEORIES

(CASE STUDY: SONGDO AND SEOUL IN SOUTH KOREA)

Hamid Doost Mohammadian

ABSTRACT

One of the characteristics of our age is urbanization and the increase in the population of cities and the development of small and large cities. In fact, the sustainable development of a city, which in its general sense means all-round growth and expansion without contradiction, with two main goals: (1) improving the quality of life in terms of the capacity of the environment and (2) responding to the needs of the present generation without restrictions for the future ones. But according to the fifth wave theory, entering the first edge of the future from 2020 to 2030, with the development of urbanization, we are witnessing instability in recent decades. Depletion of forests, thinning of the ozone layer, accumulation of greenhouse gases, noise pollution, threatening air pollution in cities are among these indicators in urban areas. Other important indicators also show the social and economic instability of the world community. The increase in poverty, economic growth without job creation, increase the gap between developed and developing countries, social disintegration, and the sense of citizenship of the residents of local communities and the class gap are among these indicators. Therefore, since 1970, with the advancement of technologies, today the realization of urban development depends on attention to the factors of innovation, digitalization, and intelligentization. In the concept of development, aspects, domains, and scales become important. This improvement of the current situation will be realized when it brings all-round intelligence without conflict and contradiction in meeting the tangible and intangible needs of this society. Growth in material and spiritual aspects, from local to global scales and in different economic, cultural, educational, social,

Smart Cities for Sustainability
Advanced Series in Management, Volume 32, 211–231

Published under exclusive licence by Emerald Publishing Limited
ISSN: 1877-6361/doi:10.1108/S1877-636120230000033013

technological, and even political fields. Therefore, paying attention to the quality makes the concept of development more effective. These qualities are aligned with the concept of sustainability. In this chapter, U-city, smart sustainable mobility through 5th Wave, i-Sustainability Plus, and doost cultural theory (DCT) will be probed as a new concept of urban living for blue-green smart city and mobility. In addition, challenges made some countries not to prefer to utilize smartness and ubiquitous idea as a tool for their problems, will be explored.

Keywords: i-sustainability plus theory; 5th wave theory; digitalization; urban mobility plan; ubiquitous city; doost cultural theory (DCT); ubiquitous mobility; environmental challenges; blue-green smart mobility; blue-green smart city

1. INTRODUCTION

In the twenty-first century, rapid and unplanned urbanization causes global challenges such as poverty, slum, unemployment, economic problems, environmental problems and climate change, infrastructure challenges, social instability, health diseases problems, etc. To maintain high livability and quality of life, it is vital to struggle with these problems. In other words, new sustainable and green urban setting with high quality of life and livability are required. Sustainable cities through digitalization, information and communication technology (ICT) could create new urban settings like smart cities and ubiquitous cities (U-cities).

Ubiquitous idea was born in South Korea in political arena in 2004. U-city is defined as a city in which ubiquitous-based services are offered at any time and in any place in urban space by realizing high-tech IT, construction, traffic application technologies to enhance urban life quality and city competitiveness. U-city is the next-generation informatization city where high sustainability, low environmental problems, and high quality of life exist. Generally, U-city creates new sustainable and green urban setting through digitalization, ICT, and smartness.

The primary goal in this project is to analyze how U-cities could create new urban living with high sustainability and quality of life. In addition, two sub-goals are proposed for this research. The first sub-goal is to indicate which indicators made U-city and smart city. The last aim of the research is to find out why high sustainability and low environmental challenges are required for new urban setting. In order to achieve aims of the research, some questions should be answered:

- Why is it vital to create new urban living such as smart city or U-city?
- How would U-cities improve livability and quality of life?
- Why are sustainability and low environmental challenges required for new urban living?
- What are the role of technology, digitalization, ICT, and smartness in achieving U-city?

In order to answer the research questions and gain aims, it is important to pursue a sound methodology. Systematic literature review (SLR), survey that is based on research, exploring case studies, and analyzing them are proposed as methodologies' tools in this research.

2. LITERATURE REVIEW

2.1 Sustainability

Nowadays, sustainability and sustainable development are remarkable and timely terms at the global level. Generally, sustainable development is declared that society should be stable and indefinitely sustainable in order to improve human situation and life. In addition, ecological stability and sustainable economy, in the long term, would be capable of satisfying basic human conditions and it could affect environmental, social, and economic situation (Feil & Schreiber, 2017). High quality of life, health, and prosperity with social justice and maintaining the earth's capacity to support life in all its diversity are the main aims of sustainable development (Hui, 1996).

Environment, economic, and social well-being are three traditional pillars of sustainability. There must be in equilibrium to achieve sustainability (Brundtland Report, 2019).

Economic development makes a maximum flow of income in terms of rational use, resource efficiency, particularly scarce resources. The aim of economic sustainability is to make a balanced and sustainable economic environment by producing goods and services to preserve manageable levels of government and external debt and avoid extreme sectoral imbalances affecting agriculture and production industrial. Environment development would be the ability to hold whiles the three basic functions of the environment: the power function of resources, waste receptor function, and the direct utility. It is said that "ecological component of sustainable development captures not only the actual economic development in relation to the environment, but also the entire development. Ecological development is closely correlated growth and interrelation with environmental laws, the ecological balance". Environmental development is the capacity to grow and bring the environment and its peculiarities, while ensuring the protection and renewal of natural resources and environmental heritage (Duran, Gogan, Artene, & Duran, 2015). Generally the main aim of environmental sustainability is to leave the Earth in as good or better shape for future (Szczuka, 2015). Social sustainability is related to social interactions, relationships, behavioral patterns, and values of humanity (Duran et al., 2015).

2.2 Digitalization, Smart Life, and Smart Cities

In recent decades, besides of sustainable development, digitalization is one of the most remarkable phenomena. Digitalization is the integration of digital technologies into everyday life that can be digitalized. Digitalization could be one of the indicators that could gain sustainability (Seele & Lock, 2017).

In fact, we can say that with the passage of time and the advancement of technology, smarter cities are created. In fact, smart cities are digital solutions to create a better life. Because as the cities become smarter, the quality of life in them becomes better. Meanwhile, cities will be able to respond more and better to their citizens.

Until several years ago, city leaders thought of smart technologies as efficient tools behind the scenes. But today, technology has directly entered the lives of the residents of cities and villages. Smartphones are known as the keys to cities, and through a short and simple search, you can get instant and practical information on a variety of topics such as transportation, traffic, health services, safety alerts, and community news.

Smart cities use data and digital technologies to make better decisions and improve quality of life. More comprehensive, real-time data gives agencies the ability to see events as they happen. In addition, it provides a better understanding of how demand patterns are changing and provides faster and less expensive solutions (Cocchia, 2014; D'Auria, Bifulco, & Tregua, 2014).

The three layers must work together to build a smart city.

The first layer consists of a massive technology base of smartphones and sensors connected by high-speed communication networks.

The second layer refers to a set of specific applications. Converting raw data into alerts, insights, perspectives, and actions requires appropriate and efficient tools. In this layer, technology providers and application developers come into play.

The third layer deals with the use of cities, companies, and general members of the society in line with the goals. Many programs succeed only when they are widely accepted and can bring about changes in the behavior and culture of society. They encourage people to avoid commuting during rush hours, how and when to change their route, use less energy and water, or better take care of their body and mind by taking special care. In general, it can be said that technology has helped to make people smarter and makes them more aware. In this way, many problems faced by cities and people in the past will be reduced (Trindade et al., 2017).

2.3 Urban Planning

Urban planning could make a city sustainable, smart or normal (Sarma, 2016). Generally, sustainable city is a city designed with environmental consideration, with minimum consumption of energy, water, and food and minimum output of waste and air, water, and soil pollutions (Deffner & Vlachopoulou, 2011). Generally, urban planning has a role in achieving sustainable development (Sarma, 2016).

Urban infrastructure planning is required to achieve urban planning and sustainability in urban setting. Fundamentally, urban infrastructure comprises essential facilities, services, and social structures for cities and communities. Urban infrastructure includes:

- Transport: rail, ports, airports, highways, roads, footpaths, and public transportation.
- Water: water supply, collection and refinement of wastewater, the removal of storm water (collectively the "three waters").
- Energy: electricity and natural gas production, transmission, and distribution.
- Telecommunications: fixed line, mobile, and internet.
- Social and community infrastructure: public recreation spaces, reserves, libraries, schools, and hospitals (Aotearoa, 2017).

Besides, urban infrastructure planning, urban design could lead cities to be sustainable. Urban design and urban planning are related. Urban design is the process of shaping the physical setting for life to deal with the three-dimensional space in cities, towns, and villages, and its objective relies in accordance with the vision of the future that they represent. It focuses on the intersections among architecture, landscape design, and planning. While, urban planning means that it is the organization of all elements of a city/town planning or other urban environment, using different tools (land use plan, the operational plan, or strategic plan, structural plan all are the parts of urban planning). Generally, urban design and planning includes not just the design of public spaces, streets, neighborhood, and homes, but the configuration of and greenway systems, regions growth patterns, transportation network, water and sewerage systems, and even industrial process (Shahreen & Voghera, 2019). Fundamentally, urban plan and design should be aligned with sustainable urban to gain sustainability in cities. Urban sustainability is the ability of cities to diminish the environmental impact of urban activities while improving social equity and livability and promoting economic condition in urban areas. Fundamentally, sustainable urban planning deals with separate issues of an urban system, such as security, livability, and social equity; productivity, personal, and public finances; and pollution levels, amount of reserve habitat, and resource consumption, respectively (Abdullahi & Pradhan, 2017).

2.4 Environmental Challenges

Generally, sustainability is an important phenomena and environmental sustainability is required to gain sustainability. Environmental challenges threaten not only environmental sustainability but also humans' being. So, it is necessary to find out environmental challenges and struggle with them. Environmental challenges are occurred by humans' activity which aren't aligned with environment. These activities could change the environment and disturb environmental sustainability (Bervar & Bertoncelj, 2016).

Besides of climate change, there are more environmental challenges such as air pollution, water pollution, water shortage, sewage, solid waste, noise pollution, soil degradation, biodiversity loss, deforestation, overexploitation, smell, infrastructure problems, agricultural and industrial challenges, and energy.

2.5 The 5th Wave/Tomorrow Age Theory or Theory of Comprehensive Everything

Prof. Dr. Hamid Doost Mohammadian's research on human life from 70,000 years ago to now began in 2010, and it led to the fifth wave theory with the issue of future forecasting in order to prevent and face shocks and crises that humans, societies, technology, businesses, and economic lawyers will face. Today we have finished the fourth wave, we are in a new stage of evolution and innovation called the Fourth Industrial Revolution. Educational technologies, such as the Internet of Things (IoT), have affected the livability and quality of human life to prepare for future concerns. A concept called Edu 5.0 has changed global education policies from traditional methods to modern, innovative, and sustainable methods and created a revolution in the education sector. Smart education has led to the creation of modern fields such as massive open online courses (MOOCs) and e-learning. On the other hand, the increasing speed of technologies, growth and many changes in the world have caused unrest in the soul, psyche, and mind of humans and created educational challenges in societies. So that these challenges require appropriate solutions to face them. Basically, this theory states that the threshold of tomorrow is from 2020 to 2030; and on the edge of tomorrow, we face shocks such as the educational challenges of tomorrow, the dangers of the modern educational style, the obstacles of new businesses and the challenges of human competencies, etc.

The figure shows the theory of wave 5/the era of tomorrow, the ages and technologies we have passed and the point where we are (Fig. 12.1).

2.6 i-Sustainability Plus Theory

The continuous changes and transformations that are happening at a high speed are one of the most obvious features of the present age that can be seen in many aspects of life and have placed the achievement of the organizations' goals in a complicated and difficult path. Today, for their sustainability and survival, organizations must be able to quickly adapt to developments in various fields in order to remain competitive and meet the needs of society. In this regard, educational policies, creativity, and innovation are considered as a determining factor in the long-term success and survival of organizations and societies.

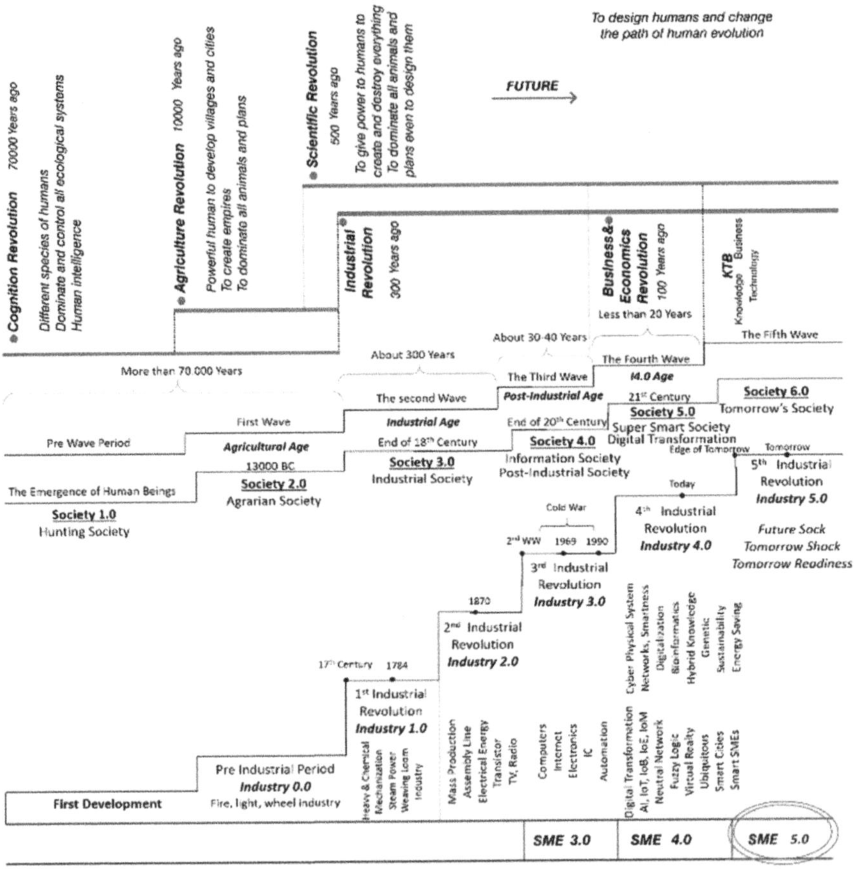

Fig. 12.1. The 5th Wave/Tomorrow Age Theory (Doost, H. 2010–17) (Author).

i-Sustainability Plus is a theory of work with high sustainability, innovation, and digitalization. The intelligence and digitality of this structure, which includes the concept of Edu 5.0 and the idea of sustainable intelligent education, is explored as a new idea of tomorrow's societies and innovative ecosystem in schools and academic society. In other words, this theory consists of a triad that, with today's rapid technological advances, makes businesses more innovative to be more sustainable and stable against the future of the fourth industry. The importance of digital learning in this theory, innovative management is innovation in business, marketing and technological, and educational systems of an organization to achieve success and best results. To achieve educational sustainability, cultural, environmental, social, economic, technological, and political dimensions should be improved almost equally. In general, social justice, livability and high quality of life, preserving the environment are the main goals of sustainable development. Sustainable development strategies based on three environmental, social, and economic indicators are a solution to maintain ecological integrity, protect the environment, achieve equality, meet basic human needs and cultural diversity, and improve the quality of life. Due to having the roadmap and theories of the fifth wave and i-Sustainability, we are able to predict and prevent educational shocks; and finally, by using open innovation, implementation, development, and use of these digital and educational technologies to provide an equal gender for sustainable leadership in engineering education through the theory of tomorrow, let's face educational crises with strength (Fig. 12.2).

2.7 Doost Cultural (DCT) Theory

According to Prof. Doost's DCT theory, education mirrors culture through the educultural sustainability. Sustainable development is the result of mainly local activities with each person's awareness of the development situation, which should be the basis of comprehensive and human development. This type of development has different environmental, economic, social, cultural, technical, educational, and sometimes political dimensions that form the process of sustainable development in relation to each other. In the meantime, the cultural dimension is very important as the foundation of this model, because every society and thought provides a definition for culture based on its worldview. Culture is not an abstract subject and the same as other social issues, but culture is a spiritual space in which a person or society walks. Therefore, the influence of culture in various fields of life, including education, is undeniable. The culture of educational technology innovation in society can be a catalyst for development and sustainability. According to the fifth wave theory, in order to identify, prevent, and deal with today's shocks and tomorrow's crises, education and culture building are essential steps in innovation and development and play an important role like a backbone. For this reason, the cultural vision and future integration of the fourth industry of small and medium-sized enterprises (SMEs) as a symbol of Western culture and the future of Society 5.0 (Society 6.0) as a symbol of non-Western culture is the basis of policy for industry and technology. Education and engineering education is culturally very sensitive. What "works" in one culture for women sustainable innovative in engineering education might be seen as negative in some other cultures which cannot work. This chapter describes and analyzes the cultural experiences of expat women from different cultures. DCT is to help societies, educators, and women leaders toward blue-green sustainable education and help women engineer leaders to deal with cultural differences and achieve cultural adaptation with a focus on recognizing, sharing, allocating, and balancing of the resources in engineering education. DCT theory is three dimensional: knowledge, cultural adaptation, and cultural differences. The first dimension, knowledge, includes tangible culture and intangible culture, for example, language and

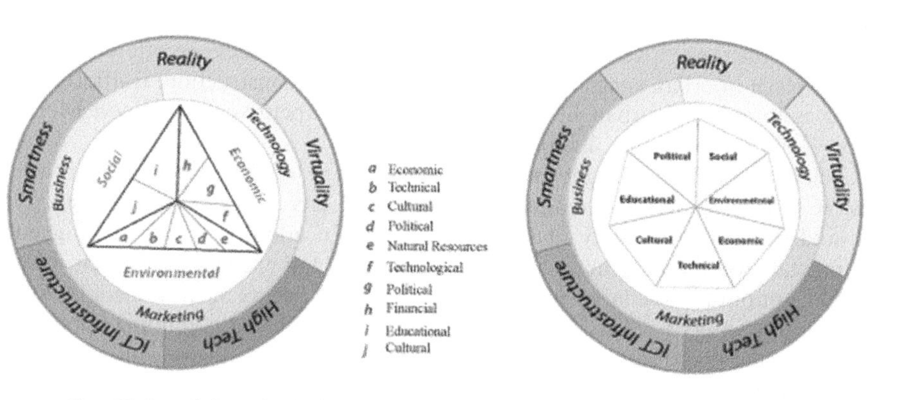

Fig. 12.2. i-Sustainability Plus Theory (Doost, H. 2010–17) (Author).

food belong to visible and tangible culture, while religion and psychology are invisible and intangible culture. When it comes to cultural differences, there are four of the most well-known models by researchers:

(1) Hofstede,
(2) Victor,
(3) Trompenaars,
(4) Hall and Globe.

The last dimension refers to cultural adaptation, all factors from dimensions of sustainability plus are included. In addition to the factors of the high sustainability model, competencies like:

- 7PS factors competencies,
- Communication competencies,
- Digital transformation,
- Legal issues,
- Intercultural transformation,
- Predisposition, and
- Cross cultural shock

should also be considered. These three quadrants together form a cube, data analysis is performed using multiple regressions, and a single-line calculation is performed at the three-dimensional level based on DCT. This theory supports leaders to manage engineering education through the 5th wave theory and to carry out gender adaptation for women leadership in innovative, technology, and entrepreneurship educationally (Fig. 12.3).

2.8 Ubiquitous Life and Ubiquitous Cities (U-Cities)

ICT has been proposed as the main axis of transformation and development of the third millennium. In this area, the city is also changing due to the evolution of the concept of urban space and the emergence of new ICT infrastructure. In other words, today the role of information technology and communication infrastructures are extremely important, the reason for this is the emergence of environmental, social, and economic opportunities and the disappearance of time and place limitations (Lee, Han, Leem, & Yigitcanlar, 2008; Jang & Suh, 2010).

U-city is defined as "a city in which ubiquitous based services are offered at any time and in any place in urban space by realizing high-tech IT, construction, traffic application technologies to enhance urban life quality and city competitiveness" (Yeon Mee, Hyun Soo, Soo Young, & So-yeon, 2009).

In U-city, any citizen can get any services anywhere and anytime through any ICT devices. ICT development has made the conventional city in terms of intelligence, innovation, and evolution to E-city and then to U-city. In addition, U-city is aligned with environmental sustainability and sustainable smart (or knowledge) city which makes the ubiquitous computing available to among the urban elements such as people, building, infrastructure, and open space. High speed networks, advanced communication services, digitalization, and smartness have an important role in U-cities. Digitalization such as ICT is used in several areas in U-cities like social science, geography, planning, and architecture (Lee et al., 2008) (Fig. 12.4).

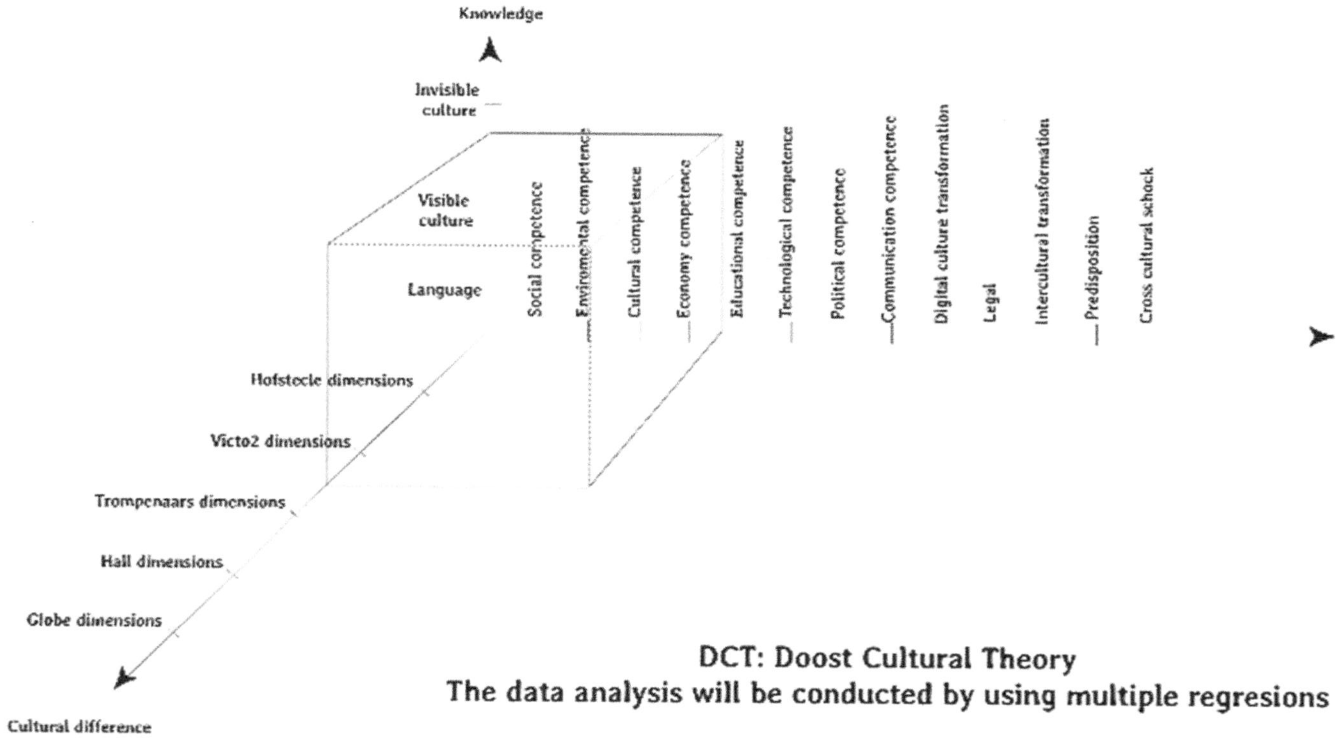

Fig. 12.3. DCT (Doost Cultural Theory, Doost. H, 2012–17) (Author).

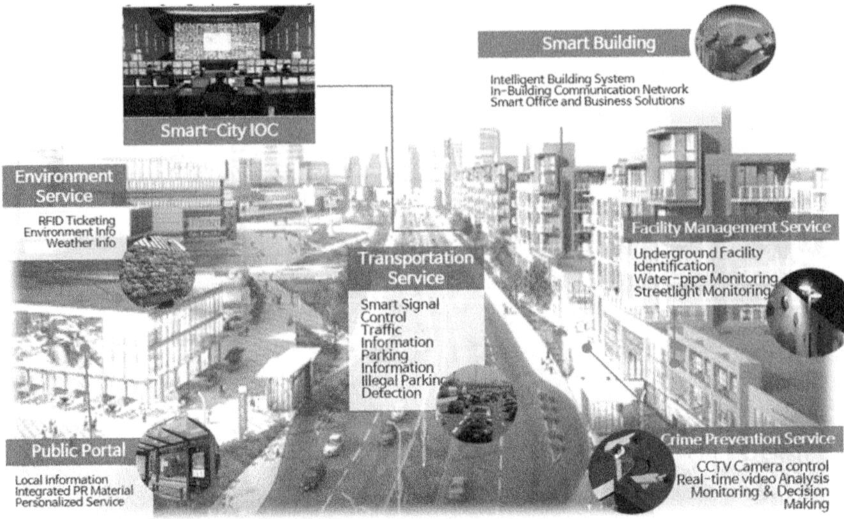

Fig. 12.4. U-City's Component.

Fundamentally, U-city is the next-generation informatization city that can innovate every function of the city such as improvement of quality of life, increase in convenience of urban life, security guaranteeing by systematic city management, new industry creation and improvement of citizen welfare (Jang & Suh, 2010).

U-life is a new idea of living encompassed U-city services such U-education, U-banking, U-home, U-healthcare, U-shopping, etc. Generally U-life and U-services have improved quality of life through producing new job, increasing citizens' knowledge, promoting security, and other advantages. Fundamentally, U-government, U-business, U-mobility, U-port, smart and intelligent building, and U-services are component of U-life (Fig. 12.5).

3. CASE STUDY

That being said, South Korea continues to lead the way in digital opportunities with its recent, innovative and inclusive urban projects. The U-city initiative in 2008 in South Korea is a national urban development project with inclusive technologies, which focuses on strengthening the role of ICT in civil planning and management (Yeon Mee et al., 2009) (Fig. 12.6).

The U-city construction and implementation plan includes construction, management, services, financing, and a system to bring forward the project. Misunderstanding about relevant technologies and services, resulting in budget waste due to multiple investments for the same technologies stipulates and error possibilities are the most important mistakes concerned on U-city construction. To minimize mistakes was the enactment of the law and the guideline for U-City infrastructure establishment (Yeon Mee et al., 2009) (Fig. 12.7).

U-technologies and ecology in cities create U-eco city and it was developed at 2007 for the first time in Korea. This project lasted from 2008 to 2013.

The important and main dimensions of the project are:

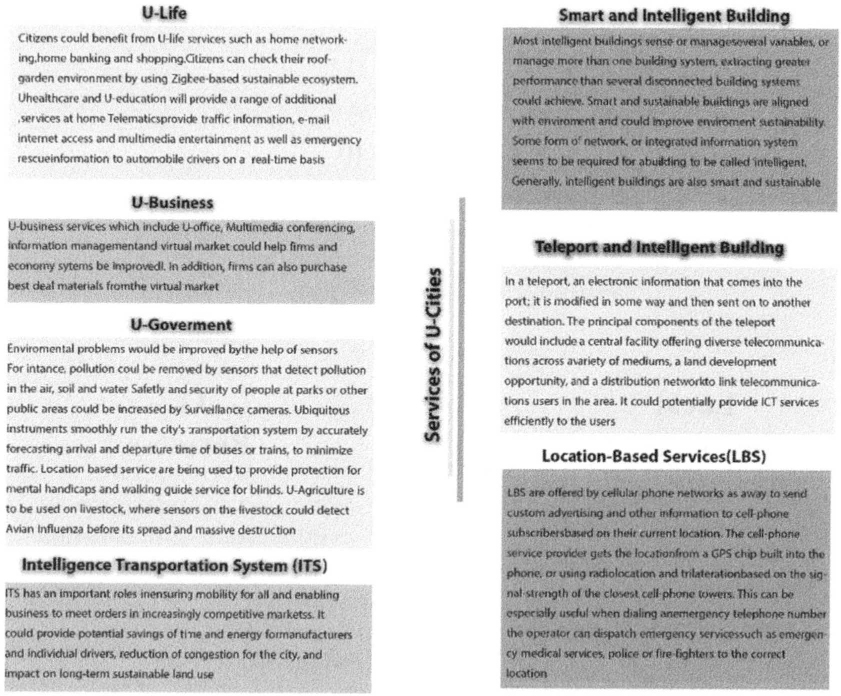

Fig. 12.5. Services of U- Life (Lee et al., 2008).

- Fusion technology of eco-city construction based on u-technology.
- Planning, design, evaluation of eco-city based on u-technology.
- Construction technology based on low energy and recycling.
- U-water water circulation system manufacturing technology.

The main and important goals of this project are:

- Future-oriented and sustainable urban design.
- Urban management technologies based on the combination of comprehensive infrastructure and ecological system.
- Creating a comfortable urban environment for urban citizens (Yeon Mee et al., 2009).

It should be said that cities are the main platform for using information technology; for this reason, the importance of information in the decision-making process of information and communication technology is increasing rapidly. As a result, the expansion of smart cities can create many opportunities for living, working, and leisure environments.

Seoul and Songdo will be explored as an example of U-city in this research.

3.1 Seoul

Currently, South Korea is considered one of the leading countries in the field of using new technologies. Now its capital city, Seoul, has gone a step further than a smart city and is

Fig. 12.6. The Main U-City Operation Project in Korea (Yeon Mee et al., 2009).

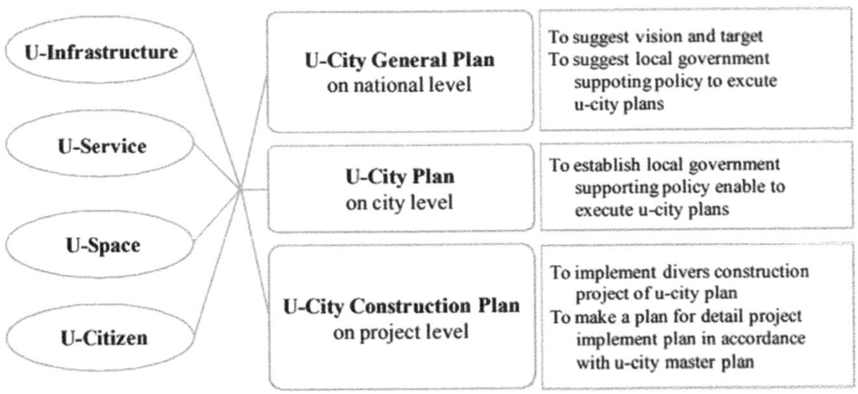

Fig. 12.7. U-City Planning System (Yeon Mee et al., 2009).

planting the seeds of a metaverse ecosystem called "Metaverse Seoul" for all its urban management areas. This effort combines the twins of augmented reality and virtual reality to improve urban service delivery as well as virtual tourism planning, management, and support.

3.2 Songdo

A smart city uses information and communication technology to create, deploy, and promote technologically integrated infrastructure to address urban challenges and improve operational efficiency, share information with citizens, and provide better quality government services, and promote economic growth. Songdo International Business District in South Korea, which is also known as the ubiquitous smart city, is considered one of South Korea's smart cities. In this area, smart strategies have been fully utilized for management so that structures, trash cans, public spaces, and all affairs are done by digital technology. And since 2000, Songdo has been considered as one of the most advanced and smartest cities in the world. This smart city is an ambitious project that will be built on an artificial island 56 kilometers west of Seoul, the capital of South Korea. In fact, this city can be seen as a glimpse of the future of Seoul and other cities in the world in the not too distant future. According to United Nations forecasts, in less than 40 years, more than 70% of the world's population will live in cities. This means 6.4 billion people out of the 9.2 billion population at that time! Currently, cities are home to half of the world's seven billion people. However, the addition of another three billion people to the population of current cities will bring their capacity to their limits, and it seems that the way out of this problem is the construction and development of smart cities. In fact, a city that interacts with its residents and the surrounding environment (Ilhan, Möhlmann, & Stock, 2015; Simon, 2014).

4. DISCUSSION

The main aim of this research is to find out novel idea and concept of urban life. Specific infrastructure, tools, and techniques are vital to achieve this goal. Telecommunication, regulation, mobility, financial stability, and generally i-Sustainability Plus could be used as tools and techniques to create new urban setting with high quality of life and livability.

As it was mentioned sustainability has three main aspects: environmental sustainability, economic sustainability, and social sustainability. In authors' viewpoint, there are four more aspects concerned on these three mains pillars: educational sustainability, cultural sustainability, political sustainability, and technical sustainability. A country needs these pillars of sustainability to gain sustainable infrastructure and generally sustainability. These seven aspects create a puzzle that is named sustainability; all these parameters are relevant and directly or indirectly could influence each other.

Education is an important issue for future of each country. Generally, the future of a country is concerned on their juveniles. Skillful, perseverance, and responsible youths are required to achieve sustainable society; so, high and sustainable education is needed to grow innovative and entrepreneur children who are able to create sustainable and successful society. Only education is not sufficient for sustainable and successful society. Besides of sustainable education, technical sustainability is required to gain sustainable society. Technical sustainability could create new technologies and techniques to make new urban setting with novel idea where high quality of life and livability are existed. In addition, democracy, freedom, equality, transparency, honesty, and political stability are components of sustainable society. Therefore, political sustainability for high quality of citizens' life is required. Cultural deviation is a significant threat for future of each country. Fundamentally, not only political and economic problems cause revolutions, but also cultural deviation is another reason for revolutions. Cultural norms could effect on social, political, and economic aspects of a society. So, it is vital to achieve cultural sustainability to create a country with high sustainability. Besides of all these aspects, economic aspect has an important role in achieving sustainability in a country. Proper management strategies, stable banking, economic stability, appropriate exports, proper strategies concerned on business and companies especially SMEs, stable communication in business are formed sustainable economic. Economic policy makes human well-being, quality of life, and equality without threatening the ecosystem's sustainability. Essentially, a balance between work and life, opportunities for job and education, welfare, wealth, social responsibility, high quality of life, certified future, and social security are vital for a sustainable country. These are indicators of social sustainability. Social responsibility and social sustainability could develop quality of life, livability, and sustainability in a country. Besides of all these parameters, environment has an important role in developing sustainability. Environment is one of the most controversial issues at recent decades that concerned on global. Environmental sustainability is required to gain all aspects of sustainability. Sustainable environment means that humans' activities are aligned with environment not to threaten the future of environment and resources. Fundamentally, these seven parameters are required to gain sustainability in a country.

Fig. 12.8 shows the seven aspects of high stability and its classification, that all these seven parameters should be improved almost equally, and the blue shape is more stable than the red one due to its regularity.

Fundamentally, sustainability makes sustainable infrastructure in cities that is necessary to achieve sustainability and green life in a country (Fig. 12.9).

Nowadays, although sustainable city is a proper key to improve quality of life and livability in urban setting, it isn't sufficient. In other words, a new concept and idea more than sustainability is required. Rapid and unplanned urbanization causes a lot of problems such as environmental problems, slum, poverty, economic problems, social instability, health problems, traffic, and infrastructure challenges in cities especially big and hub cities like capital. So, besides of sustainability, high precise, high speed, and high technology are needed in cities to gain high quality of life and livability in such cities. Furthermore, digitalization and ubiquitous are declared as tools to create new concept for urban settings. Fundamentally, combination of seven parameters of sustainability (educational, technical,

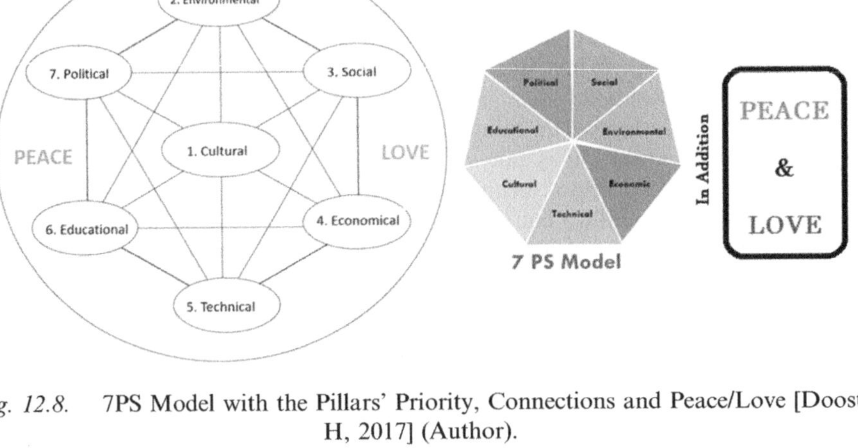

Fig. 12.8. 7PS Model with the Pillars' Priority, Connections and Peace/Love [Doost. H, 2017] (Author).

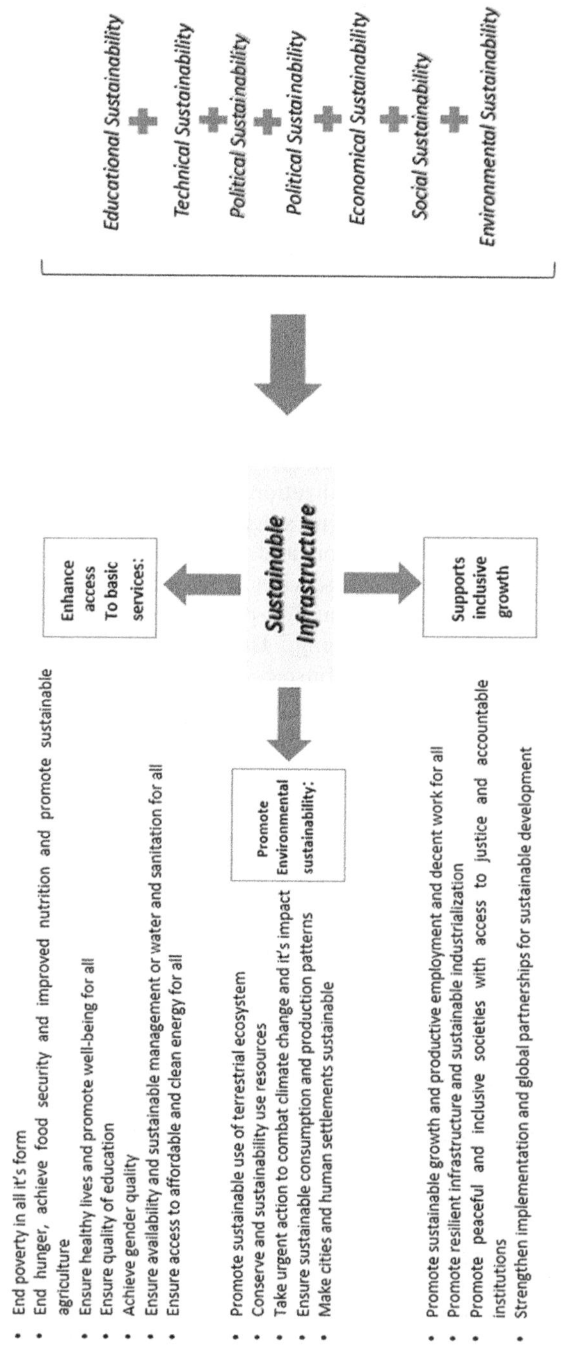

Fig. 12.9. Advantages of Sustainable Infrastructure (Author).

political, cultural, social, economic, and environmental) and ubiquitous idea included digitalization, high technology, and virtual reality create new concept named sustainability plus. Prof. Dr. Doost Mohammadian declares i-Sustainability Plus as a new concept concerned on green urban setting with high quality of life and livability (Fig. 12.10).

U-life, U-mobility, U-building, U-health, U-government, U-politic, U-education, U-security, U-airport, U- technique, U-infrastructure, U- citizen, and U-services are component of U-city. These components are performed by digitalization, high technology. U-security is one of the most remarkable aspects of U-city. For instance, cameras existed on buses, automobiles, and every building creates high security and safety in U-city. U-mobility is concerned on transportation, traffic, and accident. Generally, a cyber person besides driver or sensors showed ahead traffic could reduce accidents or application indicated the most proper transit services near a person would improve public transportation. These are concerned on U-mobility. U-health is one of the most aspects of U-city. Through U-health, health diseases, and health problems would be reduced. For example, applications related to personal health signs such as pressure, pulse, etc., will help people be healthier in less expensive way. New technology such as computers, mobiles, and virtual learning happen a revolution in education that could be named U-education. It would improve self-regulated learning, personal knowledge, and increase well-educated people. Fundamentally, technology and digitalization alter e-government to ubiquitous government. For instance, in Seoul government services for citizens are done through mobile devices. It is declared that participation and real-time information services are factors make U-government successful. U-life services would improve quality of life and livability. U-banking, U-shopping, U-airport, and U-port are existed as the most important services in U-cities. U-business involving, U-office, management, multimedia conferencing, virtual marketing, and other U-business services could promote economic situation. In addition, smart buildings and construction in U-city could improve quality of life, reduce energy consumption, and reduce environmental impact. One of the most substantial aspects of U-city is U-infrastructure. Infrastructure in concerned on power, energy, water, waste management, transportation, mobility, construction, urban plan, and U-services. For instance, pollution in soil, air, and water could be eliminated by sensors, or in Songdo combination of waste management is utilized to reduce pollution related to waste. In addition, using renewable energy could enhance infrastructure of U-cities such as Songdo and Seoul. Besides of all these parameters, U-citizens are required to develop U-city. In other words, inhabitants who will use technology, digitalization, and ICT have an important role in achieving U-city.

One of the most significant advantages of U-city is Internet of Energy (IoE). Through technology and information technology (IT), it is possible to integrate the planning, monitoring, control of power process and management of energy sharing, optimizing energy efficiency, and minimizing wastage. Generally, IoE manage energy to operate cost, minimize environmental impact, increase profitably, reduce energy consumption, reduce energy wastage, and save energy. Therefore, IoE could improve sustainability in a country. Fundamentally, digitalization and technology could influence on IoE and improve sustainability in U-city.

Digitalization and new technology not only could make U-life, but also, they could influence urban planning. In other words, urban planning and strategic plan would improve by digitalization and technology. For instance, all aspect of risks such as project risk management (PRM), financial risk management (FRM), governmentality risk management (GRM), enterprise risk management (ERM). In Seoul, a thief was arrested just in 10 minutes, because cameras of buses indicated where the thief was, so high security and

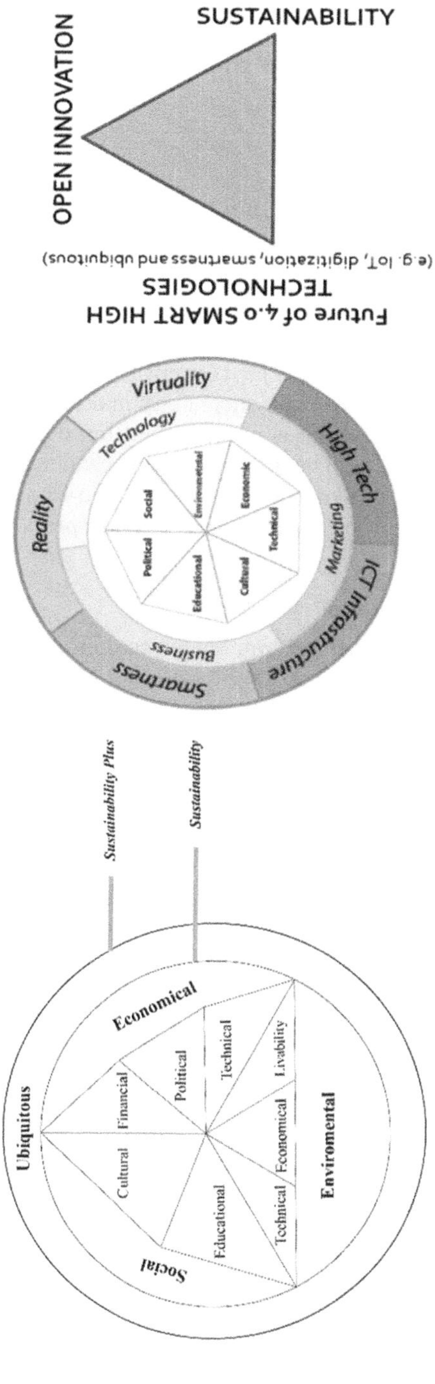

Fig. 12.10. i-Sustainability Plus Theory (Author).

safely is existed in U-city. In addition, U-business and U-government reduce risk management. Therefore, risk management will be reduced in such a city and generally, urban strategic plan could develop through ubiquitous and digitalization.

Fundamentally, sustainability plus city is a city where sustainability and ubiquitous idea are existed. The main aim of this city is to declare new green and sustainable area for living with high quality of life and livability (Fig. 12.11).

5. CHALLENGES

U-city and sustainability plus (sustainable U-city plus) are the best responses of new and green urban setting with high quality of life and livability. Generally, U-city is a proper solution for capital and huge city struggled with urbanization problems such as slum, population, poverty, unemployment, economic problems, health problems, environmental challenges, etc. Asian countries like Korea utilize this concept to deal with problems of huge city and create better area for living. Although such a city makes new urban setting with high quality of life, just few countries use this concept to confront with their problems and develop better situation for their citizens. The most important challenge concerned on this idea, is that: Why don't some countries like western countries and Iran use ubiquitous idea to struggle with their urbanization problems? Some reasons are proposed:

5.1 Cultural Norms

Through Hofstede's viewpoint, there are six indicators related to culture: power distance, individualism, masculinity, uncertainty avoidance, long-term orientation, and indulgence.

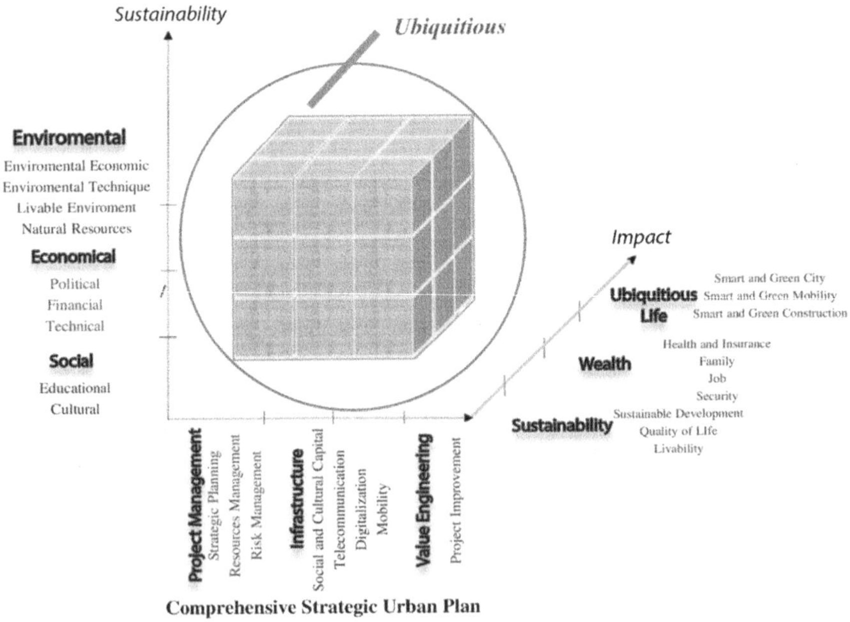

Fig. 12.11. U-DSR Model (Author).

Western countries are very different from Asian countries in these six dimensions. For instance, Western countries like America are individualism, while Asian countries like Korea are collectivism. Generally, in collectivistic society, everyone takes responsibility about group and all society, but in individualist cultures people care about themselves and their direct family. In addition, most of Western countries are masculine, hence Asian countries are feminine. In masculine society, people try to compete and be successful. Feminine society, quality of life and carrying about all citizens are important. These parameters make difference in societies and their viewpoints. Some countries prefer to utilize ubiquitous idea to make urban life better, because of their culture and quality of their life, while some other countries are sustainable through Hofstede's viewpoint and inhabitants are satisfied by their life. So, difference in culture make some countries like Korea use this concept as a new idea for better urban living.

5.2 Security Issues

In such a city, transparency is the main idea. In other words, transparency is the base of U-city. The contradictions exist between transparency, security, and open data policies. In some countries, government thinks that some security and political aspects aren't aligned with concept of transparency like military budget. In other words, some societies think that more transparency makes security worsen. So, some governments prefer not to use open data polices related to security aspect of country.

5.3 U-City Disadvantages

Although U-city has a lot of benefits, there are some disadvantages concerned on U-city. High cost and human capital familiar with technology to build U-city are the most important challenges related to U-cities. Therefore, educating inhabitants how to live in such a city, needs proper strategies. Fundamentally, cultural and educational infrastructure for citizens are required; so, it is one of the most significant challenges concerned on U-life. In addition, cultural shock, technical shock, lack of privacy, weaken of family relation, attenuate humanity, machinist life, reducing innovation because of using high technology, increasing stress and depression through high comfortability and rising physiological diseases are disadvantages of U-life.

5.4 U-City as New Idea for Sustainable Development

U-services create sustainable development in a new concept. In other words, U-services could improve all aspects of sustainability and lead the country to promote sustainable development in new era that besides of sustainability; digitalization, technology, and ubiquitous idea have roles in achieving sustainable development. Generally, U-services influence city infrastructure such as economic, social, cultural, technical, educational, political, and environmental aspects and make them be improved; while, some policies aren't aligned with benefit of U-services. For instance, U-health could reduce hospital or U-education could decrease number of universities and schools. These could cause financial problems for businesses. So, some businesses are against U-life.

6. CONCLUSION

In these decades, capitals and huge cities confront a lot of problems such as slum, population, poverty, unemployment, economic problems, health problems, environmental challenges, etc. These challenges threaten humans' being and diminish quality of life and livability in such cities. So, new concept for urban setting is required. U-city is declared as new idea for sustainable, green city with high quality of life and livability.

Generally, sustainability plus has seven indicators: educational, political, cultural, technical, social, economic, and environmental. To achieve sustainability plus in cities, these seven parameters should develop approximately equable. Sustainability plus isn't sufficient enough for huge cities to struggle with urbanizations' problem, so ICT, new technology, and digitalization are used as new tools to create new idea for urban life. These tools make U-life. Sustainability plus, digitalization, smartness, and new technology are utilized to design a new urban setting named ubiquitous city. Generally, U-city is combinational of real life, high technology, and virtual reality. The main aim of U-city is to create green and sustainable city with high quality of life, livability through sustainability plus, digitalization, and technologies.

Although U-city is the best solution for high quality of life in urban setting, few countries receive this idea to deal with their problems. Cultural issues, transparency, security aspects are the most important challenges that some counties prefer not to utilize ubiquitous idea as a solution.

Fundamentally, U-city is a novel idea for green cities to gain high quality of life.

REFERENCES

Abdullahi, A., & Pradhan, B. (2017). *Sustainable urban development*. Spatial Modeling and Assessment of Urban Form.

Aotearoa, T. (2017). *Better urban planning. Chapter 10: Urban planning and infrastructure*. The New Zealand Productivity Commission.

Bervar, M., & Bertoncelj, A. (2016). The five pillars of sustainability: Economic, social, environmental, cultural and security aspects. In *Management international conference. Pola, Croatia*.

Brundtland Report. (2019). Chapter 2: Towards sustainable development. United Nation World Commission on Enviroment and Development.

Cocchia, A. (2014). *Smart and digital city: A systematic literature review*. Springer International Publishing Switzerland.

D'Auria, A., Bifulco, F., & Tregua, M. (2014). Digital city towards smart city: A theoretical overview. In *The 2nd year of international virtual scientific conference*.

Deffner, A., & Vlachopoulou, C. H. (2011). *Creative city: A new challenge of strategic urban planning?* Regional Science Association.

Duran, D., Gogan, L., Artene, A., & Duran, V. (2015). The components of sustainable development – A possible approach. In *4th World conference on business, economics and management, WCBEM*.

Feil, A., & Schreiber, D. (2017). Sustainability and sustainable development: Unraveling overlays and scope of their meanings. *Cadernos EBAPE BR, 14*.

Hui, S. (1996). *Sustainable architecture and building design (SABD)*. Retrieved from http://www.arch.hku.hk

Ilhan, A., Möhlmann, R., & Stock, G. (2015). *Citizens' acceptance of U-life services in the Ubiquitous city Songdo*. Springer Science+Business Media Singapore. Citizen's Right to the Digital City.

Jang, M., & Suh, S. (2010). *U-city: New trends of urban planning in Korea based on pervasive and Ubiquitous geotechnology and geoinformation*. Springer-Verlag Berlin Heidelberg.

Lee, S. H., Han, J. H., Leem, Y. T., & Yigitcanlar, T. (2008). Towards ubiquitous city: Concept, planning, and experiences in the Republic of Korea. In *Knowledge-based urban development: Planning and applications in the information era* (pp. 148–170). IGI Global.

Sarma, A. (2016). *Smart city and urban sustainability*. Institution of Engineers (India). Tech Vol cum Souveni.

Seele, P., & Lock, I. (2017). The game-changing potential of digitalization for sustainability: Possibilities, perils, and pathways. *Sustainability Science Journal, 12*.

Shahreen, F., & Voghera, A. (2019). *Urban planning and design methods for sustainable development*. Retrieved April, 22.

Simon, M. (2014). Songdo, Korea: Aerotropolis, metropolis, and cyberopolis. In *102nd ACSA annual meeting, globalizing architecture/flows and disruptions*, at Washington, DC.

Szczuka, M. (2015). Social dimension of sustainability in CSR standards. In *6th international conference on applied human factors and ergonomics (AHFE 2015) and the affiliated conferences, AHFE 2015*.

Trindade, E. P., Hinnig, M. P. F., da Costa, E. M., Marques, J. S., Bastos, R. C., & Yigitcanlar, T. (2017). Sustainable development of smart cities: A systematic review of the literature. *Journal of Open Innovation: Technology, Market, and Complexity*, *3*(3), 1–14.

Yeon Mee, K., Hyun Soo, K., Soo Young, M., & So-yeon, B. (2009). *Ubiquitous eco-city planning in Korea. A project for the realization of ecological city planning and ubiquitous network society*. Real Corp 2009: Cities 3.0 – Smart, sustainable, integrative.

INDEX

Air pollution, 20–21
Air quality monitoring systems, 20
Aliveness concept, 187
America Online (AOL), 195
Amsterdam Smart City platform, 41
Anti-tourism sentiment, 28
Archeology and digitalization of cultural heritage tourism, 159
Artificial Intelligence (AI), 9, 18, 51, 152, 164, 185–186
Augmented reality (AR), 19, 152, 160–161, 164
Autonomous vehicle, 60–61

Battery electric vehicles (BEV), 60
Bibliography, 172
Bibliometrics, 172
  analysis method, 172–173
Bidirectional communications process, 51
Big data, 8, 51, 139–140, 143–144, 152, 185–186
  analytics, 24–25
  platforms, 24–25
Biodiversity, 141
Bioregionalism, 35
Blue Flag, 23–24
Blue-Green city, 88, 130
Blue-Green mobility, 94–96
Blue-Green Smart Comprehensive urban plan, 94
Blue-green smart sustainable mobility, 128–129
Blue-green smart U-cities
  blue-green city, 88
  blue-green ubiquitous cities as livable and sustainable urban settings with high quality of life and less greenhouse gas emissions, 92–96
  case study, 96–98
  challenges and solutions, 99
  Denmark, 96–97
  Finland, 97–98
  greenhouse gas emissions as significant threats for sustainable threats, 88–89
  literature review, 85–88
  restrictions, 87
  sustainability, 85
  toward sustainable cities with livability and high quality of life, 89–92
  ubiquitous city, 87–88
  urbanization challenges, 85–87
Blue-Green strategies, 95, 98–99
Blue-green sustainable mobility, 107–108
Blue-green sustainable urban plan to blue-green sustainable mobility, mobility risk management, 119–120
Blue–Green comprehensive strategic urban plan, 133
Blue–green smart city and mobility
  case study, 220–223
  challenges, 228–229
  cultural norms, 228–229
  DCT, 217–218
  digitalization, smart life, and smart cities, 213–214
  environmental challenges, 215
  5th Wave/tomorrow age theory or theory of comprehensive everything, 215
  i-sustainability plus theory, 215–217
  literature review, 212–220
  security issues, 229
  Seoul, 221–222
  Songdo, 223
  sustainability, 212–213
  U-City as new idea for sustainable development, 229
  U-City disadvantages, 229
  ubiquitous life and ubiquitous cities, 218–220
  urban planning, 214–215
Blue–Green ubiquitous cities, 94

British oil company, 56
"Broadband Village" project, 42
Building information model (BIM), 159
Built environment sustainability and quality of life (BESQol), 119, 193
Business economy, 111–118
Business life, 73

Carbon dioxide emission (CO_2 emission), 86, 88–89, 170
Challenges of urban transport solutions, 63–66
Chenggong New District, 42
Chinese language, risk, 127
Cities, 71–73, 139–140
CIVITAS ECCENTRIC, 41
Clean energy, 201, 203–204
Clean mobility, 204
Climate change, 22–23, 78
Close-range photogrammetry, 159
Cloud computing, 9, 92, 94, 143–144, 152
Communication technologies (CT), 40
Comprehensive blue-green sustainable strategic urban plan, 108, 129–130
Comprehensive strategic urban plan, 129–130
Confluence District Project, 43
Conventional vehicles (CVs), 60
Copenhagen waste management program, 78
Corona epidemic, 191
Corporate social responsibility (CSR), 190
Covid-19 epidemic, 4
Cultural deviation, 90, 223–224
Cultural diversity, 141
Cultural heritage
  digitalization of cultural heritage for sustainability of tourism, 158–161
  tourism, 158
Cultural sustainability, 90
Customer relationship management (CRM), 25–26
Cyber-physical systems (CPS), 10

Data collection process, 173
Data mining, 8, 152
Decarbonization, Decentralization, and Digitalization (D3 Revolutions), 112
Decision-making process, 221
Deep ecology, 44–45
Deep learning, 51
Department of Smart Cities, 78
Digital archiving, 159
Digital city, 5–6, 195
Digital close-range photogrammetry, 159
Digital learning, 216–217
Digital methods in archeology, 159
Digital platforms, 41–42
Digital systems, 148
Digital techniques, 159
Digital technology, 161
Digitalization, 87–88, 112, 158, 186, 202, 213–214, 218, 226, 228
Digitalization of cultural heritage, 161
  acknowledgment of digitalization for sustainable cultural heritage tourism management, 159–161
  archeology and digitalization of cultural heritage tourism, 159
  for sustainability of tourism, 158–161
  tourism, 159
Dongtan Eco-City, 42
Doost Cultural Theory (DCT), 110, 217–218
Doost Mohammadian's concept of tomorrow's SMEs, 114
Doost research method (DRM), 110, 192
  methodology, 193
Doost Risk Mitigation Method (DRMM), 107, 110
  risk with, 120
Doost Sustainability Digitalization Impact Comprehensive Plan (D-SDIC model), 110

E-commerce, 8
e-learning, 215
E-pattern, 79
E-tourism, 17
Eco-communities, 34–35
Eco-friendly practices, 16, 28
  green applications, 23–24
  green energy, 22–23

green infrastructure, 23
information communication technologies, 19–22
low carbon, 24
problems and solutions, 27–28
smart tourism and, 18–24
Eco-friendly tourism practices, 19
Eco-gastronomy, 74, 77
definition, importance, and features of smart city, 74–77
e-pattern, 79
examples of smart cities in World and in Turkey, 77
farmer information system, 79
Gaziantep ecological village project, 78
gluten free cafe online sale, 78
Izmir, 78
Konya, 79
smart city examples in Turkey, 78
smart city examples in world, 77–78
smart plant factory in city, 78
sustainability, 72–74
sustainable cities-eco cities and eco-gastronomy, 77
www.konyapedia.com, 78–79
Eco-neighborhoods, 34–35
Eco-city, 34, 36–37, 77, 146–147
Builders Foundation, 36–37
concept of sustainability and sustainable tourism, 140–142
smart cities and smart tourism destinations, 142–144
smart eco-city tourism, 145–148
standards, 146
tourism, 146–147
Ecologicalization process, 35–36, 45
Economic development, 213
Economic Growth, 141
Economic policy, 90
Economic risks, 105–106
Economic sustainability, 90
Ecophobia, 44–45
Edu 5.0 concept, 215–217
Education, 223–224
Educational sustainability, 89
Educational technologies, 215
Electric technology, 60
Electric vehicles (EVs), 60
Electricity, 21
Emerged technologies, 63
Employment, 139–140
Employment Quality, 141
Energy, 170
management, 201
monitoring systems, 21
resources, 170
Enterprise Risk Management (ERM), 87, 105, 226, 228
Environmental challenges of blue-green smart city and mobility, 215
Environmental cleaning, 141
Environmental development, 213
Environmental risks, 105–106
Environmental strategies, 125–126
Environmental sustainability, 84, 90
Environmental technologies, 40
European Fund for Regional Development, 41
European Union, 4, 6
Eco-Label, 23–24
Explanatory mixed methods, 192

Failure modes and effects analysis (FMEA, 107
Farmer information system, 79
Fifth wave theory (5TH wave theory), 109, 111, 116, 118, 187, 189
blue-green smart sustainable mobility, 128–129
calculate risk through DRMM Method, 120
case study, 122–127
challenges and solutions, 132
comprehensive blue-green sustainable strategic urban plan, 108, 129–130
DRMM, 107
Germany, 125–126
Iran, 126–127
literature review, 104–118
livable urban setting, 108–109
methodology, 118–122
mobility, 107–108
mobility risk management provide comprehensive blue-green sustainable urban plan to

blue-green sustainable mobility, 119–120
modern livable urban setting urban plan, 130
results, 127–132
risk analysis and mitigation, 105
risk management, 104–105
risk management in mobility, 105, 107, 127–128
risk management patterns, 107
risk with DRMM measurement, 120
share, control, and mitigation, 122
Fifth-generation technology (5G technology), 62
Financial Risk Management (FRM), 87, 105, 226, 228
5-Why method, 120
5th Wave/Tomorrow Age Theory, 109, 118, 186–187, 215
background, 110
business economy, 111–118
impacts, 111
main points, 109
results, 111
Food, 73
safety, 73
Fossil fuel production, 56–57
Fourth Industrial Revolution, 215
Free Wireless Internet service (Wi-Fi service), 153
Free-vehicles transportation, 201

Gas, 21
Gastronomy, 72–73
Gaziantep
Gaziantep Ecological Village Project, 78
gluten free cafe online sale, 78
smart plant factory in city, 78
Geographic Information System (GIS), 27, 62
German Federal Institute for Risk Assessment (BfR), 125
Global developments affecting transportation system, 106
Global energy situation, 54–57
Global Positioning System (GPS), 154
Global transport of dangerous goods, 106–107
Globalization, 73, 206
Govern Mentality Risk Management (GRM), 87, 105
Government Financial Organization Associations (GFOA), 129–130
Governmentality risk management (GRM), 226–228
Green applications, 23–24
Green city, 5–6
Green comprehensive strategic plan, 104
Green Destinations (GD), 23–24
Green energy, 22–23
Green Globe, 23–24
Green infrastructure, 23
Green Key, 23–24
Green mobility, 128–129
Green movement, 35
Green strategies, 94, 200–201
Greenhouse gases, 86, 88–89
emissions as significant threats for sustainable threats, 88–89
Gross Domestic Product (GDP), 16, 65

H2020 Smarter Together Project, 41
Healthy communities, 35
High quality of life, blue-green ubiquitous cities as key toward livable and sustainable urban settings with, 92–96
High sustainability model, 218
High technologies, 185–186, 198, 202
Historical building information model (HBIM), 159
Home automation systems, 24–25
Human beings, 24
Hybrid electric vehicles (HEV), 60
Hybrid SME, 110

i-comprehensive urban plan, 95
i-sustainability plus, 95, 98
i-sustainability plus theory, 110, 118, 189, 215, 217
for technological society and innovative ecosystem, 187–189
Inclusive mobility, 204
Industry 4.0, 10

Information and communication technologies (ICTs), 5, 16–17, 19, 22, 38, 51, 53, 84, 87–88, 99–100, 143–144, 152, 180, 183, 186, 212, 218
air quality monitoring systems, 20
energy monitoring systems, 21
noise monitoring systems, 21
smart lighting systems, 22
smart parking systems, 21–22
traffic control systems, 20
waste management systems, 21
Information technology (IT), 40, 183, 186, 226
AI, 9
big data and data mining, 8
cloud computing technology, 9
at foundation of smart cities, 7–10
Industry 4.0, 10
IoT, 9
Innovation, 59
Innovative, and smart mobility technologies, 191–192
Innovative ecosystem, i-sustainability plus theory for, 187–189
Innovative management, 187–189
Intelligent city, 5–6
Intelligent Transport Systems (ITS), 59, 61, 63, 108
Internal combustion engine vehicles (ICEVs), 60
International Data Corporation, 185
International Environmental Management Standard, 23–24
Internet and information, technologies based on, 183–185
Internet of Energy (IoE), 183–184, 202, 226
Internet of Things (IoT), 4, 9, 17, 24–25, 51, 53, 99–100, 112, 139–140, 143–144, 152, 183–184, 186, 202, 215
Internet services, 143–144
Ishikawa diagrams, 120
Izmir (Izmir Agriculture Mobile Application), 78

Knowle West Media Center (KWMC), 41
Knowledge, technology, and business model (KTB model), 110
Knowledge city, 5–6
Konya (Smart Tourism Guide Application), 79

Light-emitting diode (LED), 22
Livability, 181–182, 187
concept, 86
toward sustainable cities with livability and high quality of life, 89–92
Livable urban settings, 108–109
with high quality of life and less greenhouse gas emissions, blue-green ubiquitous cities as key toward, 92–96
Local Control, 141
Local welfare, 141
London's smart city approach, 41–42
Low carbon, 24
Lyon Smart Community project (LSC project), 43

M.O.V.E. project, 41
Machine learning (ML), 7, 51, 185
Maersk shipping countries, 97
Marketing strategies, 64
Massive open online courses (MOOCs), 215
Metaverse, 164
ecosystem, 221–222
Seoul, 221–222
Methane, 88–89
Ministry of Environment, 78
Ministry of Environment and Urbanization, The, 145
Mobile applications, 139–140, 152–153
Mobility, 59, 99–100, 107–108, 180, 197
blue-green sustainable mobility, 107–108
risk management in, 105–107, 127–128
risk management provide comprehensive blue-green sustainable urban plan to move to blue-green sustainable mobility, 119–120
smart blue-green mobility, 108
Mobypark (smartphone application), 10–11

Modern cities, 198
Modern livable urban Setting urban plan, 130
Modern technologies, 186
Moovit (app), 201–203

National Labels, 23–24
National Smart Cities Strategy and Action Plan (2020–2023), 38, 75
Natural resources, 16
Near field communication (NFC), 152
Near frequency communication (NFC), 21–22
Network analysis, 174, 176–177
Nitrous oxide, 88–89
Noise monitoring systems, 21
Novel technology adoption, 64
Novel transportation technologies, 51, 59–61
  autonomous vehicle, 60–61
  electric technology, 60

OKOPROFIT program, 41
Opportunities of urban transport solutions, 63–66
Overtourism, 28

Parallel algorithms, 8
Pars Oil and Gas Company (P.O.G.C Company), 128
PESTEL analysis, 63–66
  economic factors, 63–64
  environmental factors, 66
  political and legal factors, 63
  social factors, 64–65
  technical factors, 65–66
Physical integrity, 141
Plug-in electric vehicles (PHEV), 60
Political sustainability, 89
Profitable quality leader, 125–126
Programming models, 8
Project Risk Management (PRM), 87, 105, 226, 228
Public transportation systems, 5, 201
Pudong, 42

Quality of life, 181–182
  toward sustainable cities with livability and high, 89–92

Radio frequency identification (RFID), 9, 21–22, 152
REACT-EU initiative, 190
Recent developments in evaluation of renewable energy resources in tourism businesses
  conceptual framework, 170–172
  data collection process, 173
  findings of research, 173–174
  method, 172–174
  research model, 172
Renewable energy, 27, 89
Renewable energy sources, 23, 99–100, 170–171
  in tourism enterprises, 170
Research on energy, 176–177
Research process, 172
Research tool, 119
Resource Efficiency, 141
Resources, 21
Risk analysis and mitigation, 105
Risk calculation through DRMM Method, 120
Risk management, 104–105, 128
  challenges and solutions concerned on risks of road transport system of global countries, 106
  classification, 105
  global developments affecting transportation system, 106
  global transport of dangerous goods, 106–107
  in mobility, 105–107, 127–128
Risk priority number (RPN), 107
Risk with Doost Risk Mitigation Method measurement, 120
Road transport sector, 54, 66
Road transport system of global countries, challenges and solutions concerned on risks of, 106
Robotic, 186

SARS-CoV-2, 4
Self-driving vehicles (*see* Autonomous vehicle)

Semi-supervised learning, 185
Seven pillars of sustainability (7Ps model)
Seven-pillar sustainability model (7Ps sustainability model), 110, 181, 189
Sino-Singaporean Tianjin Eco-City (SSTEC), 42
Small and medium enterprises (SMEs), 111, 217–218
Smart and Sustainable Destinations (SSD), 17–18
Smart Blue-Green city, 98
Smart Blue-Green infrastructure, 95
Smart blue-green mobility, 108
Smart Cities Wheel (SCW), 6, 38
Smart city, 4, 19–20, 28, 37–38, 67, 72, 75, 142–143, 152–153, 186, 213–214, 223
  applications in world, 10–11
  characteristics of, 6
  components of, 38–39
  concept of, 4–5, 20, 51, 53–54, 75
  Copenhagen waste management program, 78
  definition, importance, and features of, 74–77
  definition of, 5–6
  examples in world, 77–78
  examples of smart cities in World and Turkey, 77
  information technologies at foundation of smart cities, 7–10
  IoT technology, 24–25
  smart economy, 39
  smart environment, 39
  smart governance, 38–39
  smart living, 39
  smart mobility, 39
  smart people, 39
Smart destinations, 15–17
  application, 26
  and characteristics of, 16–18
  concept, 29
  future of, 19–22
  problems and solutions, 27–28
  smart tourism and eco-friendly practices, 18–24
Smart eco-city tourism, 145, 147–148
  eco-city tourism, 146–147
  sustainable city tourism, 145–146
Smart eco–city, 34, 39, 43, 147–148
  Amsterdam, 41
  Bristol, 41
  Chengdu, 42
  components of smart city, 38–39
  eco-city, 34–37
  Kunming, 42
  London, 41
  Lyon, 43
  Milton Keynes, 42
  Munich, 40–41
  project examples, 40
  Shanghai, 42
  smart city, 37–38
  Tianjin, 42
Smart economy, 6, 39, 142–143
Smart education, 215
Smart energy consumption monitoring system, 21
Smart environment, 6, 39, 142–143
Smart governance, 6, 38–39, 142–143
Smart infrastructure, 182–183
Smart life, 142–143, 213–214
Smart lighting systems, 22
Smart living, 6, 39
Smart London Innovation Network, 41–42
Smart mobility, 6, 17, 39, 58, 63, 96–97, 180, 200–201
  definition, 58–59
  intelligent transportation systems, 61–63
  novel transportation technologies, 60–61
  services, 180
  systems, 202–203
Smart parking systems, 21–22
Smart people, 6, 39, 142–143
Smart sensors, 28
Smart systems, 139–140
Smart technological applications, 148
Smart technological systems, 139–140
Smart technologies, 11–12, 152
  integration into tourism destinations, 154
Smart tourism, 17–18, 24, 164
  concepts, 143, 152–153
  green applications, 23–24
  green energy, 22–23

green infrastructure, 23
information communication technologies, 19–22
low carbon, 24
Smart tourism destinations, 143–144, 153–154
significance of sustainability in, 154–158
Smart tourism technologies (STTs), 26–27
Smart transportation/mobility, 142–143
Smart waste management systems, 143–144
Smart Yangpu, 42
Smartness, 87–88, 90, 92, 94, 128–129, 186, 198
strategies, 95
Smartphones, 139–140, 213
Social ecology, 35, 44–45
Social equality, 141
Social Peace, 141
Social sustainability, 90
Software Failure Modes Effects Analysis (SFMEA), 120
Songdo International Business District, 223
Spring City (*see* Kunming)
Sustainability, 38, 72, 74, 84–85, 108–109, 112, 212–213, 215
concept, 140, 142, 174, 176
cultural heritage tourism and digitalization, 158
digitalization of cultural heritage for sustainability of tourism, 158–161
of energy, 170–171
movement, 37
plus, 228
plus city, 228
seven pillars of, 98
significance of sustainability in smart tourism destinations, 154–158
smart city and smart tourism concepts, 152–153
smart tourism destinations, 153–154
and sustainable development, 181
theory, 118
in tourism, 18
of urban tourism, 139–140
with livability and high quality of life, 89–92
sustainable cities-eco cities, 77
Sustainable city, 5–6, 212, 214, 224, 226
tourism, 145–146
Sustainable comprehensive strategic plan, 104
Sustainable cultural heritage tourism management, acknowledgment of digitalization for, 159–161
Sustainable development, 181, 203, 213, 217–218
strategies, 216–217
U-City as new idea for, 229
Sustainable Development Goals (SDGs), 36, 157
Sustainable gastronomy, 73–74
Sustainable infrastructure, 182–183
Sustainable living, 33–34
Sustainable mobility, 58, 63, 107, 119, 200–201
definition, 58–59
intelligent transportation systems, 61–63
novel transportation technologies, 60–61
systems, 202–203
Sustainable Mobility Management, 125
Sustainable smart city, 85, 181
Sustainable thinking, 35
Sustainable tourism (ST), 140–141
activities in destinations, 23–24
approach, 157
concept, 140–142
Sustainable urban settings with high quality of life and less greenhouse gas emissions, blue-green ubiquitous cities as key toward livable and, 92–96
Sustainable urban tourism, 145
System adaptation process, 181
Systematic literature review (SLR), 84, 104, 212

Talk London, 41–42
Tech Londoners, 41–42
Technical sustainability, 89, 202–203

Technological society, i-sustainability plus theory for, 187–189
Technological systems, 143–144
Technologies, 183–186
Technologies based on internet and information, 183–185
Technology mirrors culture, 198
Theory of Comprehensive Everything, 109, 118, 215
  case studies, 193–196
  5th Wave/tomorrow age theory or, 186–187
  high technologies, 185–186
  i-sustainability plus theory for technological society and innovative ecosystem, 187–189
  interviews, 197
  literature review/background, 180
  livability and quality of life, 181–182
  methodology, 191–193
  mobility, 180
  Netherlands/Amsterdam, 195–196
  questionnaire, 196
  results, 197
  smart and sustainable infrastructure, 182–183
  smartness, digitalization, and ubiquitous, 186
  South Korea/Seoul, 193–194
  sustainability and sustainable development, 181
  technologies, 183–186
  technologies based on internet and information, 183–185
  Urban 6.0 and Utopia, 189–191
3-E Houses (3-E HOUSES), 41
3D
  line drawing, 159
  recording, 159
  Socio-Eco-Environment SMEs model, 110
Tianfu New District plan, The, 42
Tomorrow Age Theory, 109–118
Tomorrow's SMEs concept, 110
Tourism, 164
  big data system, 25–26
  businesses, 171
  digitalization of cultural heritage for sustainability of, 158–161
  industry, 16, 153, 170
  sector, 18, 148, 170–171
  strategy, 16
Tourist experiences, 164
Traffic control systems, 20
Transformable energy, 143–144
Transition, 50
  process, 63, 66
  project, 64
Transmission technologies, 24–25
Transparency, 229
Transport sector general situation, 57–58
Transportation, 200–201
  global developments affecting transportation system, 106
Turkey, examples of smart cities in World and in, 77
Turkey, smart city examples in, 78
Turkish culinary culture, 72

U-airport, 226
U-building, 226
U-cities, 87–88, 212, 218, 220, 228
  construction and implementation plan, 220
  disadvantages, 229
  as new idea for sustainable development, 229
U-citizen, 226
U-democracy, 112
U-education, 226
U-governance, 112
U-government, 226
U-health, 226
U-infrastructure, 226
U-life, 226
U-mobility, 226
U-politic, 226
U-security, 226
U-services, 226, 229
U-technique, 226
Uber (app), 201
Ubiquitous (U), 186, 193–194
  blue-green cities, 84
  concept, 98
  life, 218–220
  smart city, 223
  tools, 202

United Nations Environment Conference, 140–141
United Nations Environment Program (UNEP), 36, 140–141
United Nations forecasts, 223
United Nations World Urbanization (2016), 3
Urban 6.0, 189–191
Urban areas, 99–100
Urban design, 214–215
Urban development, 198
Urban eco-village, 34–35
Urban Ecology, 34–35
Urban infrastructure planning, 214
Urban livability, 204
Urban mobility technologies, 51
Urban planning, 4, 11, 214–215
Urban settings, 50
Urban strategic planning, 108
Urban sustainability, 214–215
Urban transformation, 198
Urban transport solutions
  challenges and opportunities, 63–66
  concept of smart city, 51–54
  global energy situation, 54–57
  overview, 54–58
  sustainable and smart mobility, 58–63
  transport sector general situation, 57–58
Urbanization, 78
  challenges, 85–87
  challenges threaten livability and quality of life, 86–87
  environmental problems, 86
  system, 34–35
Utopia, Urban 6.0, 189–191

Value engineering, 98–99
Vehicle to Grid (Vehicle2Grid), 41
  application, 62
Vehicular ad hoc network (VANET), 63
Virtual display technology, 159
Virtual reality (VR), 19, 152, 160–161, 164
Virtual restoration, 159
Visitor Satisfaction, 141
VOSviewer package program, 173

Waste management systems, 21
Water, 21
  water-based concerns, 200–201
Web-based applications, 18
Wired city, 5–6
Wireless internet networks (Wi-Fi), 154
Wireless sensor networks (WSNs), 22, 51, 62
Wireless technologies, 21–22
World and in Turkey, examples of smart cities in, 77
World Health Organization (WHO), 126–127
World Tourism Organization (WTO), 77, 140–141
World Travel and Tourism Council (WTTC), 16
WoS database, 173

ZigBee network, 22

Printed and bound by CPI Group (UK) Ltd, Croydon, CR0 4YY

23/09/2026

14967326-0001